Idealist Alternatives to Materialist Philosophies of Science

Idealist Alternatives to Materialist Philosophies of Science

Edited by

Philip MacEwen

BRILL

LEIDEN | BOSTON

Cover illustration: Inspired by Philip and Maria MacEwen.

The Library of Congress Cataloging-in-Publication Data is available online at http://catalog.loc.gov
LC record available at http://lccn.loc.gov/2019954196

Typeface for the Latin, Greek, and Cyrillic scripts: "Brill". See and download: brill.com/brill-typeface.

ISBN 978-90-04-41526-3 (hardback)
ISBN 978-90-04-41527-0 (e-book)

This book is printed on acid-free paper and produced in a sustainable manner.

Contents

Foreword

William James had it right. The history of philosophy is the arena of an ongoing tug-of-war between a hard and a soft tendency of thinking. The hard-liners incline to be materialistic, receptive to surdity, and insistent on explanation by efficient causation. The soft-liners accept finalistic explanations and envision a rational order in nature's arrangements.

There has, unsurdly, been balance here. Antiquity had its Platonists and its Atomists; nineteenth century physics had its positivists and its devotees to rational mechanics. But in twentieth century philosophy of science, the hard-liners had things pretty much to themselves, save for an occasional and usually much decried voice crying in the French wilderness. (Henri Bergson and Teilhard de Chardin come to mind.)

However, the pendulum of fashion swings back and forth. In particular, recent developments in cosmology have been running decidedly against the grain of the traditional hard-liners.

The time is ripe for a foray into "Idealist Alternatives to Materialist Philosophies of Science," so, as I see it, this volume is a judicious instance of seizing upon "a tide in human affairs." Beyond this, it affords a wealth of highly interesting, instructive, and philosophically thought-provoking reflection. Few who read these stimulating pages will continue to think that idealism is a lost cause in the philosophy of science.

However, most philosophers have not yet begun to take notice.

Nicholas Rescher
Distinguished University Professor of Philosophy
Chairman of the Center for the Philosophy of Science
University of Pittsburgh

Preface

This work originated from a conversation I had with a colleague at the 2008 Congress of the Humanities and Social Sciences in Vancouver. This colleague drew to my attention the fact that something needed to be written on idealist alternatives to materialist philosophies of science. I had plenty of experience writing about idealism, some of it as an alternative to materialism, but little writing about the philosophy of science so it took me a long time to work out the project in its final form.

I am indebted to all the contributors of this book for coming to my aid in various ways and helping me bring the project to completion. I am also indebted to my colleague Stanley Tweyman who has contributed enormously to my understanding of science in the Enlightenment and Hume's view of science in particular. Last but not least, I owe a great debt of gratitude to my wife, Maria, who helped me design the cover illustration for this volume, and to our daughters, Kasia and Aletheia, who assisted with some of the technical details in preparing the manuscript for publication.

Finally, I would like to extend my appreciation to the staff at Brill, including Jennifer Pavelko, Bram Oudenampsen, and Benjamin Suchard, all of whom helped to facilitate this endeavour.

Philip MacEwen
York University

Illustrations

Notes on Contributors

Leslie Armour
was Research Professor of Philosophy at Dominican College, Ottawa, and a Fellow of the Royal Society of Canada. He published in many areas including logic, metaphysics, and the history of philosophy.

Philip MacEwen
is a Fellow of McLaughlin College, York University, where he teaches humanities and philosophy. He has published in the history of ideas, ethics, and literary theory.

Hugo Meynell
is Professor Emeritus of Religion Studies at the University of Calgary and a Fellow of the Royal Society of Canada. He has published in a number of areas including the history of philosophy, religion, philosophy of science, and music.

John D. Norton
is Professor of History and Philosophy of Science, University of Pittsburgh and Director of the Centre for Philosophy and Science. He has published extensively in the philosophy of science.

Edward L. Schoen
is Professor Emeritus of Philosophy at Western Kentucky University. His many publications focus on the interface between science and religion.

Elizabeth Trott
is Professor Emeritus of Philosophy at Ryerson University. She has published in a variety of areas including the history of philosophy, ethics, metaphysics, social and political philosophy, and the philosophy of education.

Fred Wilson
is Professor Emeritus of Philosophy at the University of Toronto and a Fellow of the Royal Society of Canada. He has published in a wide range of areas including the history of philosophy and the logic and methodology of science.

K.M. Ziebart
has taught philosophy at Loyola University Maryland and has published in the history of philosophy and philosophical idealism.

Introduction

Philip MacEwen

Is the world which scientists study independent of the scientists who study it? In one sense, the answer to this question is affirmative. How could scientists study the world unless the world was there for them to study? Thus construed, the activity of scientists is like many other human activities, practical or theoretical. How can a farmer sow the field unless the field is there to be sowed? How can a doctor examine the patient unless the patient is there to be examined? How can a mathematician calculate the square of the product of the first five prime numbers unless there are prime numbers, the square of the product of the first five of which is to be calculated?

If worlds, fields, patients, and prime numbers exist by themselves, there is no reason to think that their existence depends in any way on those who study, farm, examine, or perform calculations about them. This view makes eminent sense until we realize that, in each case, the object concerned already has the earmarks of cognitive intervention. A "world" is very different from a "field" which is very different from a "patient" who is very different from "prime numbers." None of these can be interchanged with any of the others. Each of them has a set of properties which the others do not have.

Furthermore, the set of properties which each object has is, to some extent, the result of human mediation. The world studied by science, it is true, is a system of material objects acting according to certain laws but what is "a system of material objects acting according to certain laws" precisely? The field that is farmed, it is true, is a plot of arable land but what is "a plot of arable land" precisely? The patient who is examined, it is true, is a subject receiving health care but what is "a subject receiving health care" exactly? The square to be calculated, it is true, is that of the product of the first five prime numbers but what is "the square of the product of the first five prime numbers" exactly?

The difference between the descriptions without quotation marks and their counterparts inside them is that the former are extensional while the latter are intensional.[1] The former refer to certain states of affairs in the world while

1 More precisely, the latter are intensional in one of at least four different ways: 1) as conceptions in the mind of the subject (e.g. Berkeleyan idealism); 2) as categories of the mind of the subject through which the extensional is perceived and in terms of which it is understood

 | DOI:10.1163/9789004415270_002

the latter refer to certain ideas about certain states of affairs in the world.[2] This distinction has caused enormous difficulty in the history of thought, including the history of science, and is in large part responsible for the differing world-views of materialism—or more pervasively and yet more precisely, naturalism—and philosophical idealism. According to naturalism, all language is either extensional, or at least capable of being equated with language that is so, while according to philosophical idealism, all language is intensional, or at least capable of being equated with language that is so.[3] According to naturalism, therefore, the claim that the square of the product of the first five prime numbers = 5,336,100 is a claim about the nature of the numerical world while according to philosophical idealism, the claim, "The square of the product of the first five prime numbers = 5,336,100," is a claim about the relationship between ideas, i.e., the *idea* of the square of the product of the first five prime numbers = the *idea* of 5,336,100.

Is mathematics the study of objective states of affairs in the numerical world, then, or is it the study of relationships among numerical ideas? Since opposites, unlike contraries, are not mutually exclusive, and since naturalism and philosophical idealism thus construed are opposites rather than contraries, another response to this question is that mathematics is neither the study of objective states of affairs in the numerical world nor the study of relationships among numerical ideas but the study of *both*. Accordingly, the claim that the square of the product of the first five prime numbers = 5,336,100 is both extensional and intensional. It is extensional insofar as it refers to a state of

(e.g. Kantian idealism); 3) as concepts which are grasped by the mind of the subject but exist extensionally as part of universal Mind or Spirit (e.g. Hegelian idealism), or 4) as concepts which may be grasped by the mind of the subject but exist extensionally in a world of universal Ideas or Forms (e.g. Platonic idealism). There are other types of philosophical idealism, like emergent mentalism as discussed by C.D. Broad or dualistic idealism as developed by Descartes, but these are four of its principal types. What these types have in common is the view that intensionality plays a crucial role in determining, to use Leslie Armour's expression in Chapter 1 of this volume, "the things that fill the world." Furthermore, intensionality always involves the intensionality of the subject, though it may not be limited to it. For the sake of simplicity, I speak of intensionality in terms of "ideas" above though, as we can see, philosophical idealism interprets this expression in a variety of ways.

2 This statement is true even of Berkeleyan idealism since it is God who perceives the "things" i.e., states of affairs in the world, and presents them to us in orderly ways which cause us to perceive ideas.

3 This is true of extensional forms of intensionality, such as Hegelian and Platonic idealism, since the extensional existence of Mind/Spirit (Hegel) or the Forms (Plato) is itself intensional. Hence, all language about Mind/Spirit/the Forms is intensional though it can be about intensionality in its objectivity.

affairs in the numerical world and intensional insofar as it refers to the *concept* of such a state, something which becomes explicit when we say, "The square of the product of the first five prime numbers = 5,336,100."

This gives rise to a third position between naturalism and philosophical idealism which might be called "pluralism." According to this view, the answer to the question, "Is the world which scientists study independent of the scientists who study it?," is both affirmative and negative. To the extent that the world must exist in the first place if scientists are to study it, the answer is "Yes." To the extent that "the world" which scientists study is carefully delineated from other constructs like "fields," "patients" (as distinct from "subjects"), and "prime numbers," the answer is "No." According to pluralism, science is the study of both the extensional world and the intensional "world," and much of the confusion in the history of thought can be avoided if this distinction is understood in like manner.

The papers in this volume can be taken as concerted attempts, both systematic and historical, to address these sorts of issues from a variety of perspectives. Philosophical idealism is certainly well represented, but so also are naturalism, pluralism, and other positions. While the overriding purpose of the volume is to advance idealist alternatives to what in ordinary terminology is called "materialism" (but which more pervasively, albeit more precisely, can be referred to as "naturalism") in the philosophy of science, this has not come at the expense of either downplaying naturalism or limiting the debate to the two principal interlocutors.

The historical papers are all concerned with the role of science in the thought of, the philosophy of science of, or attendant metaphysical considerations in the thought of, particular figures. Sometimes one figure has been selected for examination (e.g. Hume, Bradley, Einstein), sometimes several figures have been selected for scrutiny (e.g. Charles De Koninck and John Leslie), and sometimes many figures have been selected for comparative analysis (e.g. Anaxagoras, Socrates, Plotinus, Hume, J.S. Mill, T.H. Green, Bradley, Althusser, Russell, Blanshard, etc.). On the other hand, the systematic papers are more concerned with general problems of ontology about the world which science studies (Armour), the methods of science (Schoen), or interpretations of the *philosophy* of science (Meynell). I shall observe this delineation in commenting on each contribution.

In "The Things That Fill the World" (Chapter 1), Leslie Armour tries to determine of what "the world" consists, construed as the amalgam of objects we think we encounter directly and the "theoretical entities" of physics. Armour argues that even the objects we think we encounter directly—things like chickens, works of art, coins, account books, automobiles, etc.—require a good deal of "making" on our part if we are even to perceive them. It takes a lot of concepts and mental activity to see an automobile. As Gilbert Ryle was fond of saying, "seeing" is an achievement verb.[4] In this sense, the things that physicists talk about—electrons, protons, neutrons, quarks, etc.—are not unlike the objects we think we encounter directly since the mental apparatus required to understand both types of objects is considerable.

This raises the possibility that the best way to understand the things that fill the world is to focus on the common mental apparatus which is required to understand both the objects of ordinary experience and the objects of physics. W.V.O. Quine took such an approach by appealing to space-time regions and their successive occupation. Thus construed, the objects of ordinary experience and the objects of physics are both logico-mathematical entities which can be described using appropriate co-ordinate systems. Indeed, the space-time regions themselves can be considered as sets of real numbers. This way of understanding the things that fill the world turns the objects of ordinary experience into mathematical formulae, thereby privileging the theoretical entities of physics rather than the objects of ordinary perception, an inversion of the first view of the world where the latter were deemed to be primary and the former secondary.

If the things that fill the world are all mathematical formulae, what are mathematical formulae? The bulk of mathematicians have been mathematical Platonists—James Robert Brown in our generation and figures like Paul Dirac, Kurt Gödel , G.H. Hardy (and, as John D. Norton argues in Chapter 4 of this volume, Albert Einstein) not long before us. To some of these people, Platonism was so obvious it did not even need to be defended. We might not like (the concept of) the number two but there is nothing we can do about it. In any world, there will be countable things and more than one of them and (the concept of) the number two, i.e., "twoness" or what Russsell called the "class of all couples," is in some sense just there.

This view made a lot of sense as long as mathematics was considered to be simply a collection of eternal truths. It now seems more plausible to think of mathematics in terms of conjectures and refutations. Mathematical formulae which stand are just the ones that have not succumbed to new demonstrations of inconsistency. Despite the introduction of historicism, mathematics remains a collection of formulae, now construed as theorems which have withstood the latest round of conjectures and refutations. Mathematical Platonism

4 Gilbert Ryle, *The Concept of Mind*, London: Hutchinson, 1966, 150–152; 222.

remains as the theory behind the process of conjectures and refutations, winnowing the formulae and allowing mathematics to make progress in so doing.

Furthermore, this kind of Platonism highlights another aspect of mathematical formulae: their intelligibility. Mathematics makes progress, not by accumulating more theorems, but by selecting from the body of extant theorems those which best meet the criteria of intelligibility. This weds the best of traditional Platonism with the best of modern idealism without giving way to the mostly post-Cartesian view that concepts or ideas are simply denizens of the mind. If reality consists of physical objects and the theoretical entities of physics, and both are mathematical formulae selected according to the criteria of intelligibility, then we can understand the world and, indeed, read it as if it were a book.

So what is the relationship between the world of physical objects and the theoretical entities of physics, on the one hand, and the subject(s) which interprets them as being such, on the other? Alternatively, how do Popper's "three worlds," or something like them, fit together? The objects which physicists talk about are interpretations of mathematical formulae. Traditionally, physics is the study of movements of particles or waves or the dissemination of energy across space and time or space-time. There are theories and theorems to connect mass and energy as well as particles and waves. Physicists experience the world they study in terms of the theories and theorems they construct to interpret it. Physics is particularly germane here because it provides as direct a link as possible between the logico-mathematical entities that fill the world and the logico-mathematical methodologies physicists use to experience them.

As for the relationship between our everyday personal experiences and the world of physical objects, this too is logico-mathematical, at least on the objective side, as we saw above. Unlike the experience of the physicist with the entities physics studies, however, it is not purely so. Many of our everyday personal experiences involve sharing minds with other creatures. The fact that we discern minds in other creatures suggests that we have minds in common with them. This situation is perhaps most intelligible if the same mind appears in all of them, though different facets of it and different capacities appear in each organism. Even in everyday personal experiences with entities other than organisms, the presence of mind is often pervasive. To see an automobile or a building, for example, is to see the product of minds, though not the organic instantiation of minds. Thus, our everyday experience of the world is shared, both as perceiving subjects and in terms of the objects we experience. Like the Haidas of the Queen Charlotte Islands, it is easy to see the objects of everyday experience as extended organisms which either bring us into contact with other minds or instantiate a particular species of mind as physical entities.

The next systematic contribution is Hugo Meynell's "Idealism and the Philosophy of Science" (Chapter 3). According to Meynell, philosophical idealism is an enormously important half-truth, significantly improving upon the incoherence of materialism, which makes science impossible, but falling well short of critical realism and its standards for true judgments. The important half-truth of idealism is that it rightly emphasizes the constructive role of intelligence in our apprehension of what we take to be "reality" or "the actual world":

> The great discoveries of Kepler, Darwin, or Planck were not just a matter of soaking up sense-experiences or observable facts and recording them; they were prodigious feats of mental creativity. Certainly ... they are *more than* that ... but they are no *less*. It is one thing to conceive an hypothesis, to envisage a possibility; it is another to affirm for good reason that the possibility is a reality, that the hypothesis is certainly or probably so.[5]

Despite the constructive role it assigns to intelligence in rightly maintaining that "*the nature and structure of the mind are a crucial clue to the nature and structure of the cosmos ...*,"[6] philosophical idealism is not the correct position in epistemology and metaphysics because it has no criterion for determining under what conditions, precisely, judgments are true. It is one thing to point out that our apprehension of reality is influenced by mental creation, constitution, or construction but quite another to articulate the role of the mental, or other factors, in determining whether our judgments about the world are true. Unlike philosophical idealism, critical realism does have such a criterion. Judgments are true insofar as they are well-founded and they are well-founded insofar as they have been arrived at attentively, intelligently, and reasonably. "... attentiveness is to the relevant evidence in experience, intelligence is a matter of envisaging possibilities or hypotheses, and reasonableness consists in judging in each case as certainly or probably so that hypothesis which adequately explains the evidence."[7]

Leaving the defence of this position to the author, several questions arise, first about the relationship between critical realism and philosophical idealism, and second about whether philosophical idealism has no criterion for determining under what conditions, precisely, judgments are true. Is critical realism, as Meynell interprets it, not itself a form of philosophical idealism?

---

5 Hugo Meynell, "Idealism and Philosophy of Science," in *Idealist Alternatives to Materialist Philosophies of Science*, ed. Philip MacEwen, Leiden: Brill, 2019, Chapter 3, 77.

6 *Ibid.*, (italics in text), 75.

7 *Ibid.*, 76.

Meynell admits that the point is moot, likening it to the question whether classical sonata first-movement form is binary or ternary. Is the recapitulation the second part of the second part (historically called "the development") or is it rather a separate third part? Perhaps the special merit of sonata-allegro form, as it is otherwise know, is that it combines elements of both binary and ternary form; similarly, Meynell maintains, critical realism combines the merits of idealism, empiricism, and materialism.

It is a little surprising that Meynell includes materialism in this list since, elsewhere in Chapter 3, he argues that materialism—or at least consistent materialism—is inconsistent with science and, indeed, would make it impossible. Thus, in response to those who object that it is only a matter of time before our mental life is thoroughly explained in terms of the physics and chemistry of the brain, Meynell rejoins that it never will be for the following reasons:

> [W]e believe, and ought to believe, what scientists say on the subjects of their specialties only on the assumption that they, or their informants, have been thoroughly rational [attentive, intelligent, and reasonable] in coming to the relevant judgments. If nobody, in the last analysis, can be rational in this sense, but everyone thinks, speaks, and acts as they do simply due to the physical and chemical laws which operate on their brains, then science itself falls to the ground. It would be singular indeed if the ultimate upshot of science were that science is impossible. Yet that seems to be a valid inference from a consistent materialism.[8]

Perhaps Meynell's position is that critical realism comprises *elements* of both philosophical idealism, by acknowledging the importance of intelligence in true judgments, and materialism by maintaining that, while judgments are true if and only if they correspond to certain facts, those facts exist independently of judgments about them. "I think it is odd to deny that the … extinction of the dinosaurs between sixty and seventy million years ago … [is a] 'fact,' or that [it] … obtained prior to and independently of anyone thinking, saying, or writing what they did about [it]."[9]

There is also an element of both idealism and materialism in the process whereby critical realism arrives at true judgments. On the side of materialism, critical realism affirms independently existing *facts* to be true based, on the

8 *Ibid.*, 92.

9 *Ibid.*, 81.

side of idealism, on what is known in judgments about them "… reached at the ideal term of rational inquiry,"[10] i.e., when attentiveness to the relevant evidence in experience, intelligence in envisaging possibilities or hypotheses, and reasonableness in judging each case as certainly or probably true have been taken as far as they can go in the case at hand.

As for empiricism, not only is it part and parcel of attentiveness, but it is crucial to the epistemology of materialism as well as to certain types of idealism. Materialism is essentially a metaphysical position which claims that the ultimate constituents of reality are material or physical bodies. While the ultimate constituents of reality may not be empirically observable (e.g. Democritus' atoms or Boyle's corpuscles), they nevertheless make up the objects which are empirically observable. Thus, the metaphysics of materialism is both microscopic and macroscopic, and empirical with respect to the latter though not to the former. Idealism has been construed both metaphysically and epistemologically. Epistemologically, it is either empirical or at least has an empirical dimension since part of "the world" which it seeks to know will consist of bodies that, in certain respects, can only be accessed through the use of the senses. In the words of Descartes, "I do not in truth think that I should rashly admit all the matters which the senses seem to teach us, but, on the other hand, I do not think that I should doubt them all universally."[11] Thus, Meynell is on solid historical ground in combining idealism, materialism, and empiricism in formulating critical realism.

The third systematic contribution is "Nature, God, and Scientific Method" by Edward L. Schoen (Chapter 5). Like Meynell, Schoen challenges a strictly idealist, naturalistic, or any other monolithic approach to science, arguing that scientific methods counsel "… a tentative, rather open-ended pluralistic stance"[12] to the history of their development. After surveying the history of philosophy of science from A.J. Ayer to the present, Schoen concludes that there are only three methodological practices which characterize science typically and persistently, if not universally: 1) the drive to uncover and identify scientific laws, 2) the empirical nature of science, and 3) the methodological constitution of kinds. Even these do not stem from any distinctively metaphysical commitments about the objects that science studies, however, nor, on the subject side, does any of them preclude the possibility of supernatural agency

10 *Ibid.*

11 René Descartes, *Meditations on First Philosophy* (1641); ed. Stanley Tweyman (Ann Arbor: Caravan Books, 2002), VI:91.

12 Edward L. Schoen, "Nature, God, and Scientific Method," in *Idealist Alternatives to Materialist Philosophies of Science*, Chapter 5, 129.

or divine intervention, much as some philosophers of science might argue to the contrary.

Despite the growing evidence of indetermination in nature, and the increased scientific tolerance for slop this has induced, scientists nevertheless continue to look for patterns of regularity in nature which they might codify into statement of scientific law and, if possible, decipher whatever is responsible for these patterns. These two goals are obviously related but should not be conflated. Gregor Mendel, for example, identified various patterns of inheritance in his study of peas, regularities which were subsequently embedded in Mendel's Laws. Only now, a century after Mendel's death, is any clear understanding emerging concerning just how the patterns of inheritance, codified in Mendel's Laws, are naturally generated. Thanks to decades of biological research, Mendel's vague "factors" are slowly being replaced by a sophisticated genetic theory, replete with genes, chromosomes, the helical structure of DNA, and all the rest.[13]

Given the distinction between lawful patterns of regularity and the forces or entities responsible for producing these patterns, it is possible to envision ways in which divine agents might enter into the natural course of events without disrupting scientific laws. Suppose, for the sake of simplicity, that the patterns of the orbiting planets were to conform precisely to Kepler's Laws. In the natural course of events, this would be attributable to the operation of a natural force, the force of gravity. Imagine, however, that there is some divine agent, God, who can directly do anything, including creating or destroying the universe in whole or in part. Now suppose that God grew weary of the force of gravity and determined to destroy it. Normally, this loss of gravitational attraction among the planets would become immediately evident. The planets would fly off in all directions. But imagine that God chose to intervene so as to keep the planets spinning in their Keplerian orbits, not by recreating gravitational attraction, but by direct, divine action. In that case, the planets would continue to move across the heavens with no interruption, disruption, or violation of Kepler's Laws.[14] Other examples could be cited quite easily, the point being that there is nothing about the tight regularity of nature which excludes the possibility of God acting directly in the world.

The second methodological practice which is typical and persistent is the empirical character of science. While this might seem to entail certain metaphysical commitments about the objects science studies, typically naturalistic

13 *Ibid.*, 137–138.

14 *Ibid.*, 138.

or idealist in persuasion, this is not the case. The reason is that human perception does not demand that things appear as they actually are. All that is required is the ability to isolate items sensuously from their surrounding perceptual environments. Microbes are too small for anyone to see with the naked eye. If they have a nature or essence, perhaps a genetic one, it is reasonable to assume that the very nature or essence of microbes determines many of their features and possibly even their size. If so, then it would seem that microbes are essentially as small as they are.

If human visual capacity also were a matter of essence, so that, by their very nature, unaided human eyes could not resolve microbes as perceptual fields, none of this would have the slightest bearing on whether or not humans could see microbes. Thanks to the invention of microscopes, bright lights and, in some cases, special staining techniques, observers now regularly isolate microbes visually from their surrounding environments. With the recognition that anything that can be sensuously isolated from its surrounding background can be perceived comes the recognition that all sorts of imperceptible items, even essentially imperceptible ones, may be observed. Not only do optical microscopes make it possible to see microorganisms too small to be resolved by unaided human eyes, but electron microscopes allow the luxury of seeing things that are too small even to reflect visible light. With the aid of night vision goggles, hunters can watch the movements of deer, even when it is too dark out to see. With the aid of television and video recorders, people can watch events on the other side of the world, whether those episodes are happening now or transpired years ago.

While bereft of distinctively metaphysical commitments, the empirical nature of science does raise some interesting theological possibilities. If empirical observation does not demand that things appear as they actually are and much of empirical observation in science takes the form of observing what is essentially unobservable, then some of the tenets of theology begin to make more sense. For example, according to the Nicene Creed, God the Father is essentially incorporeal and eternal and yet Jesus is of the same essence as the Father. In that case, anyone who saw Jesus in his bodily form would have seen an essentially eternal and incorporeal God. Whatever was seen would depend upon its *identity*, not upon the way in which it appeared. Since items need not appear as they actually are, an eternal, incorporeal God might show up as a preacher in Galilee. Whatever obstacles there may be to this possibility, essential unobservability is not one of them.[15]

15 *Ibid.*, 143–144.

The third methodological practice which characterizes science typically and persistently is the constitution of kinds. According to D.Z. Phillips, the impact of methodological constraints in science, be they regularity, observability, or anything else, does not simply delimit the range of what may be studied but actually determines the kinds of realities that can be encountered. The kind of reality science studies pertains to the reality of this or that, not to the reality of kinds themselves. Since the study of God is the study of a particular kind of reality, not the study of the reality of this or that, to approach God through the methods of science would be to transform God's reality into the reality of physical objects. This would have disastrous consequences at the level of individual practice, for example, changing genuine religion into the worship of physical objects *simpliciter*.[16]

In response, Schoen argues that an examination of the history of science reveals, not that scientific methodologies determine the constitution of kinds which science studies but that the constitution of kinds which science studies determines its methodologies. For example, Kepler painstakingly chronicled in his *Astronomia Nova* the steps that led him to his revolutionary rejection of Aristotelian astronomical methods. He documented not only how each of his hypotheses fared with Tycho Brahe's data, but precisely how various aspects of those data drove him through a succession of methodological innovations. Like Galileo, Kepler discovered that instead of Aristotelian methods constituting the kinds of reality to be found in the skies, the kind of reality found there determined which methods were most effective in his quest for scientific understanding.[17]

The resilience of constitution of kinds to methodology, and any specific metaphysical commitment the latter might entail, once again raises some interesting theological possibilities. "If there is any lesson to be learned from the history of science, it is that there is no telling what the future may hold, what kinds of things scientists may repudiate or advocate, what metaphysical boundaries may be respected or violated."[18] Despite dissenting voices like Edward O. Wilson, who still believes that he can extract metaphysical clues from

16 *Ibid.*, 143–144. Even idolatry is not the worship of physical objects *simpliciter* but rather the worship of what physical objects *symbolize* or *represent*. When the children of Israel became exasperated with Moses' long sojourn on Mt. Sinai, for example, they brought their gold earrings to Aaron who made an idol in the shape of a calf. The calf was then introduced to the Israelites as "… your gods … who brought you up out of Egypt" (Exodus 32:4; cf. 32:8), something which a physical object *simpliciter* could never have done (cf. story of Elijah and the prophets of Baal in 1 Kings 18:16–40).

17 Schoen, *ibid.*, 148–149.

18 *Ibid.*, 157–158.

contemporary scientific practice, boils all of reality down to the entities of physics, and finds in science a new and better-tested ground to attain the same ends as religion,[19] the methods of science cannot determine when, where, or whether God may offer an enfolding hand. As an examination of clockwork suggests, any number of quite diverse mechanisms can produce exactly the same patterns. The hands on the face of nature might, in fact, be driven by something as radically different from gears and springs as a system of pulleys. If God is thrown into the mix, no matter how sophisticated or adequate a theory about gears and springs may be, the actual agent responsible for patterns in nature might not even be material. It might be a supernatural, incorporeal divinity. For those whose eyes of faith have seen God's hand, science offers no reason for doubt. For those who have sought God's hand, Wilson's Gilead of science offers no balm.[20]

The first historical contribution is my "Science and the Humanities in Hume's Philosophy of Religion" (Chapter 2). Unlike our "two cultures" view of science and the humanities, Hume had a "one culture" view, treating all human learning under the rubric of "the science of MAN" and dividing it into seven disciplines: mathematics, natural philosophy, natural religion, logic, morals, criticism, and politics. One can nevertheless detect a dividing line in these disciplines between the first three and the last four. The first three are in some measure dependent on the science of Man since "… they lie under the congnizance of men, and are judged by their faculties and powers" while the last four are entirely dependent on it. "The sole end of logic is to explain the principles and operations of our reasoning faculty, and the nature of our ideas. Morals and criticism regard our tastes and sentiments: And politics consider men as united in society, and dependent on each other."[21] Thus, it is possible to read our two cultures view of science and the humanities into Hume's one culture view of the sciences.

I argue that if we apply this insight to Hume's *Dialogues Concerning Natural Religion*, while continuing to respect both cultures, we can make sense of two competing strands of thought within it and also resolve the competition between them. The first strand produces Philo's hard-nosed conclusion at the end of Part 8 that "A total suspense of judgment [in matters of natural religion]

19 *Ibid.*, 158; cf. Edward O. Wilson, *Consilience: The Unity of Knowledge* (New York: Vantage Press, 1998), *passim*.

20 *Ibid.*, 150–160.

21 David Hume, *A Treatise of Human Nature* (1739–1740); ed. David Fate Norton and Mary Norton, 2 vols. (Oxford: Clarendon Press, 2007), Vol. 1, Introduction, 4.

is our only reasonable resource."[22] The second strand yields Philo's heartfelt confession at the beginning of Part 12 that "… no-one [other than myself] has a deeper sense of religion impressed on his mind, or pays more profound adoration to the divine Being, as he discovers himself to reason, in the inexplicable contrivance and artifice of nature."[23] If we put the two strands together, we get Philo's one-culture verdict, notable for both its rationality and sensitivity, at the end of Part 12:

> If the whole of rational theology … resolves itself into one simple … proposition, that the cause or causes of the order in the universe probably have some remote analogy to human intelligence … , what can the most inquisitive, contemplative, and religious man do no more than … believe that the arguments, on which it is established, exceed the objections, which lie against it?[24]

While I do not make this point in Chapter 2, the metaphysical significance of Philo's one culture verdict is considerable. The underlying metaphysic of Philo's hard-nosed conclusion at the end of Part 8 is difficult to pinpoint but it is at least consistent with naturalism to a certain extent. Of the three criticisms on which it is based (the argument from design leads to an infinite regress, produces a *reductio ad absurdum*, and is met or exceeded in plausibility by alternative cosmologies), this is probably most apparent in the case of the first criticism. There Philo argues that the argument from design is anthropomorphic because, as Cleanthes earlier notes, "… the Author of Nature is somewhat similar to the mind of man."[25] The argument from design is also idealistic because it "… trace[s] the world of objects ["the productions of human contrivance" and the "one great machine" which constitute the world] into a similar universe of ideas."[26] Having made this move, what is to stop the argument from design from going on with an infinite regress of ideal causes? "Have we not the same reason to trace the ideal world into another ideal world, or new intelligent principle? … How can we satisfy ourselves without going on *ad infinitum*?"[27] The implication is that if Cleanthes had left "the world of

22 David Hume, *Dialogues Concerning Natural Religion* (1779); ed. Stanley Tweyman (Ann Arbour: Caravan Books, 2000), 147 (hereafter *Dialogues*).
23 Hume, *Dialogues*, 172.
24 Hume, *Dialogues*, 184–185.
25 Hume, *Dialogues*, 109.
26 Hume, Dialogues, 126.
27 Hume, *Dialogues*, 125.

objects" alone, none of these logical problems would have arisen in the first place—indeed, the argument from design would never have been advanced. It is presumptuous to say that Philo's metaphysic is naturalistic as a result, but it is certainly consistent with naturalism in its overall structure.

On the other hand, the metaphysic underlying Philo's confession at the beginning of Part 12 that "… no-one [other than myself] has a deeper sense of religion impressed on his mind, or pays more profound adoration to the Divine Being …" is undoubtedly idealistic. Philo speaks of religion in terms of his "unfeigned sentiments" which he has "ever cherished and maintained."[28] For the Philo of the humanities, religion is a feeling or a passion which impresses itself on the mind and expresses itself in "profound adoration to the Divine Being." One is reminded here of Rudolph Otto's view of religion as numinous experience, an experience which may

> … at times come sweeping like a gentle tide, pervading the mind with a tranquil mood of deepest worship. It may pass over into a more set and lasting attitude of the soul, continuing, as it were, thrillingly vibrant and resonant. … It may burst into sudden eruption up from the depths of the soul with spasms and convulsions, or lead to the strangest excitements. … It may become the hushed, trembling, and speechless humility of the creature in the presence of—whom or what? In the presence of that which is a *mystery*, inexpressible and above all creatures.[29]

The metaphysic underlying the synthesis of these two strands of thought in Philo's one culture view of natural religion (science) is likely a combination of naturalism and idealism, uniting the cause or causes of the order in the universe with human intelligence in a "remote analogy" to which "the most inquisitive, contemplative, and religious man" must assent if the evidence, on balance, supports it. A fuller study of Hume would need to be done to develop this metaphysic but the possibility for combining naturalism and idealism is certainly present in Hume's philosophy, an ironic outcome to be sure for one who ended his first published philosophical work after *A Treatise of Human Nature* with a fierce castigation of traditional metaphysics:

> When we run over libraries, persuaded of these principles, what havoc must we make? If we take in our hand any volume; of divinity or school

28 Hume, *Dialogues*, 177.

29 Rudolph Otto, *Das Heilige* (1917); *The Idea of the Holy*, trans. John W. Harvey (New York: Oxford University Press, 1950), 12–13.

> metaphysics, for instance, let us ask, *Does it contain any abstract reasoning concerning quantity or number*? No. *Does it contain any experimental reasoning concerning matter of fact and existence*? No. Commit it then to the flames: For it can contain nothing but sophistry and illusion.[30]

The second historical contribution is John D. Norton's "Philosophy in Einstein's Science" (Chapter 4). Norton shows how Einstein, who read, wrote, and appreciated philosophy but was not a dogmatic philosopher, nevertheless used philosophy pragmatically to 1) legitimate an extraordinary new physical proposal concerning time in special relativity, 2) find what he called an "epistemological defect" in earlier theories which, in turn, motivated him to seek his general theory of relativity, 3) ground his theorizing in principles that distinguish the real from the unreal, 4) come to adopt a form of mathematical Platonism as the way to find new theories, such as his unified field theory, and 5) portray himself, correctly and unapologetically, as an "unscrupulous opportunist" to the systematic epistemologist by combining realism, idealism, and positivism in order to advance his theorizing.

One of the central problems Einstein faced in developing the special theory of relativity was how to deal with the apparent irreconcilability of the principle of relativity, which asserts the equivalence of all inertial motion, and the light postulate, which assigns a unique speed to light propagating *in vacuo*. For years, Einstein had tried to modify Maxwell's theory of electrodynamics, which held that light always travels at one speed, c=186,000 miles per second *in vacuo* in relation to Maxwell's ether, such that the speed of light would vary according to the speed of the emitter. After many fruitless attempts, Einstein realized that he could find no sustainable emission theory of light and that Maxwell's theory and the constancy of the speed of light must stand. So how could Einstein keep both the principle of relativity and the constancy of the speed of light? To solve his problem, he had to come to the realization that the absolute character of time, or of simultaneity, was an assumption that could be challenged. To give him the courage to mount such a challenge, he turned to the writings of David Hume and Ernst Mach. While we do not know which writings of Hume and Mach were at issue or how, precisely, they were of service, we do know that, with their aid, Einstein came to the realization that the concepts of physical theories, such as the absoluteness of time, must be properly grounded in experiences or else they are fictions:

---

30 Hume, *An Enquiry Concerning Human Understanding* (1748); ed. Tom L. Beauchamp (Oxford: Clarendon Press, 2000), 123.

> After seven years of reflection (1898–1905), the solution came to me suddenly with the thought that our concepts and laws of space and time can only claim validity insofar as they stand in a clear relation to experiences ... By a revision of the concept of simultaneity into a more malleable form, I thus arrived at the special theory of relativity.[31]

Over the ensuing decade (1905–1915), Einstein sought a new theory which would extend the principle of relativity to all motion, including accelerated motion. He believed he had achieved this in 1915 with the completion of his general theory of relativity. The motivations Einstein reported for seeking this extension of the principle of relativity are fascinating and rarely found stated explicitly in the physics literature. Einstein credited Ernst Mach for pointing out to him what Einstein called "*ein erkenntnistheoreticsher Mangel*" (an epistemological defect) in classical mechanics and, indeed, in the special theory of relativity. According to this defect, both classical physics and special relativity posit certain preferred inertial motions. These were the uniform straight-line motions followed by free bodies, unaffected by perturbing forces, which, in turn, define inertial spaces of reference—loosely speaking, the spaces carried with each set of bodies moving together inertially. So-called "inertial forces" arise if a body is constrained to accelerate or deviate from inertial forces. Newton imagined water swirling in a bucket and the resulting acceleration led the water to be hurled outward and climb up the wall of the bucket, producing a concave water surface. In the same manner, fluid spheres in rotation, such as stars and planets, bulge at their equators. What causes this bulge? Einstein categorically rejected the answer that the cause is rotation with respect to inertial spaces because these spaces were not "an observable fact of experience." Instead, he turned to the source that he eventually did accept: distant masses and their motions. In the final theory, nothing intrinsic to a space distinguishes one space from another. The discrimination of spaces into inertial and accelerating comes only by virtue of the masses distributed in them. If the masses of the universe are at rest in a space, it is an inertial space. If those masses swirl around, it is a space with inertial forces that pull water up the sides of Newton's bucket and lead fluid bodies to bulge at their equators.[32]

31 *The Collected Papers of Albert Einstein*, trans. John Stachel et al., *Volume 2, The Swiss Years: 1900–1909* (Princeton: Princeton University Press, 1989), 264; cf. Norton, "Philosophy in Einstein's Science," in *Idealist Alternatives to Materialist Philosophies of Science*, Chapter 4, 99.

32 Norton, *ibid.*, 100–105.

A perennial theme in philosophy is the separation of reality and appearance. Contemporary physics is replete with techniques which try to effect this distinction. They are associated with the notions of invariance, symmetry, and gauge transformations, whose lineage in physics goes back to Einstein's work. In Chapter 4, Norton focuses on two narrow episodes in the narrative to separate reality and appearance: the point-coincidence argument and the incompleteness of quantum theory. The point-coincidence argument was developed after Einstein had initially accepted, then rejected, only finally to accept general covariance. Spacetime theories usually label events with three spatial coordinates and one temporal coordinate. However, we can make new numerical labels for any event by adding, subtracting, or taking any combination of the more traditional choices of the event's coordinates and any rescaling of them. These manipulations create arbitrarily many more spacetime coordinate systems. If we have a physical theory that can employ any of these coordinate systems, no matter how jumbled and rescaled, then the theory is generally covariant. The central conception of Einstein's general theory of relativity is a connection between gravitation and the curvature of the space geometry. Einstein's decision to seek a generally covariant theory was pivotal since it enabled him to draw on an elaborate body of mathematical techniques emerging from the 19th century for understanding curvature.

It therefore came as a surprise to modern readers to learn that Einstein rejected general covariance in 1913. He and his mathematician friend, Marcel Grossmann, announced that they had been unable to find physically admissible generally covariant gravitational field equations. In place of these equations, they published gravitational field equations of limited covariance. Three years later, however, Einstein gave up limited covariance and returned to general covariance. What convinced him of the need for this return was the point coincidence argument, according to which the factual content of a physical theory is exhausted by the catalog of spacetime coincidences which it licenses. It thus followed that any differences between the spreadings of the fields were purely artifactual since they did not correspond to any factual differences.[33]

A better-known criterion of reality which is associated with Einstein is his critique of quantum theory. In the 1920's, physics had been drawn into the emergence of the so-called "new quantum theory" according to which, in Schrödinger's famous formulation, each particle of the new theory was associated with a wave. Einstein agreed with his colleagues that the new quantum theory was a worthy achievement which resolved a number of difficulties of

33 *Ibid.*, 105–115.

the "old quantum theory". However, Einstein resisted one aspect of the new theory resolutely: does the quantum wave associated with a particle provide a complete description of the particle or are there additional facts about each particle which are not expressed in the wave? Most physicists adopted the first view but Einstein urged the second view: the new theory was incomplete. It could not measure the position or momentum of a particle without its wave collapsing to a new state with a definite position or momentum. The old state was destroyed and the possibility that the definite measurement outcome was created by the measurement itself cannot be precluded. As a result, direct measurement does not necessarily reveal the properties of particles possessed prior to measurement. This insight gave rise to a different criterion of reality according to which, "If without in any way disturbing the system, we can predict with certainty (i.e., with probability equal to unity) the value of a physical quantity, then there exists an element of physical reality corresponding to this physical quantity."[34] This is the famous criterion of reality from the so-called "EPR" paper (after the surnames of its co-authors) of 1935.

Most of the metaphysics we have seen thusfar in Einstein's work has empiricist, positivist, or verificationist underpinnings. We might conclude that Einstein's commitments were to empiricist or verificationist principles. However strong these commitments may have been, they were subordinate to a deeper commitment—a practical loyalty to whatever metaphysic best led to new and improved theories. This loyalty even extended to mathematical Platonism, a position which Einstein endorsed in his Herbert Spencer lecture, "On the Methods of Theoretical Physics" (1933), as part of his search for a unified field theory:

> Our experience hitherto justifies us in believing that nature is the realization of the simplest conceivable mathematical ideas. I am convinced that we can discover by means of purely mathematical constructions the concepts and the laws connecting them with each other ... In a certain sense, therefore, I hold it true that pure thought can grasp reality, as the ancients dreamed.[35]

34 Albert Einstein, Boris Podolsky, and Nathan Rosen, "Can Quantum-Mechanical Description of Physical Reality Be Considered Complete?," *Physical Review*, 47 (1935), 777–780; 777; cf. Norton, "Philosophy in Einstein's Science," 118.

35 Albert Einstein, "On the Methods of Theoretical Physics" (1933); in *Ideas and Opinions* (New York: Bonanza, 1954), 270–276; cf. Norton, "Philosophy in Einstein's Science," 122–123.

The lesson to be learned from the previous examples is that Einstein was correct in describing himself as an "unscrupulous opportunist" in selecting whatever philosophy would advance his theorizing. While Einstein was certainly interested in philosophy, reading it, writing it, and discussing it with some of the leading philosophers of his time, he was not a dogmatic philosopher. Rather, he used philosophy pragmatically for other purposes. Einstein summarized his pragmatic approach to philosophy best toward the end of his life in his celebrated observation that scientists are realists insofar as they seek to describe a world that is independent of the acts of perception, idealists insofar as they look upon their concepts and theories as free inventions (not logically derivable from what is empirically given) of the human spirit, positivists insofar as they consider their concepts and theories justified only to the extent to which they furnish a logical representation of relations among sensory experiences, and may even appear as Platonists or Pythagoreans insofar as they consider the viewpoint of logical simplicity as an indispensable and effective tool of their research.[36] One is not likely to find a more candid statement of the metaphysics of science, as articulated by one of its most eminent practitioners, than that.

The third historical contribution (Chapter 6) is "Charles de Koninck, John Leslie, and The Conceptual Parameters of Science" by Elizabeth Trott. Charles De Koninck (1906–1965), a transplanted Belgium philosopher who made his career at Université Laval from 1934 onwards, was very much concerned to rethink the world in terms of the work of Einstein and other major scientists of the time. John Leslie (1940-), who emigrated from England and spent the bulk of his career at the University of Guelph, developed a lifetime interest in the extent to which principles of rationality delimit the conceptual parameters of science. His reflections on the anthropic principle and on immortality stand as conceptual limits to the truth claims of science.

De Koninck's ambition was to unite the truth claims of science with the actual affairs of human experience. In his most popular book, *The Hollow Universe* (1960), De Koninck argues that unification could only be achieved through a discussion of the values which are essential to our understanding of the complexities of the universe. If we generalize the new scientific outlook and accept it as the one true way of thinking about nature, i.e., whatever scientists *qua* scientists study, without recognizing the limits of scientific

36 Albert Einstein, "Remarks Concerning the Essays Brought Together in this Co-operative Venture," in P.A. Schilpp, ed., *Albert Einstein: Philosopher-Scientist, Library of Living Philosophers,* Vol. 7 (La Salle, Ill.: Open Court, 1970), 665–668; cf. Norton, "Philosophy in Einstein's Science," 123–124.

abstractions in explaining it, then the only wonder left is the wonder of the hollowness of the universe, both in nature and in thought. By "hollowness" De Koninck means doing scientific research without the concept of the good, i.e., what all things desire insofar as they desire their perfection. Without this concept of the good, the justification of doing scientific research would spell out as follows: scientists do research in order to understand the mechanisms of nature/the universe so that they can continue to do research to understand the mechanisms of nature/the universe. Without the concept of the good, the project of scientific research is circular and the scientific community the most self-centered profession imaginable.

De Koninck did not live to witness sociobiology attain the status of a mature science, as represented by E.O. Wilson's best-selling book, *Sociobiology: The New Synthesis* (1975), but he thought that the trends in the biology of his time were certainly headed towards the hollow universe. In particular, the application of evolutionary theory to social development, without the concepts of purpose, intent, need, choice, and decision that we employ in ordinary experience to account for the activities of biological entities, was a huge leap in that direction. For De Koninck, an organism with no purpose is no organism at all; it is not a body equipped with resources making possible the fulfilment of needs but simply a mass of functionless appendages. Attempts to apply evolutionary theory to non-human, and especially human, systems in a purely mechanistic way, accordingly, were certain to produce a skewed understanding of the biosphere.

So how can the truth claims of scientific research be unified with the actual affairs of human experience? According to De Koninck, the information that biology makes available to us must be integrated into a holistic theory which is related to other sciences and the humanities as structured by minds and their purposes in the world. This integration would view the universe/nature as a great work of art, inexhaustible in scope, where science provides knowledge about the art that fashions natures and the members they represent, but requires human experience to provide knowledge of the purposes, intents, needs, etc. they have and, where applicable, the choices and decisions they make. Ultimately, this project transcends human capacity and is sustained by God who provides unity to the world of objects science studies and which living things, to some extent, experience.

Like De Koninck, John Leslie seeks to establish a unified metaphysics of experience. Rather than dividing reality up into a world of human experience, a world of scientific experience, and a world of divine causal intervention, Leslie tries to show that they are basically the same world. There are many dimensions to Leslie's overall project but the basis of it lies in his neo-Platonic metaphysics and the ethical requiredness of the world in the inclusive sense just indicted. Indeed, neo-Platonism and the ethical requiredness of the world are intimately connected because Leslie acknowledges the Platonic world of forms as a necessary and eternal realm of abstract truths, facts, or propositions. The necessity of these propositions is established more by rational possibilities than by logic. Many propositions are meaningful in that they do not promote logical contradictions. Others stand as meaningful beyond their logical coherence. In order to preclude their inclusion in any rational position set forth, we would need to adduce arguments that the probability of their being meaningful or having reference was so low that we could dismiss them. Leslie identifies Platonic truths as highly probable truths. They could not be discredited even if there were nothing in existence by which existence they could be established.

These hypothetical conditionals that are logically linked provide a unifying structure for the world. Leslie extends these eternal, irrefutable hypothetical conditionals to include ethical truths. For example, if there were a world that was full of misery, then such a world would be undesirable and the ethical requirement that it not exist would be a necessity in the same sense that 3x4=12 is a necessity. Leslie then wants to argue that, given that there can be ethical necessities of negative value, there can also be ethical necessities of positive or intrinsic value, one being that the world exists.

Leslie discusses the possibility of a world existing out of necessity in his book, *Immortality Defended* (2007), though as Trott notes, his consideration of the idea goes back to his first published book, *Value and Existence* (1979). Ethical requiredness means that if there be any good in any sense of the word, then it ought to exist. If there is reason (requirement) for a world of misery not to exist and if the claim that nothing ought to exist is vacuous, since the proposition is contradictory (in the same way that round squares are contradictory), and hence does not qualify as an abstract truth, then there is reason (a requirement) for a good world to exist, i.e., a world which includes and sustains goodness but does not preclude evil, in the same way that there is a reason (requirement) for an evil world not to exist.

If the existence of a good world as thus defined is ethically required, then the world of science is not a "hollow universe" in De Koninck's sense but a world permeated by value. So what about the scientists who study the world? What about human experience and divine intervention, indeed? Where do they figure in Leslie's inclusive understanding of "the world?" Actually, the ethical requiredness of the world itself requires the existence of mind, not so that a good world can *exist* but so that that it can be *understood* as a world, the goodness of which is ethically required. One reason (requirement) for the scientists who study the world, as well as human experience and divine intervention, is that patterns, which enable prediction and hence further discovery in science, are hard to deny. Patterns and their recognition, however, are peculiar to the minds of persons. The existence of patterns, then, requires the existence of beings who can recognize them. From the existence of beings who recognize patterns, Leslie infers the existence of a cosmic or divine mind whose thoughts give existence to these patterns. The divine mind makes comprehensible to our minds abstract and eternal Platonic truths, including the truth that the existence of a good world is undeniably more rationally convincing than the existence of an evil world. Thus, the worlds of scientific experience, human experience, and causal divine intervention are the same world and the "hollow universe" of De Koninck has been filled out into a universe where the terms are connected by its own ethical requirements.

The fourth historical contribution (Chapter 7) is "Idealism and Naturalism: A Really Old Story Retold with Variations" by Fred Wilson. The author has written a *tour de force,* narrating a story from Anaxagoras to Brand Blanshard, and should be required reading for anyone who wants to understand the history, logic, and ontology of the debate between idealism and naturalism in the philosophy of science. I will not attempt to summarize it here but will rather confine myself to an observation which goes to the heart of Wilson's argument and, indeed, the subject of idealist alternatives to materialist/naturalist philosophies of science.

According to Wilson, naturalism cannot accept Socratic forms and necessary connections and the legacy they have bequeathed to later idealisms—the soul (Plotinus), consciousness (Green), the Absolute (Bradley), the claims of Marxism notwithstanding, the structural idealism of Marxism (as evidenced in the thought of Althusser), objective necessary connections (Blanshard), etc.—into its ontology because they all violate the Principle of Acquaintance. According to this Principle, one may justifiably introduce something into one's ontology only if one is acquainted with it in sensible experience or inner awareness, i.e., one's passions and feelings along with reflecting on what one is acquainted with through sensible experience. The problem with the Principle of Acquaintance is that it can be interpreted either extensionally or intensionally. Interpreted extensionally, i.e., in terms of the *objects* of acquaintance, it precludes the forms and necessary connections, together with the varied legacy to which these have given rise in later idealisms, because we are not acquainted with any of these in our sensible experience or inner awareness. We do not perceive such things through our senses, in which case we cannot reflect on them by virtue of being acquainted with them through our sensible experience, nor do they figure among our passions and feelings.

Interpreted intensionally, i.e., in terms of our *ideas* of the objects of acquaintance, however, there is no reason to preclude them at all. To deny them extensional existence, as the naturalist does, is one thing, but in order to do so, the naturalist must have ideas of everything (forms, necessary connections, the soul, consciousness, etc.) the existence of which is being extensionally denied. Indeed, without such ideas, the naturalist could not justifiably implicate various forms of idealism for violating the Principle of Acquaintance. This does not mean that the forms of idealism Wilson examines are extensionally true. It does mean, however, that an extensional interpretation of the Principle of Acquaintance cannot be used as the ultimate criterion to show that they are not extensionally true, given that this Principle requires an intensional interpretation to generate such a result.

This problem came up at the beginning of my Introduction where I noted that the distinction between the extensional and the intensional has caused enormous difficulty in the history of thought, including the history of science, and was in large part responsible for the differing world-views of materialism—or more pervasively and yet more precisely, naturalism—and philosophical idealism. A good way to resolve the problem, it seems to me, is to realize that meaningful language is never purely extensional but always intensional. It does not have to be *about* anything we are acquainted with in sensible experience or inner awareness (e.g. the truths of mathematics) but it must have intensional significance, otherwise it could not be *meaningful*. The claims of historical forms of idealism are certainly meaningful, as Wilson's analysis amply indicates, although they may not be extensionally true. The challenge for naturalism is to prove that they are not extensionally true without resorting to language which is intensional. Given what I have said about the nature of language, this goal would seem to be unattainable.

The last historical contribution is Chapter 8, "Bradleyan Idealism and Philosophical Materialism" by K.M. Ziebart. What is the relation of Bradley's metaphysics to philosophical materialism and to what extent is Bradley's rejection of philosophical materialism idealist? Zeibart argues that the historical polarization between materialism and idealism is not really helpful here since Bradley's metaphysics is one in which the ideal and the material coexist and even overlap. One concrete way in which this coexistence is manifested is Bradley's view of the development of the function of judgment. According to Bradley, material and ideal processes are fundamentally embodied, natural phenomena. Judgment may be what distinguishes human being from that of other creatures but it is an operation which has evolved out of other physical processes and therefore should not be regarded as a clear or essential dividing line between man and other animals. Judgment is a "late acquisition" of human beings, indicating that the possession of mind is really a matter of degree. Bradley affirms the general association of the distinction requisite of judgment with the acquisition of language but stresses that this too is a matter of degree, given the inherent difficulty of saying with certainty where language begins. Even among other animals, there is an employment to some degree of universals without which, he maintains, it is incomprehensible that an animal burnt one day at the kitchen fire will the next day shrink from a lighted match.[37] If this is so, judgment must come before language and cannot be distinctively human. Mind is thus to be regarded, not as a distinctively human faculty, but rather as something continuous with overall physical evolution.

Bradley's analysis of mental development is indicative of his overall metaphysics. Not only is it very empirical in approach but it regards ideal operations as continuous with the physical without reducing them to it. This careful balancing of the ideal and the material highlights the realism of Bradley's metaphysics as well. It is both idealist and materialist since there is nothing that falls outside reality in its idealist and materialist instantiations. Furthermore, this careful balancing of the ideal and the material as jointly constituting reality without according priority to either one highlights the realism of Bradley' metaphysics. Accordingly, Bradley's metaphysics might best be described as empirical in methodology, idealist in development, materialist in embodiment, and realist in totality.

Zeibart does not consider whether Bradley's metaphysics, thus construed, can withstand philosophical analysis or whether, assuming a positive verdict, it could provide an alternative to both idealist and materialist philosophies of science. Nevertheless, this is an exciting way to end for it heralds a possibility that may transcend idealist alternatives to materialist philosophies of science, and critical responses thereunto, which occupy most of the present volume.

References

Bradley, F.H., *The Principles of Logic*, 2nd ed., 2 vols., Oxford: Oxford University Press, 1922.

Descartes, René, *Meditations on First Philosophy* (1641); ed. Stanley Tweyman, Ann Arbor: Caravan Books, 2002.

Einstein, Albert, "On the Methods of Theoretical Physics" (1933); in *Ideas and Opinions*, New York: Bonanza, 1954, 270–276.

37 F.H. Bradley, *The Principles of Logic*, 2nd ed., 2 vols. (Oxford: Oxford University Press, 1922), 1:37; cf. Zeibart, "Bradleyan Idealism and Philosophical Materialism," in *Idealist Alternatives to Materialist Philosophies of Science*, Chapter 8, 305–306.

Einstein, Albert, "Remarks Concerning the Essays Brought Together in this Co-operative Venture," in P.A. Schilpp, ed., *Albert Einstein: Philosopher-Scientist, Library of Living Philosophers*, Vol. 7, La Salle, Ill.: Open Court, 1970, 665–668.

Einstein, Albert, *The Collected Papers of Albert Einstein*, trans. John Stachel et al., *Volume 2, The Swiss Years: 1900–1909*, Princeton: Princeton University Press, 1989.

Einstein, Albert, Boris Podolsky, and Nathan Rosen, "Can Quantum-Mechanical Description of Physical Reality Be Considered Complete?," *Physical Review*, 47 (1935), 777–780.

Hume, David, *An Essay Concerning Human Understanding* (1748); ed. Tom L. Beauchamp, Oxford: Clarendon Press, 2000.

Hume, David, *A Treatise of Human Nature* (1739–1740); ed. David Fate Norton and Mary Norton, 2 vols., Oxford: Clarendon Press, 2007.

Hume, David, *Dialogues Concerning Natural Religion* (1779); ed. Stanley Tweyman, Ann Arbour: Caravan Books, 2000.

Otto, Rudolph, *Das Heilige* (1971); *The Idea of the Holy*, trans. John W. Harvey, New York: Oxford University Press, 1950.

Ryle, Gilbert, *The Concept of Mind*, London: Hutchinson, 1996.

Wilson, Edward O. , *Consilience: The Unity of Knowledge*, New York: Vantage Press, 1998.

CHAPTER 1

The Things that Fill the World

Leslie Armour

A good guess is that much of what fills the world consists of objects described by physicists. I am deliberately, for the moment, avoiding ontological terms like "existence," "being," and "reality," and even Quine's preferred expression, "what there is," though I shall come to those in due course. "The world" is an expression which I take to denote the scope of one's focus, the things one wants to talk about and feels that one needs to talk about. The titles of magazines on the newsstands run to expressions like *Business World, Art World*, and, as Gilbert Ryle reminded us, *Poultry World*.[1] These expressions mark out overlapping areas of focus though perhaps those who find their home in *Poultry World* may not always read *Art World*. What I mean by "the world" is the most general field within which each of us finds a focus. Just as in *Poultry World* one can find glimpses of animal genetics, animal health, human health, markets, architecture, and technology—about everything short of stars and galaxies—from the point of view of those who raise chickens and like birds, so in each of our worlds one can find almost everything though, for most of us, many things have only a shadowy background existence.

Dewi Phillips argues that no reason can be given for the existence of *the* world because it is unique.[2] If we understand "world" in the recently popular sense of the "possible worlds" dispute, he may be right. But that is because "all possible worlds" denotes whatever can be spoken of without contradictions. Presumably that set of possibilities—if, indeed, it forms a set—cannot be summed or totalised. Such an assembly does not have an "explanation" in the sense that its "existence" (if that is a sensible notion) simply follows from the rules of logic and elements available to insert in a universe of discourse. Whether or not such a concept is even intelligible depends on what we think about logic and on what we think about what goes into universes of discourses.

1 Gilbert Ryle, *Dilemmas*, Cambridge: The University Press, 1954, 73. "We know that a lot of people are interested in poultry and would not be surprised to find in existence a periodical called '*The Poultry World*' ... In fact, there is such a journal and has been since 1907, not much read, perhaps, in Oxford, but there has always been a copy in the Bodleian. *Art World* and *Business World* also exist and similar remarks might be made about them."

2 See D.Z. Phillips, *Recovering Religious Concepts*, Basingstoke: Macmillan, 2000, 48–50.

 | DOI:10.1163/9789004415270_003

This may sound unlikely, but the expression "possible worlds" has had an active life in the past half century.[3]

1 The Idea of the World

By contrast, if we can understand the idea of "a world" in the way that I have suggested, we might make some progress. The world is not the universe, of which, by definition, there is only one. When Ralph Cudworth wrote The *True Intellectual System of the Universe,*[4] he meant to include everything that God could think about. Cosmologists have generally meant to sum up all the things that physicists ought to think about—stars, galaxies, gases, and odd bits of matter which structure the space-time continuum within which we identify them. The ideas people express are not among the things which populate the world they describe, nor, indeed, in some sense, are many of the activities of people. Many now doubt, however, that even all the things cosmologists study can be put in a single package. Such people talk about the "multiverse" or even "multiverses."

When we talk about "*the* world" as opposed to the poultry world or the world to which the editors of *Nature* and *The New Scientist* attend, we mean what we can bring into focus and what we think might be brought into focus if our various ways of gathering knowledge were really adequate and the results could be rendered as an intelligible whole. Trying to get an idea of *the world* is a balancing act because it means getting a picture of what various sciences have information about, together with what we make of, history, art, literature, and our own public and inner lives. But the parts of such a picture not painted by physicists have many "explanations" drawn from culture, biology, social psychology and, no doubt, faith. All such pictures may be good, bad, or indifferent and "the world" we actually speak of may be well or ill defined.

Still, most people just now are inclined to think that chickens, works of art, and the objects of the business world like coins and account books are all in some sense related to the sorts of entities which one finds mentioned in the physics journals. Electrons, protons, neutrons, and quarks figure in such mentions. Perhaps they include or, at bottom, *are* the somewhat mysterious strings which figure in arguments among theoretical physicists. There is currently no fixed list of such "physical objects." Some things now talked about

3 See David K. Lewis, *On the Plurality of Worlds,* Oxford: B. Blackwell, 1986, and Rod Girle, *Possible Worlds,* Chesham: Acumen, 2001.

4 London: R. Royston, 1678.

may disappear from the pages of *Nature* in times to come. Mysteries about dark matter and dark energy remain.

It has sometimes been thought that when we talk of knowledge there is an important line to be drawn between the things that figure in direct experience and "theoretical entities" which we know only indirectly but which serve in our explanation. But this belief is probably the residue of an inheritance which stems from a very old tradition, one which was articulated (doubtless rearticulated) by Francis Bacon in Stuart times and by Denis Diderot in the midst of a great rethinking which occupied the end of the 18th century.[5] On this view, we have three sources of knowledge: memory, experience, and imagination. From them we get history, science, and poetry. But one taking such a view might argue that theoretical entities are really creations of the imagination and so the work of poets. Indeed even hard-headed physicists like Paul Dirac insisted, sometimes very seriously, that beauty entered into theorising about mathematical physics, especially when the line between the mathematics and the physics became blurred.[6] We also know that the "experienced" world does not consist of simple atomic data but of complex structures in which our concepts and predispositions play a part.

The attempt to draw lines by reference to origins leads to a muddle. Indeed we do have memory, direct experience, and imagination. But the truth is that all three lines of descent must enter into any serious claim to "knowledge" if to know something is both to grasp it and to understand it.

2 Objects of Encounter and Objects of Inference

To start with, the line between objects that we think we encounter directly—say people, horses, houses, and stars, and things we know about by inference or "theoretical entities"—is not clear. Theoretical entities are revealed in exotic circumstances, perhaps by vapour trails in cloud chambers, or as rapidly changing numbers in machines which count the arrival of particles. In which group does one place the astronomical objects which are revealed only in computerised scans of photographic plates? And the automobile in your driveway is hardly an object of simple apprehension. It takes a lot of concepts and activity to see that what is in the driveway *is* a car.

5 See Nicholas Lash's intriguing essay, "Reason, Fools, and Rameau's Nephew," in his book, *Theology for Pilgrims,* London: Darton, Longman and Todd, 2008, 123–136.

6 Dirac's biographer, Graham Farmelo, finds this notion runs very deep in Dirac's thinking. See *The Strangest Man,* London: Faber and Faber, 2009.

The truth is that you do see an automobile in your driveway but, as Ryle would have it, "seeing" is an "achievement" verb.[7] There are visual data which you can, if you apply a complex theory to more difficult observations, trace back through a path from your brain to your optic nerve, through your cornea, and on to the reflection and refraction of light waves in interaction with something "out there." But all this is an imaginative construction. It requires not only mention of light waves and optic nerves but also the concepts which you formed in the past and which help you to group things in intelligible bundles. You create this scenario and to an important extent you create the experience of the motor car itself. So memory, immediate experience, and imagination all enter in. Sometimes we want to sort them out, especially if something goes very badly wrong and you can no longer recognise your car or perhaps any car as a car.

The things that physicists talk about, therefore, are, like everything else in the worlds we assemble, far from being "simply there," but still it is a good guess that they have a strong claim to be taken to be among the things that fill the world even if they do not do so in any way which is quite immediately obvious. The claim that they are not "simply there" does not, of course, by itself make a case for idealism. True, such objects appear in a constructed "world" and they themselves require a good deal of "making" on our part, but we need to know what sorts of objects they are, just what goes into their construction, and what we should say about their place in any world when we are not attending to them.

It is surely a good guess that "physical objects" have a place when we put all our worlds together and speak of what is likely in "the" world. It is a good guess partly because they form systems which make wide ranges of our experiences intelligible in the sense that we can say "this is just what we expected to happen" given that we are dealing with precisely defined entities and physical laws which predict their behaviour. We can, for instance, know the exact moment when there will be a full eclipse of the moon, and exactly how much of what sort of enriched uranium can be divided into masses, each less than a critical mass, but together greater than the critical mass it takes to make an atomic bomb, and just how to keep it safe and how to set it off, though this is something one may well think it would be better not to know. The calculations which are necessary have been checked many times by many people in many places and not found wanting. It is only a guess because there are still questions about the underlying nature of human experience and about the physics

7 *The Concept of Mind,* London: Hutchinson, 1966, 150–152, 222.

which underlies such predictions. Some very fundamental levels of physics involve probabilities, but we know quite a lot of about those, too. Given the facts and the theories, we understand what is going on.

We must always remember, however, that the concept of knowledge itself is in principle open. Attempts to fence it in by calling it "justified true belief" produced only paradoxes.[8] This is not an accident or the fault of the slogan, though, to begin with, the slogan failed to take account of understanding. Just being aware of the facts and being able to predict is not enough. The best snooker players can predict just about as accurately as the physicists how the balls will end up on the table. But knowing how to pot a snooker ball is not the same as knowing how the world works.[9]

The peculiarities of the cases in which we have good grounds for believing something which turns out to be true, but on grounds other than those that we are aware of, are well known. But expressions like "justified" and "true" as well as "knowledge" set the perimeters of what we are prepared to accept. They are all always open to modification as we go along, and so notions like "knowledge in physics" or "mathematical knowledge" can be given meanings, but the list of things which we put in them can always be supplemented. We ask, "Is there 'moral knowledge' or 'theological knowledge'." It may seem so, but then we meet more concepts that are in principle open and subject to philosophical development as the material they organise comes to light.

The issues this poses will become clearer once we have looked more closely at the nature of physical objects. The things that figure as objects in the physicist's world quickly turn out to be very abstract and rather peculiar, but very instructive. A good place to start is with Willard Quine whose predilection has always been to associate knowledge with logic, mathematics, and their applications to what is usually termed "the physical world."[10] There is little doubt that his natural instinct was always to equate, as nearly as possible, the list of things that fill the world that we ought to talk about in the most serious tones with the list of things physicists talk about. Yet Quine says "... our physical objects have themselves grown so tenuous that we find ourselves turning to space-time regions for something to cling to."[11]

8 See Edmund Gettier, "Is Justified True Belief Knowledge?," *Analysis* 23 (1963), 121–123.

9 The miraculous control of the balls exhibited by John Higgins or Ronnie O'Sullivan is a skill, like riding a bicycle, which does not need a degree in physics and is much harder to obtain.

10 For a general account see Peter Hylton. *Quine*, London, Routledge, 2007, 301–317.

11 "Whither Physical Objects," in *Essays in Memory of Imre Lakatos,* eds. R. Cohen, P. Feyerabend, and M. Wartofsky, Dordrecht: Reidel, 1976, 497–504. The quotation is from 499.

3 21st Century Pythagoreanism?

What we know about physical objects are mathematical formulae for predicting successive occupations of space-time regions. But Quine thinks that, given some coordinate systems, the space-time regions themselves can be considered as sets or ordered quadruples of real numbers.[12] So are we to take it that what fills the world Quine talks about is a collection of numbers or, at any rate, some entities which do duty for numbers, and, if so, what should we say about them? We can think of numbers as "abstract entities" to no avail. It simply forces us to say from what they are abstracted. Perhaps we could call them intangible entities but this would have some problems too, as would saying that they are things somehow known directly to the mind, a suggestion that no doubt moved Sir Arthur Eddington to a mentalistic idealism in a way that others found doubtful. Perhaps it is best just to call them logico-mathematical entities and wait to see what we want to say about them.

Supposing that we can make something of this term, however, what sorts of logico-mathematical entities ought we to entertain? Bertrand Russell at one time supposed that we needed real propositions.[13] Propositions are what is expressed by sentences, so that the same proposition might be expressed in French or English or Swahili. Later he thought that we could make do with mathematical functions, a form of mathematical expression which could generate other needed expressions. Quine supposed that we could make do with sets, eliminating traditional universals, numbers, and most other mathematical entities. He opposed modal logic and he was not tempted to accept real possibilities. For the most part, his objection to entities such as the possibilities, which he exiled to non-being, was that they could not easily be identified. How many possible fat men will fit into a doorway and how can we tell if there is one or more than one?[14] Peter Hylton has called this Quine's "hyper-Pythagoreanism" though one must hasten to say that Quine himself was not happy with this theory and did not ever think it quite sensible to espouse it. We will come to that shortly.[15]

12 Peter Hylton, *Quine*, 306.

13 Peter Hylton, *Propositions, Functions, and Analysis: Selected Essays on Russell's Philosophy*, Oxford: Clarendon Press; New York: Oxford University Press, 2005; 9–30 explore Russell's development.

14 *From a Logical Point of View*, Cambridge, MA: Harvard University Press, 1953, 4. Quine asks if there is a possible fat man and a possible bald man in the doorway, whether they are the same man and if there are more possible thin men than fat ones. He wants to know *how we decide*. I will approach this question as the chapter proceeds.

15 Peter Hylton, *Quine*, 306–308.

Quine, nonetheless, did examine the Pythagorean possibility, noting that real numbers could be reconstrued as sets. But this leads to Russell's paradox about the set or class of all sets or classes that are not members of themselves. We can, I think, dissolve Russell's paradox,[16] but a problem posed by Georg Cantor remains. The Cantorian absolute is so large that it cannot actually be considered as a set.[17]

16 Russell's paradox suggests that elements considered as particulars cannot be totalled. Such a totality involves a class of all classes. Some of these classes are members of themselves—the class of all unities is itself a unity, for instance, but most are not. The class of all yellow things is not a yellow thing, and the class of all uncles is not an uncle. There will therefore have to be a class of all classes which are not members of themselves. If this class is not a member of itself then it is a member of itself. But if it is a member of itself then it is not a member of itself.

One might expect that not all aggregates can be totalled. Indeed, it seems obvious that there can be no totality of the things in the world. Kant's antinomies show that experiences cannot be aggregated, and we have seen that preferences and pleasures cannot be aggregated either.

There is an answer to the Russell problem which I have suggested in *Being and Idea. Developments of Some Themes in Spinoza and Hegel*, Hildesheim: Georg Olms, 1992, 78–82. Part-whole relations are always internal in the sense that the relations make a difference to their terms. But part-part relations are always external. If this particular yellow shirt were not yellow, the class of yellow things would be different. But if this shirt were green, all the other shirts might still be yellow. Thus there can be a totality which is internally related to all the parts, insofar as these can all be regarded as expressions of its nature. But the parts do not themselves form a coherent addable class (the class of all classes which are not members of themselves) because they are not internally related to one another. Nothing about any one of them taken by itself implies any relation (other than the negative one of difference) to any other.

17 Shaughan Lavine in *Understanding the Infinite* (Cambridge, MA: Harvard University Press, 1994), 51–57, offers a detailed analysis of Cantor's position. He says "… Cantor's argument that 'the absolutely infinite sequence of numbers' is a suitable symbol of the Absolute can be turned into a mathematically precise proof that the class of all ordinal numbers is not a set. Cantor gave such a proof in a letter to Dedekind dated 3 August, 1899" (See I. Gratton-Guinness, "The Rediscovery of the Cantor-Dedekind Correspondence," *Jahresbericht der Deutschen Mathematiker-Vereinigung,* Vol. 76, 1974/75, 104–139). Since he could now obtain a contradiction from the assumption that the system of all numbers is a set, he began to call it "an *inconsistent* absolutely infinite multiplicity."

Basically, the idea embedded in the proof amounts to this: every set is a collection of distinct members. Therefore every set has a final ordinal number—a number which places its members in an order. If a set has nine members, then its members can be ordered so that there are subsets which contain all the numbers up to nine. But an absolutely infinite set cannot have an ordinal number like this. So the idea of such a set is a kind of contradiction. It is clearly not a set in the ordinary sense and cannot be exhausted because its final ordinal number cannot be stated. The rule is that if you take a number, say 3, then any set less than three forms a well-ordered subset of three. In particular there must be a set which is the number representing the set if one does not include it among its own

To repeat, Quine did not really think that this hyper-Pythagoreanism was enough. He points out that the whole concern about mathematical entities develops out of the needs of physics.[18] How shall we construe this claim? Evidently, the concern with abstract entities comes about this way: We cannot simply construe the things that physicists talk about as if they were part of a system which contains a mixed bag of electrons, protons and quarks, and also contains aeroplanes, plant pots, old dinosaur bones, and boxes of human ashes waiting to be scattered.

The supposition is that, somehow, atoms and electrons are more fundamental than plant pots and wheelbarrows. But *if* this so, then there are not "really" any plant pots and wheelbarrows; there just seem to be such. Sir Arthur Eddington liked to alarm his hearers by telling them that the table in front of him on which he had his lecture notes was really mostly composed of empty space. I do not know if this made them feel giddy or if they only laughed. But there was something there that made Susan Stebbing suspicious of what he was saying.[19]

If one pursues this quest for "what is most fundamental" and if one means by "most fundamental" what one has to talk about because statements about it cannot be analysed into statements about anything that seems even more basic, one finds, as Quine suggests, that one is left with mathematical formulae. It is the numbers that count and, indeed physicists do sometimes talk of "getting good numbers." The other stuff consists of interpretation. Wheelbarrows and plant pots emerge from still further interpretation and from images that we construct as our mind confronts what we are presented with. It was in our search for intelligibility that we were driven from our ordinary modes of thinking about what is in our world.

Quine is also keen on contingency. His general view is that, while there seem to be some necessary truths and some contingent ones, we can always regard any statement as either. Is it a contingent fact that all men are mortal?

members. That is, 0, 1, 2 is a well-ordered subset of three. It is also a subset which is smaller than 3 itself. If there are transfinite numbers in Cantor's sense, we can order these so that each has its own number—aleph-null, aleph-one, aleph-two, and so on. The smallest infinite number is the number of the integers, but there are larger infinite numbers which represent the integers and all the numbers between them. But if there are not merely transfinite numbers but absolutely infinite numbers, we cannot count them using such numbers since to do so is to put limits on the numbers, and so we cannot put them in an order. That is to say that there will be some subset, let us call it s-, which will have to be smaller than itself in order to make a well-ordered set. As Cantor says, this is impossible.

18 *Whither Physical Objects?, Essays in Memory of Imre Lakatos*, 502–503.

19 See *Philosophy and the Physicists*, London: Methuen, 1937.

Undertakers bet on it from experience. If, however, we found someone who gave every sign of being immortal—she could convince us that she was around when people started to circulate those horrible stories about Abraham and Isaac, and she shows no sign of ageing—then we might decide that she was not human. She certainly would be importantly different. So then we might take it that it was a necessary truth that being human includes being mortal. Are there any exceptions to the idea that we can shift the necessary and the contingent about as suits our purposes?

Quentin Meillassoux in *After Finitude, an Essay on the Necessity of Contingency,*[20] argues for infinite contingency. He claims it is necessary! He wants to confront the real directly—but it is a mathematical reality: "What is mathematically conceivable is absolutely possible."[21] But then what is it best to say? The claim that everything is contingent except the claim that everything is contingent is another of those paradoxes of self-reference which Russell's ramified theory of types conspicuously failed to dissolve. Meillassoux seems to have missed this. He also missed the problem that there needs to be a certain amount of minimal tidiness in the world for us to talk sense in it. If we cannot talk sense we cannot question the possibility of talking sense. The amount of tidiness in the universe required for us to do this is not inconsiderable. One has to know at the end of a sentence what the beginning meant and the stability needed for this involves quite a lot. Notice for our purposes here that talking sense of any kind requires a certain establishable order which is no doubt the foundation for at least elementary mathematics and it also involves a continuity of mind.

4 Mathematics and Worlds

This leads us to ask what mathematics is. Probably a preponderance of great mathematicians is now and always has been mathematical Platonists. In every generation mathematical Platonism is expressly defended—James Robert Brown in ours, and thinkers like Paul Dirac, Kurt Gödel and G.H. Hardy not long before us.[22] To some of these people, Platonism was so obvious that it

20 Tr. Ray Brassier, London: Continuum, 2008; published as *Après la finitude*, Paris: Éditions du Seuil, 2006.

21 *After Finitude*, 126.

22 James Robert Brown defends mathematical Platonism in *Smoke and Mirrors*, London: Routledge, 1994 (see especially 99–100) and in *Philosophy of Mathematics*, London: Routledge, 1999. Hardy was very vehement in *A Mathematician's Apology*, Cambridge: The University Press, 1940. Most striking is his use of Euclid's proof that there

hardly needed defending. One understands why. You might not like the number two, but there is nothing you can do about it. In any world there will be countable things and more than one of them, and the number two is surely in some sense just there.

But just what mathematical Platonism means is a more difficult question. When mathematics was considered simply a collection of eternal truths there seemed little doubt. The facts expressed truly in mathematical theorems were simply there to be described. But as Peggy Marchi explains, it now seems more plausible to think of mathematics in terms of conjectures and refutations.[23] A mathematical theorem that stands is just one that has not fallen before some new demonstration of inconsistency. But she does talk of progress and of adding to mathematical knowledge, and so there is some direction to the process. She builds on notions of Karl Popper and Imre Lakatos to show how it is possible for this to be true and yet for mathematics to make progress. Essentially, she sees mathematics as a collection of theorems. It is the theorem that is the datum to be explained. We can make progress by collecting theorems and then by analyzing the changes both in the way in which we express our grounds for accepting them and in the grounds themselves. Traditionally, a theorem is different from a conjecture because there is a proof for a theorem. There are not former theorems, but apparent theorems which were mistaken for real theorems. Of course yesterday's appearance might be today's truth as much as vice-versa. But this leaves everything uneasily floating about. Are the real theorems there waiting to be discovered or is something a theorem only in relation to our minds? Peggy Marchi does not really address this question.

---

is no largest prime number. Suppose we have a list of primes, and suppose that there is a largest, Q. Q is not the largest in our list but the largest anywhere in any possible list. In that case, Q=(2,3,5 ... .P)+1. Q is not divisible by any of Q, 2, 3, 5 ... .P "... for it leaves the remainder 1 when divided by any of them." But either Q is itself prime or "it is divisible by *some* prime greater than any of them (which may be Q itself)" (reprinted with a Foreword by C.P. Snow, Cambridge: The University Press, 1967, 93–94). Therefore there is no largest prime. The point is that there is always a prime greater than any we know. This prime exists and is not created by us. Gödel insisted that there was just as much reason to accept the reality of mathematical objects, especially numbers, as to accept the reality of physical objects. One is necessary to explain our perceptions, the other to explain our mathematics. (Kurt Gödel, "Russell's Mathematical Logic," in P. Benacerraf and H. Putnam (eds.), *Philosophy of Mathematics*, Cambridge: The University Press, 1983, 94 (originally published in 1944).

23 "Mathematics as a Critical Enterprise," in R.S. Cohen, P.K. Feyerabend, and M.W. Wartofsky, eds., *Essays in Memory of Imre Lakatos*, Dordrecht: D. Reid, 1976, 379–394.

One must be careful. We do not just make it *all* up; there is something there. It is important, in order not to get carried away by one of the competing theories, to distinguish between concepts and facts, though in reality there may be cases of each in which we see through to the Platonic reality.

The number two is first of all a concept. Concepts are not simply "conceptions" or items in our minds which we put to various uses when we try to think about things. They are logical devices which are elements in our universe of discourse. But they are also used to map out the world of things we want to talk about. The number two is a concept, and "twoness," as Russell preferred to put it, is the "class of all couples." Yet one cannot quite analyze it away into such a class. The number two is also an object which can be talked about, a denizen of the cluster called the integers, and another cluster called prime numbers. It is pretty much, in those contexts, a closed concept. That is to say, we can specify all its properties. I think the concept of prime number is also closed. Once one has said that a prime number is any number divisible only by itself and one, one has said about all there is to say. Some facts about prime numbers, such as that there is no largest one, are surely settled and have been since Euclid's time. But other facts about prime numbers are still in doubt and subject to conjecture. The chance that a proof of any conjecture about prime numbers will be sustained is probably proportionate to its complexity and depends on mathematical techniques. So some of what Peggy Marchi says is a little misleading while some of it is clearly sound.

It is easy to slip into a position like Quine's in which there seems to be no necessity to anything, especially when one sees that "all men are mortal" can easily be taken either as an empirical generalisation or as a necessary truth depending only on whether or not one thinks that if someone were immortal he could not be a man. But this seems to allow altogether too much fiddling with the truth for convenience unless, like Peirce, or Peirce in one mood, one simply thinks that the truth is what the community of scientists is "fated to believe." We no longer have quite that much faith in the community of scientists. It seems more plausible to say that Newtonian physics describes a possible world and, and that, within our descriptions of that world, the crucial concepts are closed. This is especially true if one takes "a world" as I have been doing as a way of designating the focus of our knowledge and interests. "The world" would then be a kind of ideal focus within which all our interests and knowledge could be presented. Mathematical systems, taken this way, are a backbone structure on which we can put various interpretations which are descriptions of one world or another.

5 Real Possibilities

Taken thus, these descriptions are sets or defined collections of possibilities. If we are willing to hold that there are real possibilities—as Quine is not—then we can take them as a kind of bedrock reality. If we do not, we are left with Quine's logical pragmatism. This allows us to take anything as true if we please and are willing to depart from the community of scientists, the community of theologians, or whatever other defining community we happen to prefer. We cannot, then, settle the battle between, say creationists and evolutionary biologists. We cannot actually claim that our worlds are nearer to what there really is than the frightening fantasies of the paranoid schizophrenic. Do we want to cure him only because we will all feel better or because we think it would be better for him to live in what we call "the real world?"

The bedrock then, we might want to say, does consist of logical possibilities, and mathematics is a basic insight into the structure of many of them. Some mathematical entities will figure in any intelligible world.

If so, how do we answer Quine who returns now to say that we do not know what to do with these possibilities, for we cannot even identify them? Quine thought that if someone said, "There are possible people in the doorway over there," and he were asked, "How many possible people are there?," there would be no answer. He also wondered if there were fewer possible fat men than possible thin ones. But if we turn the question around a little, the puzzle disappears. Suppose we ask, "How many people could possibly fit into that doorway?" Most people are a couple of feet across and doors are three feet or so, so at most two or three will fit, supposing they squeeze up a little. If we have a framework and ask how many possible x's it can accommodate, we have no difficulty. Are there more possible fat than possible thin ones? Taken as a question about *everything*, the answer is that there are equal numbers; indeed, there are infinite numbers of each. But given a framework, the answer may be different. We are told that in Burma, very many people go hungry and the average daily intake of calories is very low. One must work harder or be luckier (or unluckier if that is how one looks at it) in Burma than in Peoria, Illinois, if one is to get fat. So we might well say that the possibilities for getting fat are fewer in Burma than in Peoria. The number of calories consumed in Peoria opens the possibilities for many fat men. The frameworks are part of what I have been calling "worlds," and worlds have defining characteristics.

Such worlds then are quite different from the assemblies that David Lewis called "possible worlds." This is important. Lewis' possible worlds are indeed regions which Quine refused to include in his metaphysical geography. They are simply collections of sentences which contain no contradictions. These sentences do refer to one stratum of what I called the bedrock. But though such bedrock is a crucial element in the foundations of the world or worlds in which we live, these sentences cannot stand alone. In them we cannot say how many fat men there are or whether Barack Obama is a better president than George W. Bush, for the context which makes such determinations possible is missing. There can be any number of fat men in such a world. The number might be aleph null, Cantor's smallest infinite number. But aleph null minus one is still aleph null and so if Barack Obama disappeared from this world, it would still have the same number of men. How would we explain that to Mrs. Obama? And in some possible world, George Bush would do all the things Barack Obama would do so neither could be better than the other.

This is evidently a kind of nonsense. "Worlds" require a rich enough context to make the concept intelligible.

Where does this leave our account of physical objects? We can say physical objects are essentially sets of logical possibilities and mathematical structures. They serve in an important way to make our experiences intelligible.

We may be getting near what seemed to be our goal. But we can now see that we need to be careful. When Sir Arthur Eddington insisted that idealism was a natural outcome of a developing physics—the physics of the period from the end of the First World War to ours, he meant that mathematical entities had the best claim to "reality." Susan Stebbing rightly insisted that there was no reason to think that mathematical entities were mental.[24] They seem most likely to be what are commonly called "ideas" in what I have been calling a Platonic sense, but surely not necessarily ideas in the modern—mostly post-Cartesian—sense in which ideas are thought to be denizens of the mind.

She had a point, but she forgot that mathematical entities have another necessary property—intelligibility. If reality consists of them and we can grasp them, we can read the world—or at least the physical object part of it—as if it were a book.

6 The Idea of Physics

To see if this makes any sense, I think it is best to take Quine seriously and say that it is physics that brought us to this pass and then ask ourselves what, after all, goes into physics. Quine's own answer is that what goes into physics

24 See her essay in *Mind, Matter and Purpose*, Aristotelian Society Supplementary Volume VIII, 1928, 113–129 in the symposium on "Materialism in the Light of Modern Scientific Thought." See also *Philosophy and the Physicists*, London: Methuen, 1937.

is an interplay between concepts and experiences. We have some of each and we balance them off. As a logical pragmatist, he would say that we perform this balancing act to create a system that works for us towards whatever ends, intellectual and practical, that we have. Most physicists, though, would think that intelligibility is more than pragmatic acceptability.

So we need to see how this mixture of logical and mathematical objects and experience can be fitted together. James Robert Brown who, as we saw, is certainly a mathematical Platonist, concedes that there are various ways in which you might construe mathematical realism and let experience enter in. He notices that many great mathematical Platonists, including G.H. Hardy, thought of our knowledge of numbers and such as a kind of seeing. We actually grasp them in experience. But there are others like Karl Popper who thought numbers were items in what he called "World 2," the world of mental objects. He supposed that when we come to find things out about such objects, as Euclid did about prime numbers, we simply make a discovery. They become permanent denizens of a world of stored objects which Popper called "World 3." This, I suppose, is a way of saying that, when we know something we abstract it from time and make it available at future times, but Popper included genes and computers as storage devices. One oddity of Popper's theory is that it holds that we can *create* a prime number larger than any number anyone can think of. Yet there already *is* such a number. Much stranger still, I think, is Popper's notion that the world of physical objects is somehow primary. I believe that he thought of this "world" as being composed of rather ordinary things and therefore that "physical objects" were things like street-cars, houses, and lumps of coal. But, following Quine, we have seen that "physical objects" tend to dissolve into logico-mathematical entities. If so, his three worlds seem to dissolve into a messy soup in which they can no longer be distinguished.

Hence, to get to intelligibility, we must look more closely at the ways in which experience figures in our accounts of things. Brown is willing, but may be wrong, to settle for a sort of Lakatosian view on which our knowledge of scientific objects depends on a mixture of theories and attempts at applying them to experience and making predictions.[25] Numbers and other abstract entities enter in, but do not necessarily, one supposes, have a permanent place, for their justification depends on, and varies with, theories. Nonetheless they are somehow there.

---

25 Cf. Imre Lakatos, *The Methodology of Scientific Research Programmes, Collected Papers*, Vol. 1, Cambridge: The University Press, 1978.

This becomes confusing and we have seen that the Popperian "three worlds thesis" is very messy indeed.[26] But we can back up, look anew, and begin to sort it out. Recall that Popper used the expression "world" in much the way that I suggested, and urged that there was a world of physical objects, a world of the things of the mind, and a world of theoretical, essentially quasi-mathematical, entities. This is the common sense view, but it makes people itch because the problem is always to find the relation between the "three worlds." Popper was sometimes taken to be a mind-matter dualist who wanted to add a little (modified!) mathematical Platonism to the mix and, of course, no one has ever been able to discover how minds and material objects are related. He is not alone in seeming to fudge the issues. Mathematical Platonists, as we have seen, have sometimes thought of their favoured objects as being elements in minds or embedded in minds, and sometimes have thought of them as being, like traditional physical objects, quite independent of minds.

But if we think about the way in which I have been talking about physical objects, the situation becomes clearer. The objects which physicists talk about (say when they are talking about particles which make marks on bits of CR39, the handy plastic out of which the little lenses which are used to fix people up after their cataracts are removed are also often made) are interpretations of mathematical formulae. Physics, by tradition, is about the movements of particles or waves or the dissemination of energy across space and time or space-time. There are theories and theorems to connect mass and energy as

26 Popper laid down the "three worlds" in *The Tanner Lecture on Human Values* delivered at the University of Michigan on April 7, 1978. A simple account is available electronically (www.knowledgejump.com/knowledge/popper.html). "World 1 is the physical universe. It consists of the actual truth and reality that we try to represent, such as energy, physics, and chemistry. We may exist in this world; however, we do not always perceive it and then represent it correctly. World 2 is the world of our subjective personal perceptions, experiences, and cognitions. It is what we think about the world as we try to map, represent, and anticipate an hypothesis in order to maintain our existence in an ever changing place. Personal knowledge and memory form this world, which are based on self-regulation, cognition, consciousness, dispositions, and processes. Note that [Michael] Polanyi's theory of knowledge is based entirely within this world. World 3 is the sum total of the objective abstract products of the human mind. It consists of such artefacts as books, tools, theories, models, libraries, computers, and networks. It is quite a diverse mixture that ranges from a claw-hammer to Maslow's hierarchy to Gödel's proof of the incompleteness of arithmetic. While knowledge may be created and produced by World 2 activities, its artefacts are stored in this world. Popper also includes genetic heredity (if you think about it, genes are really nothing more than a biological artefact of instructions)."

well as particles and waves. As long ago as the eighteenth century, Kant had noticed the role of geometry. Middle-sized object—molecules, stars, and everything in between—attract one another directly as their mass and inversely as the square of the distance between them because they inhabit, or we conceptualize them best as inhabiting, three dimensional, homogenous and non-interfering spaces.[27]

More precisely, therefore, our chosen geometries define spaces through which mathematically defined entities may be said to travel, and the formulae we use provide material from which predictions can be made about what physicists might "see." Their seeing may be very indirect. Below the level of the molecule they may analyze minute scratches on bits of CR39 in order to define particles. But, if I may recall Ryle once more, seeing is an "achievement verb."[28]

Now notice that what we have is a reading of the data which enters into the experience of some physicist. Other readings are possible. The "world" of which the physicist is speaking may be made up, in the end, of strings rather than the particles and waves of many theories. The space so defined may have any number of dimensions. However it is defined, it may constitute a physical system which would be seen from another "universe" as a black hole which has imploded. The experience itself can be taken as something generated by the brain of the physicist, or the brain of the physicist may be taken as itself a reading of the experiences of some neurophysiologists (Most of us have not seen our own brains!).

There is no avoiding the fact that there are bits of mathematics and there are readings. Popper's Worlds 2 and 3 are thus accounted for more easily than his World 1. In World 1 as he imagined it, there are certainly things like those bits of CR39 which figure in scientific theories and in ordinary experience. There are houses and automobiles as well as galaxies. They may seem to be awkward. The suggestion somehow is that either there really are no tables and chairs or somehow they miraculously appear when there are only colourless waves and particles, sometimes only distinguishable by their different electric charges. But notice that the bits of CR39 are, like the particles which leave their marks on them, made up of atoms and their particles. They are not something in addition to those particles.

27 For an account of this, see G.J. Whitrow, *The Structure and Evolution of the Universe*, New York: Harper, 1959, 199–200.

28 *The Concept of Mind*, London: Hutchinson, 1966, 150–152; 222.

7 How Do Worlds Fit Together?

This is the clue to the question, "How do the worlds fit together?" Everything is in all of them. One way of putting it is that they are just different readings of the same book. But one must be careful about saying such things. Are they like different interpretations of *Hamlet* or the work of different critics who find King Lear cunning or mad or both? The answer is, oddly at first sight, both "yes" and "no."

Different readings of *Hamlet* or *King Lear* have to be justified to the extent that they must give some sensible meaning to each line of the text. But when we put worlds together to make "the world" in which we think we live, the different readings not only have to account for all the data, they have to fit together so that in some sense there is a statement which is both true in each world and corresponds to a true sentence in each of the other sub-worlds.

All our worlds, then, are constructions, but not constructions out of nothing. They are readings of experience which have an intelligible order because there is an intelligible underlying order.[29] Our best efforts at grasping this order are expressed, very likely, in mathematical physics whose entities are a kind of bedrock. It is not just "in our minds" but it needs our minds to bring it to life in the form of the world or worlds in which we live. We do live in these generated worlds, and they take not only mathematics and physics but history and poetry to bring them into intelligibility.

---

29 Thus, although I do insist that we play a crucial role in constructing our worlds, my claim is very far from the much advertised and frequently reviled "social construction of knowledge or social construction of reality." The arguments against extreme "constructivism" are marshalled by Paul Boghossian in *Fear of Knowledge,* Oxford: Clarendon, 2006. Extreme constructivism makes all knowledge relative in ways which make all claims equally good. Boghossian reviews the contradictions which arise from the claim that all knowledge is relative and explores them in claims about truth, justification and belief. I am arguing that we do construct our interpretations and assemble our worlds, but there is an underlying reality and, indeed, not all interpretations are equally good. There are defensible grounds for specific claims about truth, justification and belief. But Boghossian confines himself rather simplistically to issues which are straightforwardly logical. He seems to regard as unworthy of examination claims like those of Jacques Derrida which stem from the fact that our knowledge is expressed in words and that the relations between words and things make all our claims subject to revision. Quine, too, understood the importance of this issue. More to the immediate point, there are cases in which what is constructed by social forces is not so easy to separate from what is simply "there." The idea of a "world" is one way to confront some of these. But when we try to distinguish what the mind does and what it receives there can be difficulties, some of which are brought to light by Scott Atran and Douglas Medin in *The Native Mind and Cultural Construction of Nature,* Cambridge, MA: MIT Press, 2008. I will discuss one of their theses later.

The worlds in which we live are generated by interaction with the bedrock of order, which is the mathematical reality exploited by physicists. But each of us stands always at the centre of an expanding circle of experience. We can never leave that centre. It is not itself a point in our brains. At the centre of your brain the corpus callosum joins two hemispheres. You can sever that connection, as has been done in the treatment of epilepsy, but still you look out from the same centre.

In one sense, "what there is"—if you mean by that what you have reason to talk about and cannot analyse away into something else—consists of logical and mathematical entities which order the world you construct and the self which reads them does the constructing. But in another sense what there is is what is in the "world" which results from your best effort to combine the constructions into intelligibility. The novelist Michael Frayn argued that we are by nature readers and we read not just words but the symbols around us. Our existence depends on them and so does the world in which we live.[30]

The world, seen this way, *is* intelligible. Does that mean that it is permeated by intelligence? How are we to conceive of the minds that figure in all our readings? The worlds that can be conceived are infinite in scope. Are the minds?

The question of larger minds and more pervasive minds than ours culminating in the gods of the monotheist religions[31] intrudes at some point when we try to decide on what we should take ourselves to encounter in our worlds. The crisis of the environment has renewed thoughts of an organic view of the earth itself in which our planet might be thought to have its own aims and direction. If so, some will think again of those spirits which very old cultures have thought to animate nature, or even of the gods of the great mythologies or more subtly of the Neoplatonist world soul.

How could one decide such questions? The advance of science has played a major role in chasing the gods from the world, and attempts to introduce active intelligence into biology are met with ferocious responses from nearly all professional biologists.

30 Michael Frayn, *Constructions, Making Sense of Things*, London, Faber & Faber, 1974, 2nd ed., 2009. Also, see my article, "The World as a Work of Art," *Maritain Studies/Ėtudes Martainiennes*, Vol. 17, 2001, 3–29.

31 Each monotheistic religion, by definition, insists that there is only one God. But the hell-fire God of the old Scottish Calvinists is not much like Ralph Cudworth's God nor Allah as conceived by the Taliban. Jews tend to be circumspect when talking about God, but the God you will meet in a Reformed Synagogue is not easily recognizable as the God who knocked down the walls of Jericho and authorized the killing of everyone (although one prostitute was allowed to survive).

Notice however that no world is simply composed of the entities which compose David Lewis' possible worlds. According to my argument, Lewis' worlds fell victim to Quine's objection that their denizens cannot be identified. They are not part of "what there is." A world is a reading of a logical and symbolic structure. But it is always related to a life, and has to be lived in. Living in that world in the way that human beings and other sensing, reflecting, and reasoning beings do requires intelligence and such a world must be sufficiently permeated by intelligibility. Does that imply an over-arching intelligence? Is there an intelligence that shows itself in the logico-mathematical substructure as Arthur Eddington and James Jeans thought? Chance is intelligible mathematically, but it sounds odd—and it may well be unintelligible—to say that mathematical reality occurs by chance. If the underlying reality is not an intelligence in and of itself, it is certainly the breeding ground of intelligence.

8 Mind in the World

The thesis I have been advocating is that the basic realities are, first, logico-mathematical entities which function as symbols and constitute the basic order within which facts emerge. In the process, the minds that read them are also revealed. The readings constitute the worlds in which we situate ourselves. But we do not somehow exist *outside* these worlds; we are involved in them and understand ourselves through them. To a large extent, we obviously share the worlds which we construct, widely within the clusters of cultures which we commonly call civilizations, and less completely though still substantially with all other human beings. The worlds of animals overlap with ours, sometimes, we suppose, quite a lot as with our dogs, somewhat less so with the cows we keep, and hardly at all with lobsters.

The reading of nature and the construction of worlds involve the existence of minds. Science bears on this implication. Quantum theory suggests that states of the world change when we become conscious of them or at least that observation brings about the determination of what was indeterminate. The Russian physicist M.A. Popov in a paper entitled "In Defense of Quantum Idealism"[32] says: "In one way or another, the evolution of scientific knowledge is moving remarkably close to an idealistic picture of nature." He feels forced to this verdict though he accepts that it may seem to be a problem about

32 *Physics-Uspekhi,* Vol. 46, No. 12, 2003, 1307–1309. *Physics-Uspekhi* (Advances in Physical Sciences) is a translation of the authoritative Russian-language review journal in physics, *Uspekhi Fizicheskikh Nauk,* first published in 1918.

quantum theory. Nevertheless, he concludes: "… the traditional dominance of trivial 'materialism' in quantum physics cannot … always be tolerated."

One must tread with care. Some people have thought that the determination of quantum states causes the world to divide at each moment or with each observation so that we have very many "universes," each somehow cut off from the others. Others have supposed that the indeterminate simply becomes, as I suggested, determinate. But, if what I have been advocating is correct, what happens is that some features of the underlying logical structure become parts of the contents of our worlds when we conduct the appropriate observations. We read them into our worlds and to be a world is simply to be a reading of the logico-mathematical structure. Time in our worlds is episodic and since the events in our world follow from successive moments of observation, time appears to be an exclusion series—what is present excludes the past which is gone and the future which is to come. But this leads to the paradoxes McTaggart noticed.[33] We divide time into past, present, and future. Yet we also arrange events in a line of succession which forms a series of events from earlier to later. Looked at objectively, events form an ordered series from the far past into the distant future. Yet nothing changes in this series. Time seems to involve change. But there is a fixity in history. In *some* sense it was always true that Caesar would cross the Rubicon before the Huguenots crossed the channel and settled in England. Subjectively, however, everything is past, present, and future and things seem to change. We suppose that right now there are not literally true statements in the ordinary sense about a future that does not yet exist, though there are conditionally true statements about how things will be if present laws hold.

Literally, past, present and future are incompatible states. McTaggart reasoned that, in addition to the earlier-later and past-present-future series, there must be a third series[34] in which events succeed one another but form an inclusion series, so that the past and the future all exist in some sense in the present, but do succeed one another. If to say that an event is past, present, and future is no longer to utter a contradiction, McTaggart's problem is a massive puzzle, though it cannot be given full justice here. Nevertheless, one can see that, if what I have been saying is true, there *seems* to be an answer. The underlying logico-mathematical structure is not an infinity of atomistic possibilities but a collection of integrated systems. The evidence for this is that such a collection always manifests itself systematically—so much so that, as I said at the

33 J.M.E. McTaggart, *The Nature of Existence,* Cambridge: The University Press, Vol. II, 1927, Sections 303–351, 9–31.

34 McTaggart called the other two A and B and this one the C series.

beginning of this chapter, we can make amazingly accurate predictions about very many phenomena. The *phenomena*, of course are the manifestations of the system. When one possibility is brought into our experienced worlds, it brings with it a whole system so that our physics is law-like and our biology is structured. The underlying reality is like McTaggart's third or "C" series in that everything is contained in various sets of the possibilities, but they are possibilities *for* ordered experience. Our "free will," if it shows itself, will be the capacity to move from one ordered set of possibilities to another, a phenomenon which could be associated with what some quantum theorists regard as the splitting of the world into distinct universes.[35]

But I have introduced "minds" and here still more caution is needed. Experience is associated with minds. A "mind" is neither a logico-mathematical entity nor one of the "things that fill the world". No one claims, I think, to "be" a mind, though everyone claims to have one and supposes that other people, along with dogs, cows, and perhaps lobsters, have minds, too. It is supposed, that is, that everything that has experiences has a mind. It is not necessarily supposed that every such creature is having experiences all the time. Sometimes, perhaps, we are in dreamless sleep (though this might be disputed) and occasionally we are unconscious in some deeper sense. But to have a mind is to have a tendency to have experiences. Having a mind is thus being engaged in an activity.

The subject of such experiences is, of course, a matter of profound dispute. The first person pronoun is in a sense irreducible. I cannot replace "I am now thinking about the problem of mind" with "Leslie Armour is thinking about the problem of mind" because I might be right that I am thinking about mind while wrong in believing that I am Leslie Armour.[36] Perhaps the most we can say for certain is, as I said, that the subject is always associated with presence at the centre of a field of experience and to say that someone has a mind is to say that he or she has a tendency to have experiences.

Here, we need to focus on mind as it is associated with certain activities. It is clear that, if we are to talk sensibly about mind and its presence as a condition for the worlds which we construct, we need to look at the circumstances in

35 I will not speculate here about the possibility that this might not be compatible with McTaggart's account of volition in *The Nature of Existence*, Vol. II, Sections 444–454, 132–143.

36 Such brief analyses are only indicative. See my discussions of these questions in works such as *Being and Idea*, Hildesheim: Georg Olms, 1992, *The Rational and the Real*, The Hague: Martinus Nijhoff, 1969, and "Russell, McTaggart and 'I'," *Idealistic Studies*, January, 9:1, 1979.

which we ordinarily encounter minds in experience, for that experience is the stuff out of which we build our worlds. I obviously cannot settle such questions in an essay such as this. But I can at least sketch a theory which helps to make sense of my central thesis about physical objects.

9 Puzzling Organisms

Minds are always associated with our experiences of organisms. In one sense, of course, the minds we meet appear in the context of an array of physical objects, the objects which have been the focus of this essay. Out of our interactions with this real substratum of logico-mathematical entities we—or at least our research scientists—build our notions of the facts of biophysics, biochemistry, biology itself, and, close to home, physiology. These in turn, we suppose, take their strength from their relation to the facts about physical objects, but these sciences are all generated out of our experience. Our meetings with minds take place in the context of organisms, our own and others. People and dogs have minds, but so surely do snakes and birds. Sometimes we talk of the minds of lobsters as "rudimentary" and some people suppose it is all right to boil them alive while others do not. The claim that we do not know *at all* what it is like to be a bat or a lobster seems to be as patently absurd as the claim that we infer *adequately* the inner lives of such creatures.

Mind appears through these organisms but, if the argument in this essay is sound, we have no reason to think that it is *reducible* to them. The logico-mathematical entities which are the basis of the reality of physical objects also appear to us through our experience of organisms. There are mathematicians and physicists as well as numbers and physical objects. People think and act; logico-mathematical entities do neither. The worlds in which we live are constructed in the processes of experience from the orderly structures of this basic reality. Organisms, as we know them, are features of these constructed worlds.

Admittedly, neither the idea of an organism nor that of the relation of any organism to any mind is very clear. But we do notice some things. Importantly, living organisms, at least those of a certain complexity, have a measure of autonomy. Paul Davies remarks that, if you throw a live bird and a dead bird into the air, the dead bird will land a few feet away but you have no idea where the live bird will land.[37]

---

37 *The Fifth Miracle: The Search for the Origin and Meaning of Life,* New York, Touchstone (Simon and Schuster) 2000, 33. There is an also a British edition, London: Allen Lane, 2000.

All organisms, certainly, have contexts from which they cannot intelligibly be abstracted. The badger and its environment are a unity in the sense that, without the air it breathes, the badger cannot survive. It depends on the earth's position in the solar system for the stability of its burrow and on the fact that there are not nearby sources of deadly radiation or passing stars which might derange that system. The badger is really just one focus on the whole, the centre of a perspective on the world in which it lives.[38] The badger does of course have more autonomy than the mitochondria which power its cells. Within our experienced worlds, both badgers and mitochondria appear as descendents of once independent organisms. As the history of things appears in these worlds, the mitochondria have become so embedded in the cells or other organisms that, though they could survive on their own, they seemingly can no long reproduce themselves in isolation.

But all our experienced worlds suggest that, ultimately, in some way, reality is pervaded by intelligence, though we do not meet any minds beyond those which are associated with organisms that populate the worlds which, I have suggested, emerge out of the interactions of our minds with the underlying reality. We should notice that the worlds we generate are experienced as having embedded in them not just one organism displaying a mind, but a whole complex community. The suggestion that reality is a collection of minds, each locked into a solipsistic reality, is absurd if only for the reason that no one person could invent a language. The underlying reality is evidently so constructed that, by tapping into it, we can create worlds which we can to a significant extent share with one another.

10 Sharing Minds

Our logical constructions show us what we surely in some way know, anyway. We share minds. The fact that we discern minds in creatures suggests that we have minds in common with them. This situation is perhaps most intelligible if the same mind appears in all of them, though different facets of it and different capacities appear in each organism. If we could sensibly make out what it would be like to be God or a universal mind appearing in all things we might find it best to regard such a being as a second subject in each experience. Whatever you experience or the badger experiences, that is, would be noticed

38 See my discussion of the issues this poses in *Logic and Reality,* Assen: Royal Van Gorcum and New York: Humanities Press, 1971, Chapters 2 and 3.

equally, and at the same time, by God. The Christian God was supposed to see the sparrow fall. In these terms, God falls with the sparrow.

Such a consciousness would share in everyone's experiences. But such a consciousness would have to contain both the totality of experience and the plurality of perspectives on it held by the individuals who compose reality. It would undergo change in respect of what it shared. But if it were intelligible to itself it could not be confined to grasping the plurality as a plurality. It would also have to have a complete knowledge of things as they are apart from the variety of perspectives. It would have to have a knowledge of the objectively shared elements of experience. Such a view seems to lead to moral and political views like those of T.H. Green and some other British idealists.[39] It provides a meaning for self-realization—coming to share as perfectly as possible the eternal consciousness and it does suggest that we would have to co-operate in order to share fully in that consciousness. In religion, one can hold this view together with Christianity and, indeed, since this consciousness would be shared with us, though it would be beyond our immediate grasp, we could develop a theory of revelation to go with it.[40]

All this can be made to seem very mysterious. It is not necessarily so. But it does produce problems. The problem of "other minds," set in train by understandings or misunderstandings of Descartes, was provoked by the notion that there is no account of the idea of a "shared mind" given that each is locked into a body and has experiences which are generated by brain states. Neurophysiologists have access to these states but find no pictures or words in the brain cells which they study. It was thought for a time, at least by followers of Wittgenstein and Ryle, that claims like this hang on a misunderstanding. An exploration of the workings of language would dissolve this misunderstanding. And indeed it is easy to show that language requires a shared public experience. It was implied that this experience has to do with what people do and what they do is displayed by their bodies in acting and speaking. This may show us that there is some way of removing the mystery. It does not tell us how the mystery is to be removed unless some kind of "linguistic behaviourism" can be shown to imply the reduction of all reality to physical objects, however. And I have been arguing that that will not work.

I am suggesting that perhaps the best way of conceptualising the problem is to suppose that minds are not private in the sense that we share our minds

39 Though I have not been able to discover his reasons, T.H. Green denies what one might call the "dual subject thesis."

40 William Temple's philosophy looks rather like this though his Absolute may in other ways be closer to Bernard Bosanquet's; see Temple, *Christus Veritas*, London: MacMillan, 1924.

with others. Perhaps there is literally a Shared Mind, as theists think, but that would require more argument. There may be mind-in-general and it may show itself in many individual minds which become apparent to us when, and likely only when, we interpret reality in such a way as to generate the experience of an organism. This may sound odd, but there is nothing inherently odd in it. The self whose existence Descartes said he could not doubt is not René Descartes, but self in general. In asking us to think of the self that thinks or doubts, Descartes did not learn about himself in the way psychologists have sometimes thought that one learns by "introspection," for all the things I introspect I can easily doubt. He was simply calling attention to self-hood in general. One must add some explicit experiences to get the self of a French philosopher, or a lobster, or a cow grazing in the field. Such selves are constructions. There is always some mystery about them, especially about the minds of others. It was hard to know what went on in the minds of the cows who liked to sit on my farm and gaze towards the cars and trucks passing on Interstate 71. But we do know something because we know how cows react, what they like, and especially what they do when we leave them to exercise their autonomy. More importantly, we know how they *respond* to us.

We perceive people and cows only intermittently, as indeed, we do virtually everything that goes into our worlds. The only exception, perhaps, is the continuing centre of our experience which we tend to associate with "ourselves." But though we see George only intermittently, we do not take our individual perceptions of him to be distinct things. And farmers like to think that the cow they let into the barn to keep her out of the freezing rain is the same beast they saw in the field. For many objects, our notions of their unity has to do with features that remain unchanged. Your house is at the same address every day and it changes quite slowly. People and cows do not stand still and sometimes change quite dramatically. The kind of unity that organisms exhibit is different from that of houses and rocks. It has much to do with their autonomy. And this is reasonably construed as the appearance of minds. But this is not enough. What is more important is that our interactions with them make sense. We share with them more than the physical space that we share with stones and houses. It is conceptualizing this that poses problems at least for people in twenty-first century Western societies. What we share can be called "mind."

Most significantly, whatever the "ordinary language" philosophers were right[41] or wrong about, they were right to call attention to language. Language

41 Even diehard metaphysicians have to accept that they were often right—hence the references to Gilbert Ryle in this chapter.

is a common creation. We all play a role in its continuous creation, express ourselves through it, and find other people in their responses. But this route to a solution has evident limits. Every bit of language can, in principle, be traced to some individual user. The *Oxford English Dictionary* builds on identifiable uses from identified sources.

One way of probing a *little* further into what to twenty-first century Western minds is the rather curious idea of mind-in-general is to look at communities which still (or did until recently) continue to believe in minds which are more pervasive than our own. Members of these communities tend to suggest that they have empirical evidence for their beliefs. Scott Atran and Douglas Medin undertook very detailed and carefully analyzed studies which they have reported in *The Native Mind and Cultural Construction of Nature.*[42] There is a widespread tendency in cultures whose people live very close to nature with minimal technological separation from a shared life with plants and animals to regard nature in various forms as something with interests of its own. Such peoples also tend to have a detailed knowledge of plant and animal forms and can easily name very many tree forms and kinds of animal life. This detailed knowledge seems rather special to such societies. At any rate, the authors found astounding differences between people in such cultures and students in elite universities in their awareness of, and ability to distinguish, life forms.

The Itza group of Maya in Central America are a people who figure extensively in their book.[43] They are among the many communities in which basic elements in the environment are thought to have their own interests and to exhibit spiritual activity. Such groups are much more likely to protect common lands. A problem with common lands usually is that they are overused by people seeking to advance their individual interests. Understandably enough, where there is thought to be evidence of mind or spirit, common lands are not overused. At most, on the face it, this shows that there is a pragmatic justification for peoples' communal beliefs, though weight should be given to the fact that the more experience people have of plants and animals, the more they are likely to regard nature as "spiritualized." Some weight might also be given to the data collected by Atran and Medin about the relation between these views and the possession of a language detailed enough to come to grips with nature. The possession of a vocabulary which is adequate to make all the necessary distinctions seems to be influential. Most people living in urban environments now have lost the ability to distinguish more than a few species of trees, and

42 Cambridge, MA: MIT Press, 2008.

43 References are widespread but see, for instance, 161 ff.

Atran and Medin did a study of the frequency of tree examples in the *Oxford English Dictionary.* The frequency of these examples has declined sharply since the mid-nineteenth century, following a period which showed some recovery from the low frequencies of a hundred years earlier.[44] One does not write about trees if one does not know about them.

The peoples whose extensive knowledge of trees was investigated by Atran and Medin clearly saw their forest as an extended organism, one to be dealt with cautiously. This is true of many such peoples. The Haidas of the Queen Charlotte Islands were historically, and many still are, exceptionally careful of trees. They depended heavily on fishing and were eager and efficient traders who plied the coast in canoes which they made from the Islands' forests. They preferred not to cut down a whole tree but to extract a careful slice from it in a way that would allow the tree to grow.

These peoples' sense of the unity of things was given by a world view which is, in fact, one of the many variants of what is often called Platonism in the West. The Queen Charlotte Islanders regarded each whale as a kind if avatar of the true Great Whale, each bear as an avatar of the true Bear. When they killed an animal, they thanked its universal counterpart and asked for forgiveness. The same was true with forests—each entity instantiated a species of mind.

These are all cases of minds which appear through their finite instances. The finite instances do not exhaust them. It did not occur to such people to divide the world up into the discrete bits which Hegel described as "determinate being."

If one divides the world into isolated atomic bits, the puzzles about whether there are "other minds" and about how we know them seems to be pressing, for the manner in which mind is instantiated is understood through phenomena with which all of us are familiar. If one orders one's experience into discrete units with no necessary connection to each another, one most certainly has a problem. David Hume—a philosopher Hegel had in mind when he spoke of "determinate being"—thought experience itself was built on atomic sense data and that no datum had a logical connection to any other. But experience itself seems to be a seamless field and the question about how it should or could be analyzed into distinct parts is a troubling philosophical one. Posed simply and without a context, there is no way of answering those who raise the problem.

But if we are to understand how a world composed of logico-mathematical entities can come to be read intelligibly in the construction of our everyday

44 *The Native Mind*, 39–47.

world, we probably need to explore ideas which long predated Hume and which do not easily disappear from the history of philosophy or from the convictions of many human societies.

11 Experienced Universals

If we look at the ways in which Haidas (and other North Coast aboriginal peoples) conceptualize the world, we will see that they do not think of it in Humean terms. Specifically, the difficult part of such ideas as those of the Haidas has to do with the ways in which universals enter into our affairs. We are inclined to think of them as abstractions and, indeed, Plato suggested that one met them only when one escaped the cave in which we are trapped in our ordinary experience. But what is being suggested by peoples like the Haida is a universal experienced immediately in ordinary life.

This notion is not as obscure as one might at first suspect. We do not in fact experience the people we know as a succession of disjointed images which we somehow "put together." People change over time but we don't usually have much trouble with that. The person is not the sum of his or her appearances. The person emerges into our understanding through them. Of course, we may be wrong about the person who appears. If something strange happens we might say, "She's not the person I thought she was." We may have taken someone else to be appearing in her place. People do sometimes get bad surprises when they elect prime ministers or bishops.

The point is that what is happening is that we are experiencing universals *in things* or, as a Platonist might say, we experience forms informing things. The Haida might say that "the Great Whale is appearing through his creatures."

That one might experience God or Mind-as-Such as form informing things should not surprise Christians who suppose, precisely, that one meets God in the person of Jesus or Jesus in the bread and the wine. What is experienced, if it is God or Mind-as-Such, need not look like God or mind, for God and mind do not, in themselves, look like anything. Of course one could take a slightly weaker view. Hindus are familiar with the notion of avatars. A commonly pictured avatar of Krishna shows him fond of dancing with cow girls. It is not quite claimed that the avatar is Krishna but the avatar is a nearly-perfect instantiation of the universal Krishna. Whether or not Christians are wise to claim more than this for Jesus is another question. We can imagine an earthly triangle instantiating triangularity perfectly (though perhaps we can only *imagine* it). God is more difficult.

We can raise all sorts of questions about such discussions. Do the Central Americans studied by Atran and Medin really grasp spirit in the forest?

But we can also try to answer them. How sensible is it to liken a forest or a great swamp or the whole planetary environment to an organism? Our ideas of organism are underdeveloped. This is probably explained by the path science has taken.

We can *approach* these great questions and perhaps make some progress. Only a fool would think them solved.

References

Armour, Leslie, *Being and Idea,* Hildesheim: Georg Olms, 1992.

Armour, Leslie, *Being and Idea. Developments of Some Themes in Spinoza and Hegel,* Hildesheim: Georg Olms, 1992.

Armour, Leslie, *Logic and Reality,* Assen: Royal Van Gorcum and New York: Humanities Press, 1971.

Armour, Leslie, "Russell, McTaggart and 'I'," *Idealistic Studies,* January, 1979.

Armour, Leslie, *The Rational and the Real,* The Hague: Martinus Nijhoff, 1969.

Armour, Leslie, "The World as a Work of Art," *Maritain Studies/Études Martainiennes,* Vol. 17, 2001, 3–29.

Atran, Scott and Douglas Medin *The Native Mind and Cultural Construction of Nature,* Cambridge, MA: MIT Press, 2008.

Boghossian, Paul, *Fear of Knowledge,* Oxford: Clarendon, 2006.

Brown, James Robert, *Philosophy of Mathematics,* London: Routledge, 1999.

Brown, James Robert, *Smoke and Mirrors,* London: Routledge, 1994.

Cudworth, Ralph, *The True Intellectual System of the Universe,* London: R. Royston, 1678.

Davies, Paul, *The Fifth Miracle: The Search for the Origin and Meaning of Life,* New York: Touchstone (Simon and Schuster), 2000.

Eddington, Arthur, *Philosophy and the Physicists,* London: Methuen, 1937.

Farmelo, Graham, *The Strangest Man,* London: Faber and Faber, 2009.

Frayn, Michael, *Constructions, Making Sense of Things,* London, Faber & Faber, 1974, 2nd ed., 2009.

Gettier, Edmund, "Is Justified True Belief Knowledge?," *Analysis* 23 (1963), 121–123.

Girle, Rod, *Possible Worlds*, Chesham: Acumen, 2001.

Gödel, Kurt, "Russell's Mathematical Logic," in P. Benacerraf and H. Putnam (eds.), *Philosophy of Mathematics,* Cambridge: The University Press, 1983. (This paper was originally published in 1944.)

Gratton-Guinness, I., "The Rediscovery of the Cantor-Dedekind Correspondence," *Jahresbericht der Deutschen Mathematiker-Vereinigung,* Vol. 76, 1974/75, 104–139.

Hardy, G.H., *A Mathematician's Apology,* Cambridge: The University Press, 1940.

Hardy. G.H., *A Mathematician's Apology*, reprinted with a Foreword by C.P. Snow, Cambridge: The University Press, 1967.

Hylton, Peter, *Propositions, Functions, and Analysis: Selected Essays on Russell's Philosophy*, Oxford: Clarendon Press; New York: Oxford University Press, 2005.

Hylton, Peter, *Quine*, London, Routledge, 2007.

Lakatos, Imre, *The Methodology of Scientific Research Programmes, Collected Papers*, Vol. I, Cambridge: The University Press, 1978.

Lash, Nicholas, "Reason, Fools, and Rameau's Nephew," in his book, *Theology for Pilgrims,* London: Darton, Longman and Todd, 2008, 123–136.

Lavine, Shaughan, *Understanding the Infinite*, Cambridge, MA: Harvard University Press, 1994.

Lewis, David, *On The Plurality of Worlds*, Oxford: B. Blackwell, 1986.

Marchi, Peggy, "Mathematics as a Critical Enterprise," in R.S. Cohen, P.K. Feyerabend, and M.W. Wartofsky, eds., *Essays in Memory of Imre Lakatos,* Dordrecht: D. Reidel, 1976, 379–393.

McTaggart, J.M.E., *The Nature of Existence,* 2 vols., Cambridge: The University Press, Vol. II, 1927.

Meillassoux, Quentin, *After Finitude, an Essay on the Necessity of Contingency*, tr. Ray Brassier, London: Continuum, 2008; published as *Après la finitude*, Paris: Editions du Seuil, 2006.

Phillips, D.Z., *Recovering Religious Concepts,* Basingstoke: Macmillan, 2000.

Popov, M.A., "In Defense of Quantum Idealism," *Physics-Uspekhi,* Vol. 46, No. 12, 2003, 1307–1309 (*Physics-Uspekhi*—Advances in Physical Sciences—is a translation of the authoritative Russian-language review journal in physics, *Uspekhi Fizicheskikh Nauk*, first published in 1918.).

Popper, Karl, *The Tanner Lecture on Human Values*, delivered at the University of Michigan, April 7, 1978. A simple account is available electronically (www.knowledgejump.com/knowledge/popper.html).

Quine, W.V.O., *From a Logical Point of View,* Cambridge, MA: Harvard University Press, 1953.

Quine, W.V.O., "Whither Physical Objects," in *Essays in Memory of Imre Lakatos,* eds. R. Cohen, P. Feyerabend, and M. Wartofsky, Dordrecht: Reidel, 1976, 497–504.

Ryle, Gilbert, *Dilemmas,* Cambridge: The University Press, 1954.

Ryle, Gilbert, *The Concept of Mind,* London: Hutchinson, 1966.

Stebbing, Susan, article in *Mind, Matter and Purpose,* Aristotelian Society Supplementary Volume VIII, 1928, 113–129, in the symposium on "Materialism in the Light of Modern Scientific Thought."

Temple, William, *Christus Veritas*, London: MacMillan, 1924.

Whitrow, G.J., *The Structure and Evolution of the Universe,* New York: Harper, 1959.

CHAPTER 2

Science and the Humanities in Hume's Philosophy of Religion

Philip MacEwen

In the contemporary university, science and the humanities often operate in compartments which are hermetically sealed. This belies the name "university," derived from the Latin *universitas* ("the universal"), and lends credence to the view that the contemporary university is really a multiversity or institution where disciplines are taught and researched with little regard to their interrelationships.[1] Both "science" and "the humanities" are generic terms which denote clusters of disciplines. The former designates those disciplines which study the physical world in order to determine the principles and laws according to which it operates. The latter designates those disciplines which study what it is to be human, primarily through the examination of literary texts. Thus defined, it is easy to see how the two are closely related and yet how they can become detached. Humans live in the physical world and, as bodies, are subject to the same principles and laws as other bodies. Yet humans can and do transcend the physical world in many ways which are not always amenable to the principles and laws of science, including writing and studying literary texts.

Fruitful dialogue between science and the humanities depends upon studying their interrelationships while acknowledging their differences. This can happen in many ways. On the scientific side, science can study what it is to be human by examining the impact of humans on the natural environment and the impact of the natural environment on humans. On the humanistic side, the humanities can study the natural environment by studying literary—including scientific—texts about it. The latter venture can get testy if humanists start to relativize the methodologies of contemporary science by saying, for example, that physics is patriarchal or that science needs to revise its view of nature and understand it in a gender-neutral way.[2] Controversy is not necessarily

1 See, for example, George Grant, "Faith and the Multiversity," in *Technology and Justice* (Concord, Ontario: Anansi, 1986), 35–78.

2 There is a large literature on this subject, but a good introduction is Brain Easlea's article, "Patriarchy, Scientists, and Nuclear Warriors," in *Beyond Patriarchy: Essays By Men on Pleasure, Power, and Change*, ed. Michael Kaufman (Toronto: Oxford University Press, 1987), 195–215.

© KONINKLIJKE BRILL NV, LEIDEN, 2020 | DOI:10.1163/9789004415270_004 unfruitful, however, and can lead to enormous advances in both areas. Thus, the interest of contemporary science in global warming is, to a large extent, a product of the environmental movement, much of which, at the scholarly level, has developed within the domain of the humanities. Despite the typical distance between the humanities and science in the modern academy, then, the two spheres are not entirely removed from each other and some fruitful dialogue is taking place between them.

Another way of promoting interrelationships between science and the humanities is by studying likeminded ventures in the history of ideas. Historically, what we call "science" and "the humanities" were not regarded as "two cultures" but as one. According to this model of learning, all the disciplines which we have placed under these rubrics were sciences since they were all branches of *scientia* or knowledge. This model remained intact until the Enlightenment when the so-called "natural sciences" began to break away from the other disciplines. Due to the special kind of knowledge the natural sciences conferred, which was impartial to the investigator and repeatable subject to observation under the same conditions, they gradually assumed the name of "science" for themselves. This left the other disciplines in a sort of quasi-scientific or non-scientific limbo. Since most of these disciplines studied what it was to be human, we have here the origin of the "two cultures" view of science and the humanities that continues to dominate in the university/multiversity today.

Enlightenment thinkers were not all agreed about the "two cultures" view of the disciplines, however. While he is by no means the foremost representative of period counterculture on the subject, David Hume opposed the growing notion that science was restricted to the natural sciences and that disciplines which studied what it was to be human somehow lay beyond its borders. According to Hume, the bulk of human learning is composed of seven sciences—mathematics, natural philosophy, natural religion, logic, morals, criticism, and politics, which, in turn, are all based on what he called "the science of Man." The first three are "… in some measure dependent on the science of MAN, since they lie under the cognizance of men, and are judg'd of by their powers and faculties."[3] The other four are even more intimately related to the science of Man. "The sole end of logic is to explain the principles and operations of our reasoning faculty, and the nature of our ideas. Morals and criticism regard our

3 David Hume, *A Treatise of Human Nature*, ed. David Fate Norton and Mary Norton, 2 vols. (Oxford: Clarendon Press, 2007), Vol. 1, Introduction, 4 (hereafter *Treatise*).

tastes and sentiments: And politics consider men as united in society, and dependent on each other."[4]

Reading Hume backwards from the "two cultures" mentality of the present, mathematics and natural philosophy would now be classified as part of science while the science of man (at least in the terms Hume understood it in the *Treatise* and, later, the two *Enquiries*), natural religion, logic, morals, and criticism, would now be classified as belonging to the humanities. The only subject which seems to fall outside the "two cultures" view is politics which we would classify today as a "social science." However, Hume regarded it as a moral science, and generally treated it under the rubric of morals.[5] Thus, it is possible to read our "two cultures" view of science and the humanities into Hume's "one culture" view of the sciences. This is instructive for by studying Hume's "one culture" view of the sciences, despite the tendency of the Enlightenment to separate mathematics and natural philosophy from the other disciplines, we can identify different strands of his thought, which we would call "science" and "the humanities," dialoguing from the common platform of what Hume called "science."

I want to examine this dialogue in Hume's philosophy of religion and, specifically, his treatment of the argument from design in the appropriately entitled *Dialogues Concerning Natural Religion*. As I hope to show, the *Dialogues* is not simply about natural religion but about the interface between what we would call "science" and "the humanities" under the common rubric of science. These two strands of thought have an enormous influence on the outcome of the *Dialogues*, giving us, on the one hand, Philo's conclusion at the end of Part 8 that "A total suspense of judgment is here [in matters of natural religion] our only reasonable resource,"[6] his very different verdict at the beginning of Part 12 that "no-one [other than myself] has a deeper sense of religion impressed on his mind," and his final verdict at the end of Part 12 which brings the two strands of thought together in a unified position:

---

4 Hume, *Treatise*, Introduction, 4. More recently, the views of C.P. Snow on the two cultures are similar in many ways to Hume's. See C.P. Snow, *The Two Cultures and the Scientific Revolution* (New York: Cambridge University Press, 1959), and *The Two Cultures and a Second Look* (New York: Cambridge University Press, 1965).

5 See esp. Hume, *Treatise*, 342–364, and David Hume, *Enquiry Concerning the Principles of Morals*, ed. Tom L. Beauchamp (Oxford: Clarendon Press, 1998), Section 4, "Of Political Society," 28–32. Hume's other writings on politics are found in his *Essays* which went through multiple editions in his lifetime and have been collected by Eugene F. Miller, *David Hume: Essays Moral, Political, and Literary* (Indianapolis: Liberty*Classics*, 1985).

6 David Hume, *Dialogues Concerning Natural Religion*, ed. Stanley Tweyman (Ann Arbour: Caravan Books, 2000), 147 (hereafter *Dialogues*).

> If the whole of natural theology, as some people seem to maintain, resolves itself into one simple, though somewhat ambiguous, at least undefined proposition, *that the cause or causes of the order in the universe probably have some remote analogy to human intelligence* ..., what can the most inquisitive, contemplative, and religious man do more than give a plain, philosophical assent to the proposition, as often as it occurs; and believe that the arguments, on which it is established, exceed the objections, which lie against it?[7]

1 Science in Philo's Critique of the Argument from Design

The argument from design is originally advanced by Cleanthes in Part 2 of the *Dialogues* and runs as follows:

> Look round the world. Contemplate the whole and every part of it. You will find it to be nothing but one great machine, subdivided into an infinite number of lesser machines, which again admit of subdivisions, to a degree beyond what human senses and faculties can trace and explain. All these various machines, and even their most minute parts, are adjusted to each other with an accuracy, which ravishes into admiration all men, who have ever contemplated them. The curious adapting of means to ends, throughout all nature, resembles exactly, though it much exceeds, the productions of human contrivance; of human design, thought, wisdom, and intelligence. Since, therefore, the effects resemble each other, we are lead to infer, by all the rules of analogy, that the causes also resemble; and that the Author of Nature is somewhat similar to the mind of man, though possessed of much larger faculties, proportioned to the grandeur of the work, which he has executed.[8]

Before commencing his critique of this argument, Philo produces his own version of it. According to Cleanthes' argument, it would seem that the terms of the analogy are machines, or productions of human contrivance, and the world. However, Philo points out that resemblances between productions of human contrivance and the world are not sufficient to justify the claim that they are effects of the same kind. "But surely you will not affirm, that the universe bears

7 Hume, *Dialogues*, 172; 184–5.

8 *Dialogues*, 109.

such a resemblance to a house, that we can with the same certainty infer a similar cause, or that the analogy here is entire and perfect."[9] Cleanthes grants the point but clarifies his analogy in the argument from design to include not only adaptation of means to ends ("All these various machines, and even their most minute parts, are adjusted to each other with an accuracy, which ravishes into admiration all men, who have ever contemplated them."), but coherence of the parts ("But is the whole adjustment of means to ends in a house and in the universe so slight a resemblance? The economy of final causes? The order, proportion, and arrangement of every part?")[10] which, he claims, are present in the design of all productions of human contrivance and the world.

Furthermore, Philo notes that the principle employed in the argument from design is that "From similar effects we infer similar causes."[11] According to Philo's version of Cleanthes' argument, then, the analogy in it "… is between the means to ends relations and coherence of parts present in the design of machines and the world, and not simply between machines and the world,"[12] in which case the argument is trying to prove that the causes of the former, not simply the causes of the latter, "must be resembling."[13]

Having given his own version of the argument from design, and obtained Cleanthes' assent that it is a "fair representation"[14] thereof. Philo proceeds to make seven criticisms of it in Part 2. Due to space considerations, I shall not examine these criticisms but shall pass on immediately to consider Philo's main objections to the argument from design as they appear in Parts 4, 5, and 6–8.[15] These objections are that the argument from Design (1) involves an infinite regress; (2) is susceptible to a reduction *ad absurdum*, and (3) utilizes data which are equally supportive of a multiplicity of alternative hypotheses to a Designer of the world who is intelligent and external, and thus not supportive of Cleanthes' hypothesis at all.

Philo's first objection arises out of a dispute between Cleanthes and Demea. Cleanthes cannot understand how mystics like Demea, "… who maintain the

9 *Dialogues*, 110.

10 *Dialogues*, 110.

11 *Dialogues*, 112.

12 Stanley Tweyman, ed., Introduction, *Dialogues*, 5.

13 Hume, *Dialogues*, 112. For a fuller discussion of Philo's reconstruction of Cleanthes' argument form design, see Stanley Tweyman, ed., Introduction, *Dialogues*, 4–8; cf. Stanley Tweyman, *Scepticism and Belief in Hume's Dialogues Concerning Natural Religion* (Dordrecht: Martinus Nijoff, 1986), 36–42.

14 Hume, *Dialogues*, 112.

15 For a consideration of Philo's seven criticisms in Part 2, see Tweyman, ed., Introduction, *Dialogues*, 8–11; cf. Tweyman, *Skepticism and Belief*, 42–46.

absolute incomprehensibility of the deity, differ from sceptics or atheists, who assert that the first cause of all is unknown and unintelligible."[16] Demea responds in kind by noting that to be labelled an "anthropomorphite," like Cleanthes, is just as bad as being labelled a "mystic" like himself and others.[17] Philo picks up on the allusion to anthropomorphism and "… endeavour[s] to show you [Cleanthes], a little more distinctly [than Demea has], the inconveniences of that anthropomorphism, which you have embraced; and shall prove, that there is no ground to suppose a plan of the world to be formed in the divine mind, consisting of distinct ideas, differently arranged; in the same manner as an architect forms in his head the plan of a house which he intends to execute."[18] In other words, Philo uses Demea's attack on anthropomorphites as an opportunity to return to his critique of the argument from design, which has been in suspension since the end of Part 2.

According to Philo, anthropomorphism maintains, in Cleanthes' words, "… his [the Deity's] similarity to human mind and intelligence."[19] Thus defined, anthropomorphism is rooted in idealism which "… trace[s] the world of objects into a similar universe of ideas."[20] Having made this move, however, what is to stop us from going on with an infinite regress of ideal causes? "Have we not the same reason to trace the ideal world into another ideal world, or new intelligent principle? … How can we satisfy ourselves without going on *ad infinitum*?"[21] The argument from design fails because the analogy it draws between the means to ends relationships and coherence of parts present in the design of machines and the design of the world to prove that the cause of the former resembles the cause of the latter leads us into an infinite regress of causes in the case of the cause of the world. Philo concludes that "… it were … wise in us, to limit all our inquiries to the present world, without looking farther. No satisfaction can ever be attained by these speculations, which so far exceed the narrow bounds of human understanding."[22]

Philo's second criticism of the argument from design is again aimed at Cleanthes' anthropomorphism and adduces "more [of its] inconveniences."[23] Taking the principle, like effects prove like causes, which Cleanthes invoked in

16 Hume, *Dialogues*, 122.
17 *Dialogues*, 122.
18 *Dialogues*, 124.
19 *Dialogues*, 109.
20 *Dialogues*, 126.
21 *Dialogues*, 125.
22 *Dialogues*, 126.
23 *Dialogues*, 128.

the argument from design and to which Cleanthes assented in Philo's reformulation of the argument from design later in Part 2, Philo uses this principle to divulge the finite view of the deity which is implicit in anthropomorphism. If like causes produce like effects, then "the liker the causes are, which are seen, and the liker the causes, which are inferred, the stronger is the argument."[24] This being the case, the cause of the world cannot be unlike the finite causes of human contrivances but like them; indeed, "... the liker the better."[25] Despite claiming in the argument from design that "... the Author of Nature is somewhat similar to the mind of man, though possessed of much larger faculties,"[26] Cleanthes is forced to concede Philo's point which promptly sends Philo off on a series of consequences designed to show the absurdity of Cleanthes' anthropomorphism.

First, Cleanthes has to renounce any claim to infinity in any of the attributes of the deity. The effects we know through experience are all finite in which case it would follow, given this principle, that the causes are finite too. Second, there is no reason "... for ascribing perfection to the deity, even in his finite capacity, or for supposing him free from error, mistake, or incoherence in his undertakings."[27] Given the "many inexplicable difficulties in the works of nature,"[28] these difficulties are now simply "... new instances of likeness to human art and contrivance."[29] Third, even if the world were perfect, this does not necessarily mean that all its excellencies could be ascribed to the workman. "If we survey a ship, what exalted an idea must we form of the ingenuity of the carpenter, who framed so complicated, useful, and beautiful a machine? And what surprise must we entertain when we find him a stupid mechanic who imitated others ..."[30] Fourth, the unity of the deity can no longer be rationally sustained. "A great number of men join in building a house or ship, in rearing a city, in framing a commonwealth. Why may not several deities combine in contriving and framing a world?"[31] Fifth, if several deities contrived and framed the world, and men renew their species by generation, as do all living creatures, why must this principle be excluded from "those numerous and limited deities?,"[32] in which case we are back, like the ancients, to accounting for the

24 *Dialogues*, 128.
25 *Dialogues*, 129.
26 *Dialogues*, 109.
27 *Dialogues*, 129.
28 *Dialogues*, 129.
29 *Dialogues*, 129.
30 *Dialogues*, 130.
31 *Dialogues*, 130.
32 *Dialogues*, 131.

generations of deities, or theogony.[33] Lastly, "why not become a perfect anthropomorphite?," Philo asks, and "… assert the deity or deities to be corporeal, and to have eyes, a nose, mouth, ears, etc."[34]

In short, Philo takes Cleanthes' anthropomorphism and, using the principle, "like effects prove like causes," which Cleanthes invoked in the argument from design, reduces anthropomorphism, and, to that extent, the argument from design, to absurdity.

Philo's third criticism of the argument from design is that it utilizes data which are equally, or even more, supportive of a multiplicity of alternative hypotheses to a Designer of the world who is intelligent and external, and thus is not supportive of Cleanthes' hypothesis at all. Like the second criticism, this one is also derived from Cleanthes' commitment to the principle, like effects arise from like causes. Unlike criticisms one and two, however, it zeroes in on Cleanthes' commitment to empiricism in natural religion, as evidenced by his claim, having formulated the argument from design in Part 2, that "By this argument *a posteriori,* and by this argument alone, do we prove at once the existence of a Deity, and his similarity to human mind and intelligence."[35] At no point does Cleanthes develop his empiricism as an epistemology, but Philo correctly deigns that he is committed to the view that our limited experience is an equal, i.e., the sole and sufficient, standard by which to judge of the unlimited extent of nature.[36] In Parts 6, 7, and 8, Philo uses Cleanthes' commitment to the principle, like effects arise from like causes, as well as his commitment to *a posteriori* reasoning and to the correlative principle, "… derived from the same source of experience, That where several known circumstances are observed to be similar, the unknown will also be found similar"[37] to continue to "… argue

33 *Dialogues*, 131.

34 *Dialogues*, 131.

35 *Dialogues*, 109.

36 cf. *Dialogues*, 134.

37 *Dialogues*, 133. Philo gives two examples of this principle: "Thus, if we see the limbs of a human body, we conclude that it is also attended with a human head, though hid from us. Thus, if we see, through a chink in a wall a small part of the sun, we conclude that, were the wall removed, we should see the whole body" (133). Summing up this principle, he observes: "In short, this method of reasoning is so obvious and familiar, that no scruple can ever be made with regard to its solidity" (133). Cleanthes never professes this principle *per se,* but he does presuppose it in the argument from design. Reasoning from what we see to be true of human contrivances and the universe, such as they come under our experience, and given the fact that they resemble each other in terms of means and ends relations and coherence of parts, therefore their causes also resemble each other. Furthermore, in Part 6, Philo attributes this principle to Cleanthes ("What we see in the parts, we may infer in the whole; at least, that is the method of reasoning, on which you rest your whole theory"—136) and Cleanthes does not disavow it.

with Cleanthes in his own way" and show him "… the dangerous consequences of his tenets"[38] in natural religion.

Thus interpreted, Philo's third criticism, while extremely complex, can be stated quite simply: working solely with the data furnished by our limited experience, along with the two principles just mentioned, there are any number of alternative cosmogonies we can give which meet or exceed the plausibility of the argument from design, in which case there is no reason to privilege that argument over the competition. Despite throwing out numbers like "a hundred,"[39] Philo essentially presents two such alternatives. According to part one of the first, "… if we survey the universe, so far as it falls under our knowledge, it bears a great resemblance to an animal or organized body, and seems actuated with a like [internal] principle of life and motion."[40] Philo modifies this claim, observing that "… though the world does, in many circumstances, resemble an animal body, yet is the analogy also defective in many circumstances, the most material. No organs of sense; no seat of thought or reason; no-one precise origin of motion and action. In short, it seems to bear a stronger resemblance to a vegetable than to an animal."[41] In Part 7, Philo combines these two accounts to give his first complete alternative to the argument from design:

> Our experience, so imperfect in itself, … can afford us no probable conjecture concerning the whole of things. But if we must needs fix on some hypothesis; by what rule, pray, ought we to determine our choice? Is there any other rule than the greater similarity of the objects compared? And does not a plant or an animal, which springs from vegetation or generation, bear a stronger resemblance to the world, than does any artificial machine, which arises from reason and design?[42]

Unlike the argument from design, this first alternative has no need of an external Designer since the structure and design of plants and animals are internal to them. Quoting some of Cleanthes' opening words from the argument from design ("Look round the world."), Philo invites his empirical friend to "look around"[43] and compare the evidence for this cosmology with that for the argument from design: "A tree bestows order and organization on that tree, which

38 *Dialogues* 111.

39 *Dialogues*, 140; 143.

40 *Dialogues*, 133.

41 *Dialogues*, 135.

42 *Dialogues*, 139.

43 *Dialogues*, 140.

springs from it, without knowing the order. An animal, in the same manner, on its offspring: A bird on its nest. And instances of this kind are even more frequent in the world, than those of order, which arise from reason and contrivance."[44]

Philo's second alternative to the argument from design is presented in Part 8. Unlike the previous alternative, which Philo traces back to Hesiod's *Theogony* and Plato's *Timaeus*,[45] the second alternative is derived from "the old Epircurean hypothesis."[46] Philo modifies it by assuming that matter is finite, rather than infinite as Epicurus did,[47] and devises "… a new hypothesis of cosmogony":[48]

> … the universe goes on for many ages in a continued succession of chaos and disorder. But is it not possible that it may settle at last, so as not to lose its motion and active force (for that we have supposed inherent in it) yet so as to preserve an uniformity of appearance, amidst the continual motion and fluctuation of its parts? This we find to be the case with the universe at present. Every individual is perpetually changing, and every part of every individual, and yet the whole remains, in appearance, the same. May we not hope for such a position, or rather be assured of it, from the eternal revolutions of unguided matter, and may not this account for all the appearing wisdom and contrivance, which is in the universe? Let us contemplate the subject a little, and we shall find, that this adjustment, if attained by matter, of a seeming stability in the forms, with a real and perpetual revolution, or motion of parts, affords a plausible, if not a true solution of the difficulty.[49]

Here again, we have another cosmogony which adheres to Cleanthes' threefold commitment to empiricism, the principle that like effects prove like causes, and the principle that what we see in the parts we may infer in the whole. Given these criteria, this cosmogony is just as plausible, if not moreso, than the argument from design. Unlike the latter, however, it does not involve an external Designer but, as Philo notes elsewhere, "… an eternal, inherent principle of the

44 *Dialogues*, 141.
45 "Hesiod and all the ancient mythologists, were so struck with this analogy, that they universally explained the origin of nature from an animal birth and copulation. Plato, too, so far as he is intelligible, seems to have adopted some such notion in his *Timeus*" (*Dialogues*, 142).
46 *Dialogues*, 143.
47 *Dialogues*, 143.
48 *Dialogues*, 144.
49 *Dialogues*, 145–6.

order of the world."[50] Given this parity, or even superiority, we might expect Philo to recommend one of his alternative cosmogonies. Yet that would belie his whole purpose in Parts 2 and 4–8 of "… argu[ing] with Cleanthes in his own way; [to] show him the dangerous consequences of his tenets."[51] Philo's cosmogonies are not an attempt to establish alternatives to Cleanthes' argument from design, but rather to show that "All religious systems … are subject to great and insuperable difficulties. … [A]ll of them … prepare a complete triumph for the sceptic, who tells them, that no system ought ever to be embraced with regard to such subjects."[52] Accordingly, Part 8 ends with Philo endorsing scepticism in natural religion where "A total suspense of judgment is here our only reasonable resource."[53]

Philo's three criticisms of the argument from design are all based on formal considerations which would be accepted by modern science. His first criticism rejected the argument from design because the analogy it draws between the means to ends relationships and coherence of parts present in the design of machines and the design of the world to prove that the cause of the former resembles the cause of the latter leads us into an infinite regress of causes in the case of the cause of the world. This would be a serious objection for modern science because it recognizes efficient causality as the only kind of causality appropriate for explaining phenomena in the physical world. If an argument tries to account for certain features of the physical world, but is unable to do so because the series of efficient causes it invokes lacks a first cause *due to that very explanation,* therefore, this would be sufficient for rejecting it as a scientific account of its subject.

Philo's second criticism rejected the argument from design because it reduces to absurdity. It does this by taking the principle, like causes produce like effects, and uses it to show that the cause of the world cannot be unlike the finite causes of human contrivances, which then sets off a series of consequences revealing the absurdity of Cleanthes' anthropomorphism. Again, modern science would find this a serious objection to the argument from design because it simply traces the ramifications of one of its own principles of inference in the case at hand. Modern science infers that effects like high tides and low tides, summer and winter solstices, solar and lunar eclipses, and so on are the same because they have the same efficient causes. Since that is exactly the

50 *Dialogues*, 157.
51 *Dialogues*, 111.
52 *Dialogues*, 147.
53 *Dialogues*, 147.

principle that Philo establishes to generate his *reductio ad absurdum*, modern science would have to take his second criticism seriously.

Philo's third criticism is probably the one that modern science would find most compelling because it relies on an explanatory principle which modern science frequently employs itself, i.e., if two or more theories are able to explain some phenomenon and equally account for its properties and characteristics, there is no sufficient reason for preferring one alternative over the other; indeed, a suspension of explanatory judgment is the only appropriate response in this situation. Thus, if modern science were able to equally account for light in terms of the wave and plenum theories, neither theory would have won the day. If these were the only explanations available, the appropriate response would have been to suspend judgment on the nature of light. The reason the wave theory was eventually accepted over the plenum theory (and other theories) was that it was better able to account for certain properties of light (e.g. how the speed of light is affected by the media through which it travels) than the plenum theory (and other theories). Since this is exactly the explanatory principle which Philo employs in his third criticism, modern science would have to take it seriously. Thus, Philo's conclusion at the end of Part 8 that "A total suspense of judgement is here our only reasonable resource" is a product of the "scientific" strand in Hume's thought and would be impossible without it.

2 The Humanities in Philo's Critique of the Argument from Design

Philo's criticism of the argument from design does not resume until Part 12. As noted earlier, his conclusion in Part 12 runs as follows:

> If the whole of natural theology, as some people seem to maintain, resolves itself into one simple, though somewhat ambiguous, at least undefined proposition, *that the cause or causes of order in the universe probably bear some remote analogy to human intelligence* ... , what can the most inquisitive, contemplative, and religious man do more than give a plain, philosophical assent to the proposition, as often as it occurs, and believe, that the arguments, on which it is established, exceed the objections, which lie against it?[54]

54 *Dialogues*, 184–5.

Whatever this conclusion may mean exactly, it is not a total suspension of judgment concerning natural religion. Rather, it *is* a judgment concerning natural religion. Furthermore, it is a judgment which draws, however tentatively and detachedly, *some* analogy between the cause or causes of the universe and human intelligence which is precisely what Philo's three criticisms in Parts 4, 5 and 6–8 preclude. The question thus arises, "Why does Philo's conclusion in Part 12 change so much?" My answer is that it is the humanistic strand in Hume's thought, combined with the scientific strand we have seen operative to date, which account for the difference.

In the second paragraph of Part 12, Philo admits to being "… less cautious on the subject of natural religion than on any other."[55] He gives two reasons for his relative abandon: first, "I know that I can never, on that head, corrupt the principles of any man of common sense" and, second, "… because no-one, I am confident, in whose eyes I appear a man of common sense, will ever mistake my intentions."[56] Philo then proceeds to state his "… unfeigned sentiments on this subject; and these sentiments, you [Cleanthes] know, I have ever cherished and maintained."[57] In many respects, Philo's unfeigned sentiments on natural religion are very similar to Cleanthes' position on natural religion as presented in his argument from design in Part 2 (with one critical difference which will emerge in a moment):

> That the works of nature bear a great analogy to the productions of art is evident; and according to all the rules of good reasoning, we ought to infer, if we argue at all concerning them, that their causes have a proportional analogy. But as there are also considerable differences, we have to reason to suppose a proportional difference in the causes; and in particular ought to attribute a much higher degree of power and energy to the supreme cause than any we have ever observed in mankind. Here then the existence of a DEITY is plainly ascertained by reason; and if we can make it a question, whether, on account of these analogies, we can properly call him a *mind* or *intelligence,* notwithstanding the vast difference, which may reasonably be supposed between him and human minds, what is that but a mere verbal controversy? No man can deny the analogies between the effects: To restrain ourselves from enquiring concerning the causes is scarcely possible. From this enquiry, the legitimate conclusion is, that the causes also have an analogy: And if we are not contented

55 *Dialogues*, 172.
56 *Dialogues*, 172.
57 *Dialogues*, 177.

> with calling the first and supreme cause a GOD or DEITY, but desire to vary the expression, what can we call him but *mind* or *thought*, to which he is justly supposed to bear a considerable resemblance.[58]

The critical difference between Philo's unfeigned sentiments on the subject of natural religion and Cleanthes' argument from design is that Philo now holds that disputes in natural religion are "verbal."[59] The reason for this is that "… there is a species of controversy which, from the very nature of language and of human ideas, is involved in perpetual ambiguity, and can never, by an precaution or any definitions, be able to reach a reasonable certainty or precision. These are the controversies concerning the degrees of any quality or circumstance."[60] While many people are intrigued by controversies like whether Hannibal was great, very great, or superlatively great, what degree of beauty Cleopatra possessed, or how much praise Livy and Thucydides deserve,[61] these kinds of disputes admit of no resolution. Questions like this are interesting but answers, or at least precise and definite answers, are not forthcoming because of the nature of the questions themselves.

Philo contrasts such disputes with those involving quantity or number:

> The disputants may here [in verbal disputes] agree in their sense and differ in the terms, or *vice versa*, yet never be able to define their terms, so as to enter into each other's meaning. Because the degrees of these qualities are not, like quantity or number, susceptible of any exact mensuration, which may be the standard in the controversy.[62]

In contemporary terms, the distinction Philo is drawing here is between the humanities and science. The humanities are those disciplines which study what it is to be human through the study, primarily, of literary texts. These disciplines raise all sorts of questions about what it is to be human, none of which has a precise and definite answer. The reason they do not have precise and definite answers is because they cannot be settled by appealing to numerical and quantitative criteria. They are disputes which typically arise in the study of literary texts or, in Philo's language, "controversies concerning the degrees of any quality or circumstance."[63] Science, on the other hand, designates those

58 *Dialogues*, 174–175.

59 *Dialogues*, 175.

60 *Dialogues*, 175.

61 *Dialogues*, 175.

62 *Dialogues*, 175.

63 *Dialogues*, 175.

disciplines which study the physical world in order to determine the principles and laws according to which it operates. Because the subject of science is physical, rather than literary, it can be measured and, at least in principle, precise and definite answers can be given to the questions it raises. What Philo is doing here, then, is placing natural religion squarely in the cluster of subjects we call "the humanities" today. Despite his unfeigned sentiments on the subject, which are very similar to Cleanthes' position in the argument from design, Philo places natural religion under the broad rubric of "verbal disputes." These disputes have an ongoing fascination for human beings and, as the *Dialogues* certainly attests, are legitimate areas of inquiry, but they do not produce precise and definite results.

A further dimension of our distinction today between the humanities and science is evidenced in Philo's "unfeigned sentiments" on natural religion in Part 12 as opposed to his scientific rigour in critiquing the argument from design in Parts 4–8. Prior to his critique of "vulgar superstitions"[64] beginning at paragraph nine, Part 12 reads like a litany of personal confessions. Indeed, Philo uses the words "I … confess"[65] in prefacing his revelation of his unfeigned sentiments: "I must confess … that I am less cautious on the subject of natural religion than on any other."[66] What follows is a shocking revelation of the feelings of a man who, according to his last pronouncement on natural religion, held that "A total suspense of judgment is here our only reasonable resort."[67] Instead of this, we hear Philo divulge the innermost secrets of his heart:

> You, in particular, CLEANTHES, with whom I live in unreserved intimacy; you are sensible, that, notwithstanding the freedom of my conversation, and my love of singular argument, no-one has a deeper sense of religion impressed on his mind, or pays more profound adoration to the divine Being, as he discovers himself to reason, in the inexplicable contrivance and artifice of nature.[68]

Needless to say, this kind of language, while it has implications for modern science as we shall see in a moment, is not itself scientific or science-based. Rather, it is humanistic in the sense that it is concerned with what it is to be human as expressed though the medium of a literary text. Thus, it would be

64 *Dialogues*, 177.
65 *Dialogues*, 172.
66 *Dialogues*, 172.
67 *Dialogues*, 147.
68 *Dialogues*, 172.

of great interest to the humanities, as we understand this cluster of subjects today, but of no interest to modern science *per se*.

While Philo's confessions of his unfeigned sentiments on natural religion are not scientific, they nevertheless have implications for science. Concurring with the maxim "*That nature does nothing in vain,*"[69] Philo notes that "… merely from the contemplation of the works of nature, without any religious purpose; and from a firm conviction of its truth, an anatomist, who had observed a new organ or canal, would never be satisfied, till he had also discovered its use and intention."[70] He gives two other examples, one involving astronomy and the second involving anatomy again, to show that, while science discovers purposiveness in nature, it never brings us to a belief in purposiveness.[71] Rather, by observing the adaptation of means to ends and coherence of parts throughout nature, something which we all do as *human beings*, we (including scientists) are struck with a belief in purposiveness. Thus, our belief in the maxim that nature does nothing in vain is not a product of scientific investigation but of our lived experience of the world as human beings. In terms of our contemporary distinction between science and the humanities, then, Philo's confessions are distinctly humanistic though they do have implications for science.[72]

Having expressed his unfeigned sentiments on natural religion, Philo turns to a critique of "vulgar superstitions."[73] This material occupies over half of Section 12 and is very similar to some of Hume's other writings on religion, especially the essay "Of Superstition and Enthusiasm" (1741) and Sections XIII and XIV of "The Natural History of Religion" (in *Four Dissertations*—1757), entitled "Impious Conceptions of the Divine Nature in Popular Religions of Both Kinds" and "Bad Influences of Popular Religion on Morality" respectively.[74] What does Philo's prolonged critique of vulgar superstitions have to do with natural religion? Briefly, Philo uses the results of his critique to reassess his appraisal of the analogy in the argument from design, reached earlier in the *Dialogues* by arguing with Cleanthes "in his own way."[75] Looking at Cleanthes' analogy

69 *Dialogues*, 172.

70 *Dialogues*, 172.

71 *Dialogues*, 172–173.

72 For more on the implications of Philo's unfeigned sentiments for science, see Tweyman, ed., *Dialogues*, 85–87; cf. Tweyman, *Scepticism and Belief*, 130–132.

73 Hume, *Dialogues*, 177.

74 For "Of Superstition and Enthusiasm," see Miller, ed., *David Hume, Essays Moral, Political, and Literary*, 73–79; for Sections XIII and XIV of Hume's "The Natural History of Religion," see Antony Flew, ed., *David Hume: Writings on Religion* (La Salle, Illinois: Open Court, 1992), 168–174; 175–179.

75 Hume, *Dialogues*, 111.

from this vantage point, the results of his critique of vulgar superstition are twofold: first, given Philo's lofty idea of the Deity, as revealed in his unfeigned sentiments on natural religion, it is absurd to believe that the Deity has human passions including, as vulgar superstition maintains, "… a restless appetite for [human] applause."[76] Second, it is inconsistent to believe that since the Deity has the passion for [human] applause, he also disregards the opinions we maintain of him, such as being degraded "… to the low condition of mankind, who are delighted with entreaty, solicitation, presents, and flattery",[77] which again vulgar superstition maintains.

These results are intended to weaken the analogy in the argument from design between the causes of human contrivances and the cause of the world. They do no constitute a scientific objection in our sense of science, however, because they object to the analogy on purely humanistic grounds, i.e., our idea of God as having human qualities and such a God disregarding the opinions we have of him. Even though Philo may appear to be resuming his scientific critique of the argument from design, then, his two objections to it in Part 12 fall entirely within the purview of the humanities.

So how does the humanistic strand in Hume's thought, as evidenced by Philo's unfeigned sentiments on natural religion and his reappraisal of the argument from design, contribute to Philo's conclusion at the end of Part 12? If science alone appraises the argument from design, we reach the conclusion of Part 8: "A total suspense of judgment is here our only reasonable resource." If the humanities alone appraises this argument, we get a passionate affirmation of it in terms of unfeigned sentiment coupled with some serious reservations about the analogy, given its humanistic conception of the deity. When the verdict of science is filtered through the prism of the humanities, we get a third result which is the one given at the end of Part 12: an extremely tentative and detached affirmation of the analogy in the argument from design, to which the "most inquisitive, contemplative, and religious man" can only assent as often as the proposition affirming that analogy occurs.[78]

3 Conclusion

Whatever we think of Philo's ultimate pronouncement on natural religion, it is not the same as his conclusion in Part 8. Given the evidence I have presented,

76 *Dialogues*, 184.
77 *Dialogues*, 184.
78 *Dialogues*, 185.

the difference results from a fruitful dialogue between science and the humanities. At the end of the day, neither one wins a "complete triumph;"[79] rather, by pooling their resources, they jointly contribute to Philo's final verdict on the subject. What makes this outcome possible is not simply Philo's position on natural religion, much less Cleanthes', but Hume's overriding commitment to a "one culture" view of the disciplines which guides and directs the exchange of ideas in the *Dialogues*. Hume did not distinguish between the science and the humanities, as we do today, but we can read our "two culture" view of the disciplines into his "one culture" view. By so doing, we are not only able to *understand* the *Dialogues* better but to *listen* to it as a model case of fruitful dialogue between science and the humanities in the history of ideas.

References

Easlea, Brian, "Patriarchy, Scientists, and Nuclear Warriors," in *Beyond Patriarchy: Essays By Men on Pleasure, Power, and Change*, ed. Michael Kaufman, Toronto: Oxford University Press, 1987.

Grant, George, "Faith and the Multiversity," in *Technology and Justice*, Concord, Ontario: Anansi, 1986, 35–78.

Hume, David, *A Treatise of Human Nature*, ed. David Fate Norton and Mary Norton, 2 vols., Oxford: Clarendon Press, 2007.

Hume, David, *Dialogues Concerning Natural Religion*, ed. Stanley Tweyman, Ann Arbour: Caravan Books, 2000.

Hume, David, *Enquiry Concerning the Principles of Morals*, ed. Tom L. Beauchamp, Oxford: Clarendon Press, 1998.

Hume, David, "Of Superstition and Enthusiasm," in *David Hume, Essays Moral, Political, and Literary*, ed. Eugene F. Miller, Indianapolis: Liberty*Classics*, 1985, 73–79.

Hume, David, "The Natural History of Religion," in *David Hume: Writings on Religion*, ed. Antony Flew, La Salle: Open Court, 1992, 107–182.

Snow, C.P., *The Two Cultures and a Second Look*, New York: Cambridge University Press, 1965.

Snow, C.P., *The Two Cultures and the Scientific Revolution*, New York: Cambridge University Press, 1959.

Tweyman, Stanley, *Scepticism and Belief in Hume's Dialogues Concerning Natural Religion*, Dordrecht: Martinus Nijoff, 1986.

79 *Dialogues*, 147.

CHAPTER 3

Idealism and Philosophy of Science

Hugo Meynell

I will begin from some general epistemological and metaphysical considerations; which tend to establish, as I believe, that philosophical idealism, as generally understood, is an enormously important half-truth. It has been said that what may be called "critical realism," which I regard as the truth in epistemological and metaphysical matters, lies beyond idealism in rather the same way that idealism lies beyond naïve realism. This has important implications, as I shall try to show, for the philosophy of science. For naïve realism, what is real is simply the object of extroverted consciousness. Idealism rightly stresses the constructive role of intelligence in our apprehension of what we take to be "reality" or "the actual world," but, in many of its forms at least, underestimates the role of judgment in establishing which of our intellectual constructions are certainly or probably true and which are certainly or probably false. Critical realism has in common with naïve realism the view that there is a world of facts which exists prior to and independently of our intellectual constructions; nevertheless, we can only come to knowledge of this world by the use of such intellectual constructions. I want to stress, in contrast with such authors as Thomas Kuhn and Paul Feyerabend, the continuity between the methods of science (so far as a sympathetic outsider can understand them), on the one hand, and our ordinary ways of finding out the truth about things, on the other.

What is idealism? For Antony Flew, "... 'idealism' is [a] name given to a group of philosophical theories, that have in common the view that what would normally be called 'the external world' is somehow created by the mind."[1] (The "somehow" in this and subsequent quotations will prove crucial in the sequel, if I am right.) According to Timothy Sprigge, in his masterly survey of the subject, "Idealism is now usually understood in philosophy as the view that mind is the most basic reality and that the physical world exists only as an appearance to or expression of mind, or as *somehow* (my italics) 'mental in its inner essence'."[2] As G. Vesey and P. Foulkes see it, idealism is "... a doctrine that holds the world to be essentially a mental vision, as against the common-sense

1 Antony Flew, *A Dictionary of Philosophy* (New York: St Martin's Press, 1984), 160.
2 T.L.S. Sprigge, "Idealism," in *The Routledge Encyclopedia of Philosophy*, Volume 5, 662.

© KONINKLIJKE BRILL NV, LEIDEN, 2020 | DOI:10.1163/9789004415270_005

realist view that it is something 'out there' and independent of us." They add for good measure that "... idealism has never been a convincing point of view" in the natural sciences.[3] D.W. Hamlyn writes that philosophical idealism "... maintains in general that what is real is *in some way* confined to or at least related to the contents of our own minds."[4] Within the heterogeneous collection of philosophical views that have been termed "idealist," what I myself want to stress is that feature which draws attention to the role of mental creation, constitution or construction in our apprehension of what we call "the real world," and the latter's consequent seeming dependence on the "mental." To put it in another way, *the nature and structure of the mind are a crucial clue to the nature and structure of the cosmos and whatever may be its cause, causes, or ground.* The curious way in which mind and the material world are wrapped round one another may seem to be evaded or repressed by empiricists and materialists.

What, briefly, do I mean by the "critical realism" which I take to be the correct position in epistemology and metaphysics?[5] It is often remarked that some statements are self-defeating, without being exactly self-contradictory; the most famous is the statement, "I am lying," which gives rise to the "liar paradox." If the person who makes the statement is speaking the truth, then he is lying; but lying is incompatible with speaking the truth. The statement, "I never make a true judgment," is similarly self-defeating. Now reality, in the last analysis, is nothing other than what true judgments are about. How do we make true judgments? So far as our judgments are well-founded, they tend to be true; for all that, by way of exception, a judgment may be as well-founded as is possible for the person who makes it in the situation in which she makes it, and yet turn out to be false. (The most careful and thorough possible deliberations of a court of law may fail to come to the true answer about who committed a murder; but Grandpa may chance to hit on the right answer after consulting his tea-leaves.[6] Yet it remains that one could only find out that Grandpa was right by further deliberations of the kind carried out by the best courts of law.)

---

3 G. Vesey and P. Foulkes, *Collins Dictionary of Philosophy* (London and Glasgow: Collins, 1990), 146–147.

4 D.W. Hamlyn, "Idealism," in *Oxford Companion to Philosophy* (Oxford and New York: Oxford University Press, 1995), 386.

5 The most complete exposition and justification of critical realism is Bernard Lonergan's *Insight: A Study of Human Understanding* (*Collected Works of Bernard Lonergan*, Volume 3; Toronto: University of Toronto Press, 1992; first published 1957). Of the great works of philosophy composed during the twentieth century, this is the most underestimated.

6 I am tempted to call this "the tealeaves paradox."

That judgments tend to be true so far as they are well-founded makes it self-defeating, also, to deny that one ever makes well-founded judgments, or to make judgments which imply such denial. What is it for a judgment to be well-founded? It is for it to have been arrived at attentively, intelligently, and reasonably—where attentiveness is to the relevant evidence in experience, intelligence is a matter of envisaging possibilities or hypotheses, and reasonableness consists in judging in each case as certainly or probably so that hypothesis which most adequately explains the evidence. Such are the elements of critical realism. It accounts for the fact that, by means of our thought, we can get to know about facts which obtain prior to and independently of our thought—for example, that the Battle of Hastings occurred in 1066 or that there is a giant planet in the solar system outside the orbit of Uranus. It also accounts at once for the common-sense judgments which seem to support naïve realism, and for the psychological or historical or scientific-theoretical judgments where naïve realism seems palpably to break down. I can judge for good reason that those states of affairs obtain which I perceive directly, like that there are a computer and a yellow scrap of paper within a foot of my body as I compose the first draft of this sentence. But I can also reasonably judge, on the basis of evidence available to me, that Henry VIII had six wives,[7] that my neighbor is feeling cross at what I have just told him, and that there are over ninety naturally-occurring chemical elements. Reality as a whole is indeed, on a critical realist view, "in some way" related to the contents of our minds, as what our minds would come to know at the ideal term of attentive, intelligent and reasonable (let us say "rational" for short) inquiry.

Whether critical realism is itself a kind of idealism is moot. It is rather like the question whether classical sonata first-movement form is binary or ternary. Is the recapitulation the second part of the second part or is it rather a

7 I recently read a collection of articles, in which it was claimed by several authors that historical knowledge was impossible (*The Philosophy of History*, ed. A.L. Macfie, Basingstoke and New York: Palgrave Macmillan, 2006). Yet it appears to me that we can know with virtual certainty that Horatio Nelson died in 1805 and that Hitler was the leader of Germany during the Second World War. Are we really to believe that lawyers who get clients released from death row, on the grounds that the evidence shows that they did not commit the crimes with which they have been charged, have simply been wasting their time? One wonders whether the authors of this egregious work were implicitly contrasting the kind of "knowledge" we can have of the past with our knowledge of the medium-sized physical objects in our immediate environment, or even of our present feelings and sensations. If not, why did their scepticism not extend (perhaps it did) to judgments about other minds or about theoretical science? If postmodernism leads to this kind of scepticism, might this not give us some grounds, if the suggestion is not tantamount to blasphemy, for doubting postmodernism?

separate third part? Perhaps the special merit of this form is due to its having elements of both binary and ternary; similarly, one might suggest, critical realism combines the merits of idealism, empiricism, and materialism. Certainly, critical realism takes into account, in a way that the last two metaphysical theories may well be supposed not to do, of what I should like to call "the idealist point," which I made briefly at the end of the last paragraph but two. The great discoveries of Kepler, Darwin, or Planck were not just a matter of soaking up sense-experiences or observable facts and recording them; they were prodigious feats of mental creativity. Certainly, as we shall see, they are *more than* that on the critical realist view, but they are no *less*. It is one thing to conceive an hypothesis, to envisage a possibility; it is another to affirm for good reason that the possibility is a reality, that the hypothesis is certainly or probably so. In Bernard Lonergan's critical realist terms, intelligence is one kind of habit of mental operation and reasonableness another. Intelligence and reasonableness may be distinguished as answers to two types of questions: the former asks of a phenomenon or range of phenomena *what* or *why* they may be (Could that track in a bubble chamber be evidence of the existence of a previously unsuspected sort of fundamental particle? Could that shifty look on the bookie's face be a sign that he is trying to swindle me?), while the latter asks whether these suppositions are or are not probably or certainly so. (Evidently I cannot answer "Yes" or "No," except perhaps facetiously, to questions like, "What is the best treatment of a viral sore throat?," or "How many children had Lady Macbeth?")

Bertrand Russell wrote that common sense leads to physics, but that physics shows that common sense is wrong. How is one to understand this profound if cryptic judgment? It seems worthwhile to tease out the various conceptions involved of "common sense" and of the "science" which is exemplified by physics. To apply in the affairs of ordinary life the mental operations of attentiveness, intelligence, and reasonableness is only common sense; to identify them is something else; to work out the consequences of their place in knowledge and the ramifications of this for our overall conception of "reality" or "the world," which is nothing other than what is to be known, yet another. The theory of general relativity is quite hard to understand, or at least I find it so. But any fool can grasp, in practice if not in theory, that when you have two conflicting accounts, and observed circumstances confirm account A but conflict with what is to be expected on account B, this is good reason for preferring account A.

When Einstein first formulated his general theory of relativity in 1917, it was noted that, due to the bending of light-rays in the neighbourhood of large gravitational objects, which is a consequence of this theory as opposed to Newton's, stars in nearly the same direction as the Sun would appear, during a solar eclipse, to be in a different position than was otherwise to be expected. Such an eclipse occurred shortly afterwards, and the observations corroborated Einstein's theory. The moral is that mental procedures whose *application*, though not *articulation*, is a matter of common sense, lead to science when applied in a thoroughgoing way. In the case of the mature physical sciences, this has been done over many generations. The continuity between common sense and science is reminiscent of what apparently was Plato's view that we could not recognize the truth when we found it as the result of inquiry unless, in a sense at least, we already knew it.[8]

Closely allied to Lonergan's "act of understanding" or "intelligence" is what C.S. Peirce called "abduction" which can be described as follows: the surprising fact C is observed; but if A were true, C would be a matter of course; hence, there is reason to suspect that A is true.[9] It is among the many great philosophical merits of Peirce that he sets off the looseness involved in abductive inference so well. Suppose p is a theory, and q and r are putative observable facts. "If p," he seems to tell us, "then q would be a matter of course, but you would not expect r. But we observe q, and that it is not the case that r. Therefore, it is probably the case that p." There seems no way of making the link between theory and observations tight as a matter of deductive logic, which has led some to deny that scientific theories are supported by observations that do not themselves presuppose them. Some have inferred further that this does not matter, since science needs no support. I think it does need it, and badly, with the various types of irrationalism so prevalent at the present time. Abductions may also commend themselves by predicting facts which turn out to be the case. His "periodic table" of chemical elements enabled Dmitri Mendeleev to anticipate successfully that three previously-unsuspected elements would be found to exist.

On the subject of "induction," Antony Flew writes that it is "A method of reasoning by which a general law or principle is inferred from observed particular instances." He adds that "The term is employed to cover all arguments in which the truth of the premise, or premises, while not entailing the truth of the conclusion, or conclusions, nevertheless purports to constitute good reason for

8 *Euthydemus* 276e–277c; *Meno* 80e–86c; cf. A.E. Taylor, *Plato. The Man and His Work* (London: Methuen, 1960), 92; 136–138.

9 See Thomas A. Goudge, *The Thought of C.S. Peirce* (Toronto: University of Toronto Press, 1950), 195; cf. *Collected Papers of Charles Sanders Peirce*, ed. Charles Hartshorne and Paul Weiss, 6 vols. (Cambridge, Mass.: Harvard University Press, 1931–35), 2:511n.; 5:189; 5:590. Peirce writes also, in reference to the same sort of mental operation, of "Presumption." "Retroduction," and "Hypothesis" (Goudge, *Peirce*, 1950). *A propos* of Aquinas and his interpreters, Lonergan writes of terminological primness as the solitary achievement of lesser minds (*Verbum: Word and Idea in Aquinas*, London: Darton, Longman and Todd, 1968, 21–25).

accepting it, or them. … Induction by simple enumeration is often thought to be the fundamental form."[10] I believe this thought to be importantly mistaken; Kepler's prodigious intellectual feat sheds more light on what stupid people do every day by way of "induction by simple enumeration" than *vice versa*. Why does the planet Mars describe these peculiar loops, which seem explainable neither on the geocentric theory nor on the simple Copernican one where the planets go round the sun in perfect circles? Oh, perhaps it is because the planets go round the Sun in ellipses, such that a line drawn from the Sun to each planet covers an equal area in equal times. Why are so many of the ravens that I have observed black? Oh, perhaps it is that all ravens are black. In both cases—the everyday as well as the epoch-making, there is a creative act of understanding which goes beyond the data involved.

Of course it has been clear, at least since Pythagoras, that it is one thing to *conceive* a straight line or a circle in geometrical terms, another to *imagine* it as, say, drawn by a staff in the sand or by a pencil on paper. The same applies to terms in classical science such as Newtonian force, mass, and acceleration; it is symptomatic that they are interdefinable ("Force equals mass times acceleration."), that they are subject to mathematical analysis, and that they cannot directly be perceived (though of course they explain what is perceived and to be perceived). The "criticism of the faculties of sense and imagination," inaugurated by Rene Descartes, is said to be taken further by Nicolas Malebranche in whose *Inquiry Concerning Truth*, "The imagination appears not as a way to the truth but as the source of all the delusions to which the human mind is exposed. … To keep the imagination in check and to regulate it deliberately is the highest goal of all philosophical criticism."[11] The point is made emphatically by Robert Boyle, father of chemistry and brother of the Earl of Cork, who wrote:

> Because the use of imagining, whenever we would conceive things, is so stubborn an impediment to the free actings of the mind, in cases that require pure intellection, it will be very useful, if not necessary, to accustom ourselves not to be startled or frighted with every thing that exceeds or confounds the imagination, but by degrees to train up the mind to consider notions that surpass the imagination and yet are demonstrable by reason.[12]

10 Flew, *Dictionary*, 171–172.

11 Ernst Cassirer, *The Philosophy of the Enlightenment* (Boston: Little, Brown, 1962), 282–3; cited by Morton D. Paley, *Energy and the Imagination. A Study of the Development of Blake's Thought* (Oxford: Clarendon Press, 1970), 216.

12 Quoted by Paley, *Energy*, 216.

It is by replacing descriptions in terms of the sensible and imaginable with conceptions which are not thus perceivable and imaginable that, in accordance at once with critical realism and with popular views of science, we move progressively from knowledge of things related merely to our senses toward that of things as really related to one another. What is at issue has become especially clear with the advent of quantum mechanics, where reality has to be described in terms of unimaginable somewhats ("wavicles") which are neither waves nor particles, but somehow partake of the nature of both.

The "externality" of the "external world" is to be conceived differently on the critical realist than on the naïve realist or materialist view. For these last two views, the mind is inside us as external reality is outside us; isn't it obvious that the mind is the brain? From the point of view of critical realism, this view leads to intolerable paradoxes. It is not clear how the lumps of stuff which are your brain and mine can come to know about the other lumps of stuff supposed to constitute the universe. V.I. Lenin and D.M. Armstrong maintained respectively that the brain represents just in the same way that pictures and maps represent. I do not see how this view can long satisfy the critical mind. Does not representation of one material object by another—for example an area of countryside by a map or the state of a bank account by figures in the output of a computer—*presuppose* mind? If so, it cannot be used as *a reductive explanation of* mind.[13] The map has been drawn by a person or persons for that very purpose, and the computer similarly programmed. It might be the case, by a remarkable coincidence, that your treading with muddy boots at random on a piece of paper resulted in a pattern of marks which could be used by someone as a map of Medicine Hat, Alberta. But I do not see what sense it would make to claim that it *was* such a map, till someone had hit on the idea of using it for this purpose. (Perhaps one could rescue the view that our brains represented, at this rate, by saying that God as creator had designed them to do so; but I do not think that a materialist would thank me much for this solution to the problem.)[14]

---

13 For a valiant but, to my mind, wholly unsuccessful attempt to resolve this and related problems from the point of view of scientific materialism, see John R. Searle, *Minds, Brains and Science* (Cambridge, MA.: Harvard University Press, 1985).

14 One may compare Richard Taylor's argument for the existence of God in his *Metaphysics* (Englewood Cliffs, N.J.: Prentice Hall, 1963). We take it for granted, and cannot help doing so, that by the use of our minds we can come to know the real world. But, to take a parallel case, one cannot coherently take the sign, "You are entering Wales," as an indication that you are entering Wales, except on the assumption that the sign has been erected for that purpose.

J.L. Austin was a believer in the correspondence view of truth; this is also a consequence of critical realism. True judgments, according to Austin, correspond to the facts. P.F. Strawson, to the contrary, repudiated the correspondence theory of truth, and accordingly denied the existence of facts which obtained prior to and independently of our discourse about them. In a similar vein, Alasdair MacIntyre declared that "facts" were a human invention which came in with wigs for gentlemen.[15] I, for my part, agree with Austin. I think it is odd to deny that the existence of two giant planets in the Solar System outside the orbit of Saturn, that the extinction of the dinosaurs between sixty and seventy million years ago, and that the majority of the present population of Melbourne speak English as their first language are "facts," or that they obtain or obtained prior to and independently of anyone thinking, saying, or writing what they did about them. Idealism tends to dissolve into thin air facts as conceived by the naïve realist; but critical realism can affirm them after all as what tend to be known in judgments reached at the ideal term of rational inquiry. It is of interest that both Schelling and Lonergan accused Hegel, whom many would think of as the arch-idealist, of treating the world as though it had no facts in it.[16] This point about facts illustrates the general observation that critical realism restores common sense to you in some respects, but at a price; and the price, as we shall see, is to make theism highly plausible, perhaps even rationally inevitable.

"Facts," declared J.L. Austin, "are what true statements state."[17] Austin's lapidary maxim does not at first sight take one very far; but it turns out, I would argue, to be a surprisingly good start. One is inclined to protest: "Unless we can somehow independently get at the facts, how do we know what the true statements are? To what are we to resort but simple extraversion? Does that not do very well?" Someone tells me, when I am in the interior of a large building, that it is snowing, or when I am sitting in my office, that my Head of Department is in hers. To find out whether his statement is true, and whether these facts are the case, I go and take a look. But what of statements about other minds, or the past, or the theoretical entities of science? I cannot take a look at your thoughts and feelings, or Julius Caesar's campaigns in Gaul, or a quark or a lepton. If I am to get to know about such things, I have not only to attend to

15 P.F. Strawson, 'Truth," in *Logico-Linguistic Papers* (Aldershot: Ashgate, 2004); Alasdair MacIntyre, *Whose Justice? Which Rationality?* (Notre Dame, IN: University of Notre Dame Press, 1988), 357–358.

16 Lonergan, *Insight*, 398.

17 "Truth," *Proceedings of the Aristotelian Society*, Supplementary Volume, 1950.

my experience, but hypothesize and judge in a manner that goes beyond my experience.

This account turns out to suit the simpler cases too. I hypothesize that my Head of Department may be, or may not be, in her office; and I carry out the experiment which, short of special and extraordinary circumstances, establishes the fact of the matter one way or the other. (Such special and extraordinary circumstances might be that I had only dreamed that I had gone and taken a look or that, in order to deceive me about the matter, an actress had been hired to impersonate my Head of Department.) I would state the truth about the facts at the ideal term of rational inquiry; but they are generally what they are prior to and independently of myself or any other conscious subject actually carrying out such an inquiry.

Judgments about the facts of the world of our common-sense experience, established by ordinary people's rationality, are not apt to need revision in the manner of the judgments of scientific theory. If I, at quarter to five in the morning on a certain date, am really composing the first draft of this sentence in a senior's facility in Calgary, Alberta, then no person at any point in space-time can truly deny that I did. If we are rightly informed that the ten thousand Greeks made famous by Xenophon really saw the sea, then no one for the rest of time and over the rest of space can truly deny that thy did. On the other hand, Galileo's law of falling bodies is "true" only near the surface of the earth, as a special consequence of theories due to Einstein which themselves may be subject to similar revision, though Galileo's account is much "truer" than that of his Aristotelian rivals.[18] When we say or imply that an accepted scientific theory is true, perhaps we should say, more strictly, that it is more rational than its rivals, and therefore very likely to be closer to the truth than those rivals—as Galilean physics and astronomy are closer to the truth than Aristotelian, the Einsteinian account of cosmology to the Newtonian, and so on. As Popper says, so far as Einstein's theories are true, Newton's are "an excellent approximation." Presumably something very like our concepts of barium and sulphur would still figure at the ideal term of scientific explanation, as similar equivalents of phlogiston, caloric, or the luminiferous ether would not.

It has been characteristic of empiricism to stress the importance for science of experience, through observation and experiment, and of falsificationism, as

18 Aristotelian metaphysics comes out better than Aristotelian science. In fact, Aristotle strikingly anticipated the central elements of the generalized empirical method constitutive of critical realism, including the two kinds of question to be put to experience if we are to come to know. See *Posterior Analytics* II, 89b 32–35; Lonergan, *Verbum: Word and Idea in Aquinas*, 12–13.

championed by the late Sir Karl Popper, to insist further that a truly scientific theory is one that, while it could be falsified by such empirical tests, in fact turns out not to be so, and is thus, as Popper expresses the matter, corroborated. When Augustin Fresnel expounded his version of the wave theory of light to the French Academy, it was pointed out by S.D. Poisson that it would have the highly counterintuitive consequence that, when two beams of light crossed one another, there would be pockets of darkness where the waves cancelled out one another. This actually turned out to be the case; which was regarded, evidently rightly, as strongly supportive of Fresnel's proposal.

Both empiricism and falsificationism have been subjected to influential attack, largely on historical grounds—these are not the ways in which acknowledged advances in science have in fact been made. As Thomas Kuhn saw the matter in *The Structure of Scientific Revolutions*, inquiries into a range of phenomena tend to be more or less haphazard, until a genius like Newton, Lavoisier, Darwin, or Einstein turns up and establishes what Kuhn calls a "paradigm"—a set of conceptions which prevail over a discipline for some length of time, perhaps for centuries. There are no paradigms without anomalies or observations and experimental results which don't fit the paradigm, but these are not taken seriously so long as the paradigm is in place. Rather, efforts are made to force nature to conform to the paradigm. Eventually the anomalies may build up to such an extent that specialists cease to have confidence in the extant paradigm and a situation of confusion and uncertainty prevails until a new paradigm is established. Such a situation of "revolution" is to be contrasted with "normal science" where a paradigm reigns.

For Paul Feyerabend, the notion that scientists do their work by following a method is wholly mistaken; anarchy prevails and ought to prevail. Galileo, for example, resorted to any sort of argument he thought he could get away with to defeat his Aristotelian opponents. Empiricism and falsificationism are both highly distorting idealizations, if one looks at the history of science. Sir Karl Popper, the champion of falsificationism in science, admits that it is often appropriate to hang on, at least for a while, to a theory which has been falsified. But he provides no criteria for how much empirical evidence should accumulate against a theory before one abandons it, and it is to be concluded that the theory of falsificationism is empty, nothing more than a verbal ornament with which one adorns theories that have in fact proved successful. Kuhn and Feyerabend agree on the historical facts about the development of science, but are largely opposite in their evaluations. While Kuhn regards the reign of paradigms as proper and inevitable, for Feyerabend the scientist who sticks to a paradigm is to be denounced as a timid and intellectually slavish person.

It is a curious fact that both Kuhn and Feyerabend follow what amounts to a broadly empiricist or falsificationist approach—applying what amounts to the hypothetico-deductive method—in the historical studies with which they buttress their accounts of science. Moreover, their arguments seem to carry conviction only insofar as their authors do so. Their own conclusions, they seem to imply, are supported by objective scrutiny of the available evidence, while those of their rivals who give empiricist and falsificationist accounts of science are not. It is as though they were saying, "If the views of our opponents were right, the historical evidence that we have on the development of science would be rather different. But, if we were right, it would be a matter of course. So there is reason to think that we, and not our opponents, are correct about the matter." If one assumes that such an account must be applied if one wants to come to the truth, or at least closer to the truth, in historical matters, however, why should not the same apply when it comes to investigating the truth about things in physical nature?

The formulations of Peirce as cited above suggest a certain looseness of logical connection between the truth of theories and the evidence that supports them, and I believe that this is very important. The expectation of a tighter logical connection than in fact obtains is apt to lead first to disappointment, then to scepticism. The actual connection between each accepted scientific theory and the evidence for it obtainable by observation of experiment seems to be somewhat as follows: if A and not B, you would expect evidence W and X but not Y and Z; if B and not A, you would expect, to the contrary, evidence Y and Z, but not W and X. But W and X are found to obtain, and not Y and Z. Therefore, probably A is true and not B. This leaves it open that, if there is enough pressure elsewhere, "A is true and not B" will have rationally to be abandoned or revised.

Is it, perhaps, the disappointed expectation of a tighter logical connection between theory and evidence which has led Kuhn, Feyerabend and others to abandon the empirical and the falsificationist approaches to scientific progress? It is this that seems to have led W.V.O. Quine to his "holism" or the view that scientific theory is justified in relation to experience as a whole rather than each theory to a particular range of experience. From there it is easy to take the next step, which in fact has been done by Richard Rorty and postmodernists generally; to the effect that the claim that the prevailing theories of science, as opposed to their contradictories, are supported by experience, is a sham.[19]

19 There is an interesting parallel here with what happened to Christian theology in the fourteenth century; where the syntheses achieved by thinkers like Bonaventure, Aquinas and Duns Scotus were impugned by the ideal of a certain kind of logical rigour, which led

It appears to me that if the points which I have mentioned are properly taken into account, paradigms in science are after all quite satisfactorily to be threaded, along what one might call the "metaparadigm" of rational (attentive, intelligent and reasonable) inquiry, with outmoded paradigms as rational accounts of how things were until empirical evidence was found, and possible explanations thought out, such as only became available later. It was by no means stupid of Aristotle in *De Caelo* to suppose that the "fixed stars" were embedded in a crystal sphere that whirled round the earth. The only alternative was that the earth itself was rapidly spinning, in which case one would have expected a constant gale to be blowing in one direction, and the oceans to be pouring incontinently over the whole surface of the earth.

From the point of view of the "generalized empirical method"[20] that is part and parcel of critical realism, Galileo was not only establishing a new paradigm in a situation of revolutionary science, but he was following the metaparadigm in doing so. The famous experiment at the Leaning Tower of Pisa in 1612 was carried out, not by Galileo himself, but by one of his Aristotelian opponents, in order to refute Galileo's claim that different weights fall at the same speed. The weights hit the ground at very nearly the same time but not exactly at the same moment, which the Peripatetics seized on as evidence that Galileo was wrong. He was withering in his response: "Aristotle says that a hundred-pound ball falling from a height of one hundred cubits hits the ground before a one-pound ball has fallen one cubit. I say they arrive at the same time. You find, on making the test, that the larger ball beats the smaller one by two inches. Now, behind those two inches you want to hide Aristotle's ninety-nine cubits and, speaking only of my tiny error, remain silent about his enormous mistake."[21] There was a small discrepancy in the results from what Galileo had predicted—one would have thought well within the range of experimental error, particularly for those days—as opposed to the colossal difference from what was to be expected on the basis of Aristotelian physics. From the point of view of the metaparadigm

first to a burst of scepticism and then to decadence. Yet if the orthodoxy of revealed religion could not be defended on grounds of pure reason, it could not be thus diluted either. So the *via moderna* of William of Occam could be defended by theological conservatives like Jean Gerson. Just the same pattern of reasoning is to be found in twentieth-century Protestant theology, notably in the work of Karl Barth.

20 It is "generalized" in that, like John Locke as opposed to David Hume or Bertrand Russell, it takes account not only of our experience in the narrow sense, but of our awareness of the operations of our minds—in questioning, hypothesizing, marshalling evidence, judging, etc. on that experience; cf. Lonergan, *Insight*, 95–96, 268.

21 See John Gribbin, *Science. A History: 1543–1901* (London and New York: The Penguin Press, 2002), 76–77.

of rational inquiry as such, Galileo's sarcasm was abundantly justified, and his opponents were clearly convicted of being at least half-consciously dishonest.

While it may be true that no professional scientist will bother to justify a paradigm while it prevails, it seems to me quite wrong to suppose that no such justification can be provided. That water is not a chemical element, but rather a chemical compound of hydrogen and oxygen is a part of the accepted paradigm in chemistry, if anything is. If a schoolboy asks for the justification of it, he can be shown the experiment where a sample of water is electrolysed; he can see the quantity of water diminishing as an electrical current is passed through it, and bubbles accumulating at the anode and the cathode which respond, each in a characteristic manner. He can probably grasp that you wouldn't expect such a result if water were a chemical element, and the same applies to thousands of other observations and experiments. The fact that you could, by a prodigious feat of mental gymnastics, preserve the thesis that water is after all a chemical element, in face of the mass of *prima facie* evidence to the contrary—that there is no strict logical deduction from the existence of the evidence to the truth of the theory—is neither here nor there, if only one bears in mind Peirce's inspired vagueness.

For a critical realist, the judgments constitutive of science commend themselves as provisionally true, or at least as approximating to the truth, due to the method of attentiveness, intelligence and reasonableness by means of which they have been arrived at; if they are to be modified or even overthrown, it will be due to the method being applied further than it has been up to now. So far as Einstein is right, Newton is not absolutely wrong, but has arrived, as Popper put it, at an "excellent approximation" to the truth. The attitudes of Kuhn and Feyerabend toward truth seem to me cavalier to an extent that shocks me. Toward the end of *The Structure of Scientific Revolutions*, Kuhn remarks that the term "truth" and its cognates have only appeared once so far in the book, and that in a quotation from Francis Bacon, while Feyerabend will have it that "truth" is a mere bogey which is apt to distract inquirers from the real business of science—perhaps we should value scientific theories over one another for their aesthetic appeal.[22] If my arguments here have been along the right lines, it is a main merit of science—perhaps *the* main merit—that it follows with

22 See Kuhn, *The Structure of Scientific Revolutions* (Chicago: Chicago University Press, 1970); Feyerabend, *Against Method* (London: New Left Books, 1975); "Consolations for the Specialist," in I. Lakatos and A. Musgrave, eds., *Criticism and the Growth of Knowledge* (Cambridge: Cambridge University Press, 1970), 197–208; cf. Hugo A. Meynell, *Redirecting Philosophy. Reflections on the Nature of Knowledge from Plato to Lonergan* (Toronto, University of Toronto Press, 1998), Chapters 6 and 7.

exemplary rigour the methods by which the human mind is equipped to find out the truth about things. Once this principle is abandoned, you may be at the top of a slippery slope, a little farther down from which, as they say *a propos* historical judgments: "If you're a loyal communist, that is what you will believe. If you hold the contradictory belief, that is because you're a filthy capitalist."

In some instances of performance of the famous and crucial Michelson-Morley experiment, the results came out positive such that beams of light traveling in the same direction as the spinning of the earth were slower in relation to a point on its surface than those traveling in the opposite direction. But these results of the experiment turned out to be anomalous and exceptional. So far as it was not at first clear that they were so, it will have been proper for investigators to say that the well-known and highly-paradoxical conclusion now generally inferred from the results of the experiment—that the speed of light was the same in relation to any body whatever, however rapidly moving in the direction or the opposite direction to a beam of light—was a matter subject to reasonable doubt. Again, Proust's Law is to the effect that the atomic weight of each chemical element is an integral multiple of that of hydrogen. But the element chlorine seemed for some time to constitute an obstinate exception.[23] So long as it did so, it seems to me that it would have been natural and proper for a theoretical chemist to say something like the following: "I admit that chlorine constitutes an exception, which I cannot explain. But Proust's Law seems to be confirmed in the case of so many elements, that I cannot believe this to be mere coincidence."[24] Another theorist might, not wholly unreasonably, have regarded the apparently anomalous atomic weight of chlorine as providing adequate grounds for abandoning Proust's Law, at least tentatively. (It is not always noticed by philosophers of science, so far as an outsider can judge, that scientific statements do not have merely to be affirmed or denied, but can be asserted or negated with more or less confidence.) Ultimately, it turned out that there were two naturally-occurring isotopes of chlorine, commonly found together, each of which conformed to Proust's Law. (The slight difference in mass between protons and neutrons implies, of course, in terms of our present knowledge, that Proust's Law is only approximately true, since the nucleus of every atom but hydrogen includes neutrons as well as protons; and the masses of the two types of particle are very slightly different.)

23 The significance of Proust's Law in this connection was first pointed out, I believe, by Imre Lakatos.

24 Cf. Bode's Law, which states that planets are to be expected at certain distances from the sun. That it is not mere coincidence is suggested by the fact that it has led to discoveries. Yet I understand that no reason *why* it should obtain has yet been found.

Sir Karl Popper wrote: "To me, idealism appears absurd, for it ... implies *something like*[25] this: that it is my mind which creates this beautiful world. But I know I am not its creator."[26] What one might call "the idealist point" is, as I have already suggested, that the structure of our minds, as is presupposed by the whole of science, is a vital clue to the nature and structure of the world as a whole, and whatever or whoever is at the bottom of it. If it is, as of course it is, grossly implausible to suppose that "I," as Fichte at least came near to thinking, or "we," as social and linguistic relativists appear to suppose, are actually responsible for the existence and nature of the universe, then the idealist point, if one accepts it, gives reason to suppose that something importantly analogous to us and our minds is so. This seems to be a plausible, and perhaps inevitable, inference from the nature of knowledge, and the basic nature and structure of the world as that which is to be known. God may be deemed to be in infinite act what we are in highly limited potentiality; understanding all possible worlds, and willing the one which actually exists.

To revert to Popper's comment, perhaps it is not so absurd to suggest that something in some ways analogous to my mind would appear, from the nature of knowledge and the real world as what is to be known, to be at least an aspect of the ultimate cause, or among the ultimate causes, of the world. Hume, or rather Hume's character Philo,[27] said: "[T]he whole of Natural Theology ... resolves itself into one simple ... proposition, *That the cause or causes of order in the universe probably bear some remote analogy to human intelligence.*" This can only give rise to "... a longing desire and expectation, that heaven would be pleased to dissipate, at least alleviate, this profound ignorance, by affording some more particular revelation to mankind ..."[28] Can we take this a little further, if the inference is not from the otherwise unexplained order in the world, as from its nature as intelligible? This will apply, it should be noted,

25 My italics; cf. my argument in what follows above.

26 Karl Popper, *Objective Knowledge: An Evolutionary Approach* (Oxford: Clarendon Press, 1972), 72.

27 Of the characters in the *Dialogues*, Philo seems to represent Hume's own views most closely, though I think Hume himself was not sure about the matter. His request to Sir Gilbert Elliot, to supply him with more support for the argument from order to design, suggests that he feared he might rather have undersold Cleanthes, the defender of that argument in the *Dialogues*. Pamphilus, as his name implies, represents a position which was perhaps quite usual for thoughtful people at the time—that while Philo's scepticism was an improvement on the rigid religious orthodoxy of Demea, Cleanthes came nearest to the truth.

28 Hume, *Dialogues Concerning Natural Religion*, Section XII; in *Hume. Selections*, ed. Charles W. Hendel (New York: Scribner, 1955), 400; 401.

just as much to whatever may turn out to be at the bottom of the ontological heap—quarks and leptons perhaps—as to the higher levels of organization of this basic material. "[I]nsofar as one considers in man ... his intelligent and rational consciousness, one cannot but deal with what is related intimately to the universe and to its ultimate ground. For what is the universe and its ground but the objective of man's detached, disinterested, unrestricted desire to know?"[29] This ground may well be conceived as an intelligent will that "... freely grounds the universe in much the same fashion as the conditioned intelligent and rational consciousness of man grounds freely his own actions and products"[30] For better or worse, critical realism is at least favourable to theism, and some would maintain is soundly inferable from it.[31] Lonergan, like Hume's Philo, would look for additional revelation by this Creator of the divine nature and purposes for humankind.

The most common modern manifestation of idealism is linguistic idealism according to which we "create" the world we inhabit by employing mind-dependent linguistic and social categories. The difficulty is to give a literal form to this view that does not conflict with the obvious fact that we do not create worlds, but find ourselves in one.[32] The answer is that, while the world is nothing but what is to be got at progressively by our quasi-creative theorizing, it exists and largely is as it is prior to and independently of such theorizing. Much the same is to be said of the social idealism attributed to the later Wittgenstein as represented by *Philosophical Investigations* and *On Certainty*.

When it comes to the relation between the natural and the human sciences, the critical realist sets store by one crucial and ineluctable difference. In the case of the human sciences, the object as well as the subject of investigation is to be known as more or less attentive, intelligent, and reasonable. One may use in illustration Immanuel Velikovsky's explanation of the ferocious human sacrifices carried out by the Aztecs, and the phenomenon of the sacred cow.[33] (Whether Velikovsky was right or wrong in his main contentions is not immediately to the point.)[34] Terrifying and regularly-recurring celestial

29 Lonergan, *Insight*, 680.

30 Lonergan, *Insight*, Chapter XIX.

31 *Oxford Dictionary of Philosophy*, ed. Simon Blackburn, 2nd ed. (Oxford: Oxford University Press, 2005), 177.

32 Cf. Sprigge, "Idealism," 668.

33 See Immanuel Velikovsky, *Worlds in Collision* (London: Gollancz, 1969), 43; 123; 180–182; 245; 259.

34 I am aware that the scientific establishment confidently maintains that Velikovsky was obviously and contemptibly wrong. I myself would venture to agree with Albert Einstein in begging to differ. For a very thorough discussion of the issue, see Dale Ann Pearlman,

phenomena, accompanied by widespread agonizing destruction of human life, made people suppose that the planetary gods wanted to kill human beings in a very painful and unpleasant manner. Perhaps, if enough human sacrifices of a sufficiently cruel nature were performed, the gods would be appeased. Sure enough, when the experiment was tried, the disasters[35] ceased. Unless the sacrifices were continued, assuredly the gods would be angry again, as the Aztecs told the Spaniards. But the Spaniards, in their arrogant and insensitive colonialist way, put a stop to the sacrifices and the disasters did not return. As to the sacred cow, it appears in records from diverse continents that the comet which so ferociously tossed the earth had two horns, and left a milky edible deposit (nectar or manna). If her earthly representatives were not treated with the greatest respect and deference, the chances were that the goddess would return to chastise the earth once more.

The reader may note that the Aztecs were engaging in a perfectly good form of abductive reasoning. If the gods wanted violent human death and would get it for themselves if human beings did not provide it, they might cease to do so if human beings obliged. If this was so, the cessation of the disasters when human beings provided the required sacrifices was a matter of course. Since they did cease, the theory of what the gods wanted, and what human beings could do about it, was probably true. The abductions underlying contemporary Western science make it reasonable to suppose that the disasters are to be explained in another way, and Western anthropology can render comprehensible the less comprehensive application of rationality illustrated by the beliefs and practices of the Aztecs.

Can critical realism resolve what has been called the "science wars" issue?[36] It can. Physicists, chemists, zoologists and astronomers generally assume that, by following the scientific method, they tend to come to know the truth about the real world; so far as they are now in error about it, this will be corrected by further application of the same method. But sociologists and anthropologists of science may point out that scientific method, with its associated conceptions of truth, valid argument, accurate observation, and so on, is nothing more than a mode of procedure which happens to be highly regarded in one

ed., *Stephen J. Gould and Immanuel Velikovsky. Essays in the Continuing Velikovsky Affair* (Forest Hills, NY: Ivy Press Books, 1996).

35 The etymology of the word "disaster" (from Italian *disastro* or 'ill-starred event') is perhaps worth pondering in this connection.

36 See James Livesey, "Intellectual History and the History of Science," in Richard Whatmore and Brian Young, eds., *Intellectual History* (Basingstoke: Palgrave Macmillan, 2006), 130–132.

human society. Other societies value other procedures in accordance with their own worldviews. Suppose there is a society which attributes illness and misfortune in some of its members to the malicious intent of others who possess a diabolical substance. To say that such a society is "irrational" is nothing but an imperialist imposition of Western conceptions of rationality on its beliefs and activities in accordance with those beliefs. This society might just as well turn the tables and condemn our own methods of rational and objective investigation by comparison with its own.[37]

In accordance with critical realism and the generalized empirical method associated with it, zoologists and astronomers are quite correct that, by following a rational method (a thoroughgoing application of attentiveness, intelligence and reasonableness), they progressively get to know more and more about a world which exists, and largely is as it is, prior to and independently of ourselves and our investigations. (The planet Neptune was where it is, and had the properties that it did, before Adams and Leverrier discovered it in 1844.) A part or aspect of this world is human societies with beliefs and practices different from our own; these are studied by sociologists whose method is basically the same as that followed by natural scientists, with the differences that I have already pointed out. In coming to their conclusions, sociologists hypothesize and judge on the basis of the relevant evidence available to their experience. On some matters—say, our interactions with the spirits of the recently dead, or the healing power of certain incantatory formulae within a liturgical setting—it could turn out that their beliefs are closer to the truth than those conventionally accepted by ourselves or our scientists. One important task of sociologists is to study those social factors which promote genuine rationality in the sense in which I have defined it, and those which impugn it. The story is told that in darkest Ruritania, a young schoolboy spent his weekends helping visitors from another culture in their household chores. He had been taught in school that insects had eight legs, and told this to his employers. They invited him to examine an insect with them, and to count its six legs. When he returned to school and told his teacher about this, she slapped him on the head and told him to write down twenty times, "Insects have eight legs."

"But is it not obvious," some will object, "that it is only a matter of time before our mental life is thoroughly explained in terms of the physics and chemistry of the brain?" The answer is that it is not, never will be, and cannot

37 A view like this has been defended with forcefulness and elegance by Peter Winch, in "Understanding a Primitive Society," in D.Z. Phillips, ed., *Religion and Understanding* (Oxford: Blackwell, 1967).

be, for the following reasons: we believe, and ought to believe, what scientists say on the subjects of their specialties only on the assumption that they, or their informants, have been thoroughly rational in coming to the relevant judgments. If nobody, in the last analysis, can be rational in this sense, but everyone thinks, speaks, and acts as they do simply due to the physical and chemical laws which operate in their brains, then science itself falls to the ground. It would be singular indeed if the ultimate upshot of science were that science is impossible. Yet that seems to be a valid inference from a consistent materialism.

If what I have argued so far is on the right track, there seem to be important consequences for the philosophy of value.[38] It is not too much to say that the ethics of emotivism and prescriptivism have no firmer foundation than an error in the philosophy of logic. (It is not to be denied that to call something "good" is usually, *inter alia,* to express a positive emotion about it, or to commend a positive attitude towards it on the part of oneself and others.) The moral goodness of an action might be a matter of its actual and intended contribution to happiness, and to the fairness of its result, without there being any tight deductive connection between the former and the latter. In a similar fashion, the goodness or greatness of a work of art might be a matter of its capability of giving a certain kind of satisfaction—whatever it is that makes it worthwhile for some people to look at Michelangelo's *Pieta* or listen to a performance of Schubert's C-major Quintet, for example, without its strictly being implied that any particular person on any particular occasion would derive such satisfaction from it.

References

Austin, J.L., "Truth," *Proceedings of the Aristotelian Society,* Supplementary Volume, 1950.

Blackburn, Simon, ed., *Oxford Dictionary of Philosophy,* 2nd ed., Oxford: Oxford University Press, 2005.

Cassirer, Ernst, *The Philosophy of the Enlightenment,* Boston: Little, Brown, 1962.

Darwell, Stephen, Gibbard, Alan, and Railton, Peter, "Towards *Fin de Siècle* Ethics," *The Philosophical Review,* Vol. 101: No. 1 (January, 1992), 115–189.

Feyerabend, Paul, *Against Method,* London: New Left Books, 1975.

38 For a masterly survey of recent philosophy of value among philosophers of the analytic tradition, see Stephen Darwall, Alan Gibbard, and Peter Railton, "Towards *Fin de Siècle* Ethics," *The Philosophical Review,* Vol. 101: No. 1, January, 1992, 115–189.

Feyerabend, Paul, "Consolations for the Specialist," in I. Lakatos and A. Musgrave, eds., *Criticism and the Growth of Knowledge*, Cambridge: Cambridge University Press, 1970, 197–208.

Flew, Antony, ed., *A Dictionary of Philosophy*, New York: St. Martin's Press, 1984.

Goudge, Thomas A., *The Thought of C.S. Pierce*, Toronto: University of Toronto Press, 1950.

Gribbin, John, *Science: A History: 1543–1901*, London and New York: The Penguin Press, 2002.

Hamlyn, D.W., "Idealism," in *Oxford Companion to Philosophy*, Oxford and New York: Oxford University Press, 1995.

Hartshorne, Charles and Weiss, Paul, eds., *The Collected Papers of C.S. Pierce*, Cambridge, MA.: Harvard University Press, 1931–1935.

Hume, David, *Dialogues Concerning Natural Religion*, in *Hume Selections*, ed. Charles W. Hendel, New York: Scribner, 1955.

Kuhn, Thomas, *The Structure of Scientific Revolutions*, Chicago: University of Chicago Press, 1970.

Livesay, James, "Intellectual History and the History of Science," in Richard Whatmore and Brian Young, eds., *Intellectual History*, Basingstoke: Palgrave Macmillan, 2006, 130–146.

Lonergan, Bernard, *A Study of Human Understanding*, in *Collected Works of Bernard Lonergan*, Toronto: University of Toronto Press, 1992; first published in 1957.

Lonergan, Bernard, *Verbum: Word and Idea in Aquinas*, London: Darton, Longman, and Todd, 1968.

Macfie, A.L., ed., *The Philosophy of History*, Basingstoke and New York: Palgrave Macmillan, 2006.

MacIntyre, Alasdair, *Whose Justice, Which Rationality?*, Notre Dame, IN: University of Notre Dame Press, 1988.

Meynell, Hugo, *Redirecting Philosophy: Reflections on the Nature of Knowledge from Plato to Lonergan*, Toronto: University of Toronto Press, 1998.

Paley, Morton D., *Energy and Imagination: A Study of the Development of Blake's Thought*, Oxford: Clarendon Press, 1970.

Pearlman, Dale Ann, ed., *Stephen J. Gould and Immanuel Velikovsky: Essays in the Continuing Velikovsky Affair*, Forest Hills, NY: Ivy Press Books, 1996.

Popper, Karl, *Objective Knowledge: An Evolutionary Approach*, Oxford: Clarendon Press, 1972.

Searle, John R., *Minds, Brains and Science*, Cambridge, MA.: Harvard University Press, 1985.

Sprigge, T.L.S., "Idealism," in *The Routledge Encyclopedia of Philosophy*, Volume 5, 662–669.

Strawson, P.F., "Truth," in *Logico-Lingusitic Papers*, Aldershot: Ashgate, 2004, 147–164.

Taylor, A.E., *Plato: The Man and His Work*, London: Metheun, 1960.

Taylor, Richard, *Metaphysics*, Englewood Cliffs, NJ: Prentice Hall, 1963.

Velikovsky, Immanuel, *Worlds in Collision*, London: Gollancz, 1969.

Vesey, G. and Foulkes, P., eds., *Collins Dictionary of Philosophy*, London and Glasgow: Collins, 1990.

Winch, Peter, "Understanding a Primitive Society," in *Religion and Understanding*, ed. D.Z. Phillips, Oxford: Blackwell, 1967, 307–324.

CHAPTER 4

Philosophy in Einstein's Science

John D. Norton

1 Introduction

Albert Einstein read philosophy. It was not an affectation of a celebrity-physicist trying to show his adoring public that he was no mere technician, but a cultured thinker. It was an interest in evidence from the start.

In 1902, Einstein was a poorly paid patent examiner in Bern seeking to make a few extra Francs by offering tutorials in physics. Maurice Solovine answered the advertisement. The tutorials quickly vanished when they discovered their common fascinations in reading and talking. They were soon joined in their raucous meetings by Conrad Habicht, completing what they dubbed their "Olympia Academy." Their explorations where wide-ranging, devouring texts and sausages with gusto. They read the philosophers and philosophically-minded scientists of the day, including Pearson, Mach, Mill, Hume, Spinoza, Avenarius, Clifford and Poincaré.[1]

The philosophical interest endured. In the late 1920s, there were three portraits on the walls of Einstein's study in Berlin. Two were unsurprising: the great English physicists Michael Faraday and James Clerk Maxwell. And the third? It was not one most people would predict. It was Arthur Schopenhauer. However, as Don Howard[2] has shown in detail, once one is alerted to look, the imprint of Einstein's reading of Schopenhauer is clearly visible in his writing and thought.

Einstein's own philosophical writings have in turn attracted considerable attention.[3] In the years following his discovery of the general theory of relativity,

1 As collected and reported by the editors of the Einstein Papers. John Stachel *et al.*, *The Collected Papers of Albert Einstein, Volume 2, The Swiss Years: Writings 1900–1909* (Princeton: Princeton University Press, 1989), xxiv-xxv.

2 Don Howard, "A Peek behind the Veil of Maya: Einstein, Schopenhauer, and the Historical Background of the Conception of Space as a Ground for the Individuation of Physical Systems," in *The Cosmos of Science: Essays of Exploration*, John Earman and John D. Norton, eds., *Pittsburgh-Konstanz Series in the Philosophy and History of Science*, vol. 6. (Pittsburgh: University of Pittsburgh Press; Konstanz: Universitätsverlag, 1997), 87–150.

3 For a good introduction to his philosophical writing and thought, see Don Howard, "Albert Einstein as a Philosopher of Science," *Physics Today*, 58, No. 12 (Dec. 2005), 34–40; Don Howard, "Einstein's Philosophy of Science," *The Stanford Encyclopedia of Philosophy*, Edward N. Zalta (ed.), URL = https://plato.stanford.edu/entries/einstein-philscience/.

 | DOI:10.1163/9789004415270_006

Einstein was read and consulted by many philosophers, including Rudolf Carnap, Hans Reichenbach, and Moritz Schlick. As a result, Einstein's writings in physics and philosophy enjoyed a commanding presence in the new movements in modern philosophy that gained strength in the 1920s and 1930s.[4]

My purpose in this Chapter is not to attempt a synthetic portrait of Einstein's philosophy. For reasons I will indicate later, I am not sure how useful that would be. Rather I want to draw attention to what I believe is the most important aspect of Einstein's interest in philosophy. Einstein quite consciously integrated philosophical analysis into his physical theorizing. Its explicit use was part of how Einstein found his way to new theories and defended them. Here I will sketch a few episodes in Einstein's physics in which philosophical analysis played an important role. I will try to explain at a level relatively free of the technicalities of physical theories just what role the philosophical analysis played.

The first episode, recounted in Section 2, is Einstein's adopting an empiricist theory of concepts in order to legitimate an extraordinary new physical proposal concerning time in special relativity. Section 3 will recount what Einstein described as the "epistemological defect" in earlier theories that motivated him to seek his general theory of relativity. Section 4 will describe how Einstein twice grounded his theorizing in principles that distinguish the real from the unreal. One pertains to the completion of the theory of general relativity and the other grounds his co-authored efforts to prove the incompleteness of quantum theory. Finally, in Section 5, I will describe how Einstein came to adopt a form of mathematical Platonism as the way to find new theories, such as his unified field theory. In the conclusion (Section 6), I will explain why I believe Einstein was correct and appropriately unapologetic in portraying himself as an "unscrupulous opportunist" in the view of a systematic epistemologist.

2 An Empiricist Theory of Concepts

In June 1905, Einstein[5] sent the journal *Annalen der Physik* the manuscript of his paper, "On the Electrodynamics of Moving Bodies." In it, he laid out his special theory of relativity. Its first "Kinematical Part" is both a brilliant departure from earlier thinking and a deceptively easy analysis to read. In order to

4 See, for example, Thomas Ryckman, *Reign of Relativity: Philosophy in Physics 1915–1925* (Oxford: Oxford University Press, 2007).

5 Albert Einstein, "Zur Elektrodynamik bewegter Körper," *Annalen der Physik*, 17 (1905), 891–921; translated as "On the Electrodynamics of Moving Bodies," in Albert Einstein *et al.*, *The Principle of Relativity*, trans. W. Perrett and G. B Jeffrey (New York: Dover, 1952), 37–65.

solve certain problems in electrodynamics, Einstein tells us, he will posit two principles: the principle of relativity and the light postulate. The first asserts the equivalence of all inertial motion. The second assigns a unique speed to light propagating *in vacuo*.

The two principles are "apparently irreconcilable," Einstein mentions in passing on the paper's first page. The reader is left to imagine why. It is not hard to do. When the two principles are combined, they entail that all inertially moving observers will find the same speed for the one beam of light. Imagine that I measure the speed of a light beam and find some value, "c." If I am chasing rapidly after that same beam at the great speed of c/2, should I not find it to propagate at c-c/2=c/2? No, the two principles say. I must find the same value, c, and that just does not seem right.

Einstein turned immediately to a simple explanation of why this irreconcilability is only apparent. Implicit in our judgments of the speed of light are further assumptions about space and time. In concrete terms, an observer measures the speed of light by timing how long light takes to traverse a known distance, and that time difference is measured by clocks placed at either end of the distance. The procedure requires that the two clocks be properly synchronized. Each must read "12 o'clock" at precisely the same instant. It is easy to assume that, if one observer judges the clocks to be properly synchronized, then so also will another observer in relative motion. That, Einstein proceeded to demonstrate, is incorrect. His famous demonstration involved an ingenious thought experiment with clocks and light signals and drew on his theory's two principles. The essential outcome is that the two observers will not agree on which spatially separated events are simultaneous; and thus they will not agree on whether the two clocks are properly synchronized. In their attempts to measure the speed of light, the two observers will use clocks synchronized differently. The resulting differences turn out to be exactly sufficient to ensure that both recover the same value c for the speed of light.

This effect, "the relativity of simultaneity," was the first of the novel results of the new special theory of relativity. It led Einstein immediately to argue that observers in relative motion will, in general, not agree on the lengths of objects and the time durations measured for processes. The analysis is so crisp and simple that it is hard to suppress the image of impish Einstein casually tossing off the analysis from the comfort of an armchair one sunny afternoon.

The reality of the discovery was quite different. As I have recounted in some detail elsewhere,[6] Einstein had become convinced years before that the

6 John D. Norton, "Einstein's Investigations of Galilean Covariant Electrodynamics prior to 1905," *Archive for History of Exact Sciences* 59 (2004), 45–105.

principle of relativity must hold in Maxwell's theory of electrodynamics, even though that theory was based on an ether in which there was a preferred state of rest. Worse, Maxwell's theory asserted that light always travels at just one speed, c=186,000 miles per second *in vacuo* in relation to this ether. It seemed to Einstein that the principle of relativity would force him to give up this constancy. He struggled to find a modification of Maxwell's theory in which the speed of light would vary according to speed of the emitter. After many fruitless attempts, Einstein finally realized that he could find no sustainable emission theory of light. Maxwell's theory and the constancy of the speed of light must stand. It was a point of desperation for him. How could he keep both the principle of relativity and the constancy of the speed of light?

Einstein later recalled in his *Autobiographical Notes* how he finally solved the problem:[7]

> Today everyone knows, of course, that all attempts to clarify this paradox satisfactorily were condemned to failure as long as the axiom of the absolute character of time, or of simultaneity, was rooted unrecognized in the unconscious. To recognize clearly this axiom and its arbitrary character already implies the essentials of the solution of the problem.

To solve his problem, Einstein had to see what everyone before him had missed: that the absoluteness of simultaneity is an assumption that can be challenged. Furthermore, he needed something to give him the courage to mount that challenge and abandon the assumption. Einstein continued the above remarks by noting that this essential support came from his reading in philosophy:

> The type of critical reasoning required for the discovery of this central point was decisively furthered, in my case, especially by the reading of David Hume's and Ernst Mach's philosophical writings.

Einstein affirms here *that* reading Hume and Mach's philosophical writings were decisive. However, he does not tell us *how* they were decisive or even *which* writings were at issue. It has been natural to assume that it was Hume's and Mach's writings, specifically in philosophy of space and time, that made the difference.

7 Albert Einstein, "Autobiographical Notes," in P.A. Schilpp, ed., *Albert Einstein: Philosopher-Scientist* (Evanston, IL: Library of Living Philosophers, 1949), 2–95; 53.

When we begin to explore Einstein's other writings and remarks, another possibility emerges. In 1924, Einstein remarked:[8]

> After seven years of reflection in vain (1898–1905), the solution came to me suddenly with the thought that our concepts and laws of space and time can only claim validity insofar as they stand in a clear relation to experiences; and that experience could very well lead to the alteration of the concepts and laws. By a revision of the concept of simultaneity into a more malleable form, I thus arrived at the special theory of relativity.

The same idea is given more succinctly in a remark from Einstein's 1917 popular account of relativity theory:[9] "The concept [of simultaneity] does not exist for the physicist until he has the possibility of discovering whether or not it is fulfilled in an actual case." The breakthrough was not grounded in some novel philosophical insight into space and time specifically. Rather it was a general view about how concepts are properly employed in physical theories.

The concepts of physical theories must, Einstein here asserts, be properly grounded in experience, else they are fictions. Once one has this clue, one recalls immediately that just this sort of empiricist approach to concepts is fundamental to the thought of Hume and Mach and one can see that it is to this aspect of their writing that Einstein referred. The analysis of David Hume's *Treatise* depends on just this simple grounding of concepts ("ideas") in experience ("impressions"). The introductory section concludes with the synoptic assertion:[10] "… all our simple ideas proceed either mediately or immediately, from their correspondent impressions. This then is the first principle I establish in the science of human nature …"

Later Hume makes clear that concepts cannot extend beyond this grounding in experience without introducing a fiction. For example, he writes:[11] "Ideas

8 The remark is in a voice recording, transcribed and presented in the German in F. Herneck, "Zwei Tondokumente Einsteins zur Relativitätstheorie," *Forschungen und Fortschritte* 40 (1966), 133–135; translated in John Stachel *et al.*, *The Collected Papers of Albert Einstein,* Volume 2, *The Swiss Years: Writings, 1900–1909* (Princeton: Princeton University Press, 1989), 264.

9 Albert Einstein, *Über die spezielle and die allgemeine Relativitätstheorie* (*Gemeinverständlich*), Braunschweig: Friedr. Vieweg & Sohn, 1917; 15th expanded edition translated by R.W. Lawson as *Relativity: the Special and the General Theory* (London: Methuen, 1954), Section VIII.

10 David Hume, *A Treatise of Human Nature*. ed. P.H. Nidditch, 2nd ed. (Oxford: Clarendon, 1978), Book 1, Part 1, Section 1.

11 David Hume, *A Treatise of Human Nature*. Book 1, Part II, Section III.

always represent the objects or impressions from which they are deriv'd, and can never without a fiction represent or be appl'd to any other ..."

One finds a similar empiricist approach to concepts in the writings of Ernst Mach. More relevantly, we know that Einstein found it in Mach, for Einstein tells us just this in his obituary for Mach:[12]

> Science is, according to Mach, nothing but the comparison and orderly arrangement of factually given contents of consciousness, in accord with certain gradually acquired points of view and methods. ... concepts have meaning only in so far as they can be found in things, just as they are the points of view according to which these things are organized. (Analysis of concepts)

What is important is that this empiricist approach to concepts is quite general. It is not limited to the analysis of space and time, but applies to all concepts. Most famously, Hume applied it to causation.

All this is only the beginning of a fascinating tale. Einstein elsewhere averred that it was Hume "still much more" than Mach who guided him; and we find some differences in the way Einstein was willing to accept fictional concepts not properly grounded in experience in his theorizing as conventions.[13]

3 "An (Inherent) Epistemological Defect"

In his analysis of 1905, Einstein eliminated the ether state of rest from physics and reinstated the relativity of motion only as far as inertial motion, that is, uniform motion in a straight line. Over the ensuing decade, Einstein sought a new theory that would extend the principle of relativity to all motion, including accelerated motion. Einstein believed that he had achieved this in 1915 with the completion of his general theory of relativity.

What is important for our purposes are the motivations Einstein reported for seeking this extension of the principle of relativity. In 1916, Einstein published a definitive review article of the completed theory. In an early section, "§2 The Need for an Extension of the Postulate of Relativity," Einstein gives

12 Albert Einstein, "Ernst Mach," *Physikalische Zeitschrift*, 17 (1916), 101–104.

13 For elaboration of these issues, see John D. Norton, "How Hume and Mach Helped Einstein Find Special Relativity," in M. Dickson and M. Domski, eds., *Discourse on a New Method: Reinvigorating the Marriage of History and Philosophy of Science* (Chicago: Open Court, 2010), 359–386.

what is surely a type of reason that is rarely found stated explicitly in the physics literature:[14]

> In classical mechanics, and no less in the special theory of relativity, there is an inherent epistemological defect which was, perhaps for the first time, clearly pointed out by Ernst Mach.

Einstein's German—*ein erkenntnistheoretischer Mangel*—was a little weaker than the standard Perrett and Jeffrey translation given here and is captured more literally merely as "epistemological defect." However I have used the stronger Perrett and Jeffrey translation since it has been in the standard edition of Einstein's paper since its 1923 translation, and I like to pretend that it captures the passionate energy of its author, a barely 37-year old Einstein at the moment of his greatest scientific creativity.

Either way, it is an extraordinary idea. Our best theory of gravity and Einstein's greatest contribution to modern physics is motivated in part by the need to remedy an epistemological defect of earlier theories!

Einstein proceeded to explain the problem. Both classical physics and special relativity posit certain preferred inertial motions. These were the uniform straight-line motions followed by free bodies, unaffected by perturbing forces. These motions in turn define inertial spaces of reference; they are, loosely speaking, the spaces carried with each set of bodies moving together inertially. So-called "inertial forces" arise if a body is constrained to accelerate, that is, to deviate, from inertial motions. Newton[15] imagined water swirling in a bucket and the resulting acceleration led the water to be hurled outward and climb up the wall of the bucket, producing a concave water surface. Analogously, fluid spheres in rotation, such as stars and planets, bulge at their equators.

"What causes this bulge?," Einstein asked. We are, he noted, inclined to answer that the cause is rotation with respect to inertial spaces. This answer is rejected thunderously:

> No answer can be admitted as epistemologically satisfactory, unless the reason given is an *observable fact of experience*. The law of causality has

14 Albert Einstein, "Die Grundlage der allgemeinen Relativitätstheorie," *Annalen der Physik*, 49 (1916), 769–822; translated as "The Foundation of the General Theory of Relativity," in Albert Einstein *et al.*, *The Principle of Relativity*, 111–164; from §2.

15 Isaac Newton, *Mathematical Principles of Natural Philosophy and his System of the World*, 1729; tr. Andew Motte, revised Florian Cajori (Berkeley: University of California Press, 1934). Vol. 1:10–11.

> not the significance of a statement as to the world of experience, except when *observable facts* ultimately appear as causes and effects. [Einstein's emphasis]

He continued a few sentences later:

> … the privileged [inertial] space of … Galileo thus introduced, is merely a *factitious* ["bloss fingierte"=ad hoc, JDN] cause, and not a thing that can be observed. (Einstein's emphasis)

Then Einstein turned to the cause that he would accept: distant masses and their motions. He thereby foreshadowed the form that he hoped his final theory would take. In it, nothing intrinsic to a space distinguishes one space from another. The discrimination of spaces into inertial and accelerating comes only by virtue of the masses distributed in them. If the masses of the universe are at rest in a space, it is an inertial space. If those masses swirl around, it is a space with inertial forces that pull water up the sides of Newton's bucket and lead fluid bodies to bulge at their equators.

The analysis is driven by Einstein's conception of an epistemological defect. In his popular account of relativity, written at the end of 1916, Einstein gave a more prosaic and visceral illustration of it:[16]

> I am standing in front of a gas range. Standing alongside of each other on the range are two pans so much alike that one may be mistaken for the other. Both are half full of water. I notice that steam is being emitted continuously from the one pan, but not from the other. I am surprised at this, even if I have never seen either a gas range or a pan before. But if I now notice a luminous something of bluish colour under the first pan but not under the other, I cease to be astonished, even if I have never before seen a gas flame. For I can only say that this bluish something will cause the emission of the steam, or at least possibly it may do so. If, however, I notice the bluish something in neither case, and if I observe that the one continuously emits steam whilst the other does not, then I shall remain astonished and dissatisfied until I have discovered some circumstance to which I can attribute the different behaviour of the two pans.

16 Albert Einstein, *Über die spezielle and die allgemeine Relativiätstheorie* (*Gemeinverständlich*), Ch. XXI.

It is hard for a philosopher to read this and not see an account here of the violation of a venerable principle, Leibniz' principle of sufficient reason.[17] This, however, was not Einstein's reading. He proceeded to assert that it was "E. Mach [who] recognised [the epistemological defect of prior theories] most clearly of all ..."

Einstein clearly had in mind Mach's celebrated analysis of Newton's notions of absolute space and time in his *Science of Mechanics*, including his famous remark on Newton's bucket:[18]

> Newton's experiment with the rotating vessel of water simply informs us, that the relative rotation of the water with respect to the sides of the vessel produces no noticeable centrifugal forces, but that such forces are produced by its relative rotation with respect to the mass of the earth and the other celestial bodies. No one is competent to say how the experiment would turn out if the sides of the vessel increased in thickness and mass till they were ultimately several leagues thick. The one experiment lies before us, and our business is, to bring it into accord with the other facts known to us, and not with the arbitrary fictions of our imagination.

Einstein has left us in little doubt as to how he read Mach's critique. His first published statement of what he later dubbed "Mach's Principle" came in 1912 when Einstein had developed only a rudimentary forerunner to his general theory of relativity. The statement asserts:[19] "... the entire inertia of a point mass is the effect of the presence of all other masses, deriving from a kind of interaction with the latter."

Lest there be any doubt as to the origin of the idea, Einstein—notorious for his meagre citation habits—appended a footnote to the section of *Science of Mechanics* in which the above bucket quote appears:

> This is exactly the point of view which E. Mach urged in his acute investigations on the subject. (E. Mach, The Development of the Principles of Dynamics. Second Chapter. Newton's Views of Time, Space and Motion.)

---

17 Perhaps I need not add that philosophers now a century removed from Mach's positivism will find the insistence on the direct observability of causes excessive.

18 Ernst Mach, *The Science of Mechanics: A Critical and Historical Account of Its Development*, 6th ed., trans. T.J. McCormach (LaSalle, Illinois: Open Court, 1960), 284.

19 Albert Einstein, "Gibt es eine Gravitationswirkung, die der elektrodynamischen Induktionswirkung analog ist?," *Vierteljahrsschrift für gerichtliche Medizin und öffentliches Sanitätswesen*, 44 (1912): 37–40; 39.

Einstein's reading of Mach's remark is curious. Mach asserts (my emphasis), "*No one is competent to say* how the experiment would turn out if the sides of the vessel increased in thickness ..." Yet Einstein took this as a license to say just what would happen. Were the walls of the bucket so enlarged and set into rotation, they would drag the water in the bucket slightly. This dragging would be a massively weakened version of what Einstein believed happens when all the masses of the universe rotated around the bucket.

Einstein sought to derive these "Machian" effects in his developing theories of gravity prior to the completion of general relativity.[20] They are recovered in various forms in the final theory as well, as Einstein explains in his *Meaning of Relativity*, the closest Einstein came to writing a textbook for his theory.[21]

Matters did not continue as one might expect. Einstein later came to renounce his fascination with Mach's critique. Writing in 1946 in his "Autobiographical Notes," he reflected:[22]

> ... in my younger years, however, Mach's epistemological position also influenced me very greatly, a position that today appears to me to be essentially untenable. For he did not place in the correct light the essentially constructive and speculative nature of all thinking and more especially of scientific thinking; in consequence, he condemned theory precisely at those points where its constructive-speculative character comes to light unmistakably, such as in the kinetic theory of atoms.

It was also never clear that the general theory of relativity did meet the Machian-inspired demands concerning the origin of inertial. Eventually Einstein withdrew his support for these demands, as he noted again later in his "Autobiographical Notes":[23]

> Mach conjectures that in a truly reasonable theory inertia would have to depend upon the interaction of the masses, precisely as was true for Newton's other forces, a conception that for a long time I considered in principle the correct one. It presupposes implicitly, however, that the

20 Albert Einstein, "Gibt es eine Gravitationswirkung, die der elektrodynamischen Induktionswirkung analog ist?,"; Albert Einstein, "Zum gegenwärtigen Stande des Gravitationsproblems," *Physikalische Zeitschrift*, 14 (1913): 1249–1262.

21 Albert Einstein, *The Meaning of Relativity,* (1922); 5th Expanded Edition (Princeton: Princeton University Press, 1956), 101–103.

22 Albert Einstein, "Autobiographical Notes," 21.

23 Albert Einstein, "Autobiographical Notes," 29.

> basic theory should be of the general type of Newton's mechanics: masses and their interaction as the original concepts. Such an attempt at a resolution does not fit into a consistent field theory, as will be immediately recognized.

Finally it remains unclear that the critique of absolute space Einstein read in Mach's writings is the one Mach intended. Einstein found the critique as authorizing the search for a new theory of inertia, whereas Mach may merely have intended it to support an austere formulation of an otherwise unaltered classical physics, everywhere purged of the mention of metaphysical notions, such as Newton's absolute space.[24]

4 The Real

A perennial theme in philosophy is the separation of reality from appearance. Present day physics is replete with techniques that effect this separation. They are associated with the notions of invariance, symmetry and gauge transformations, whose lineage in physics traces back to Einstein's work. A century ago, his theories of relativity demonstrated the same reality can have very different appearances in different frames of reference. However the idea of using group theory and distinguishing the real as the invariants of the transformations of groups is a nineteenth century notion. It was a commonplace of geometry before it was brought into twentieth century physics, in large measure through the stimulus of Einstein's theories of relativity.

These are broad themes. My concern in this section, however, is two narrow episodes concerning the real. In them, Einstein sought to resolve a pressing problem in physics by positing what we might call a reality principle; that is, a principle that separates reality from appearance.

4.1 *The Point-Coincidence Argument*

The first of these episodes arose with the completion of the general theory of relativity. We saw that Einstein's initial concern was to implement a generalized principle of relativity that extended to accelerated motion. By 1916, that demand had evolved into a requirement of general covariance. To see what it amounts to, we should recall that spacetime theories label events with

24 For an investigation of this issue, see John D. Norton, "Mach's Principle before Einstein," in J. Barbour and H. Pfister, eds., *Mach's Principle: From Newton's Bucket to Quantum Gravity: Einstein Studies*, Vol. 6 (Boston: Birkhäuser, 1995), 9–57.

four numbers. They are usually three spatial coordinates and one time coordinate. However one can make new numerical labels for any event by adding, subtracting, or taking any combination of the more traditional choices of the event's coordinates, and any rescaling of them. These manipulations create arbitrarily many more spacetime coordinate systems. If one has a physical theory that can employ any of these coordinate systems, no matter how jumbled and rescaled, then the theory is generally covariant.

The central conception of Einstein's general theory is a connection between gravitation and the curvature of the spacetime geometry. His decision to seek a generally covariant theory was pivotal. It enabled Einstein to draw on the elaborate body of mathematical techniques emerging from the nineteenth century for understanding curvature. As long as he kept his equations generally covariant, this body of mathematics admitted remarkably few possibilities for the implementation of his theory. That fact is routinely used today in motivating Einstein's theory.

Hence it can come as a surprise to modern readers to learn that Einstein considered and rejected general covariance in 1913. Then he and his mathematician friend, Marcel Grossmann, published a sketch of what was the general theory of relativity in all its parts, excepting its most essential part.[25] That was its gravitational field equations, the theory's analog of Newton's inverse square law of gravitation. They announced that they had been unable to find physically admissible generally covariant gravitational field equations. In place of these equations, they published gravitational field equations of limited covariance.[26]

In a little over two years, Einstein would recognize this rejection of general covariance for the catastrophe it was. However, before then, Einstein turned his powers towards making a bad situation worse. If he could not find admissible generally covariant gravitational field equations, then he would demonstrate

25 Albert Einstein and Marcel Grossmann, *Entwurf einer verallgemeinerten Relativitätstheorie und einer Theorie der Gravitation* (Leipzig: B.G.Teubner, 1913), (separatum); with addendum by Einstein in *Zeitschrift für Mathematik und Physik*, 63 (1914), 225–61.

26 What precisely went wrong? We have a rather complete record of the computations Einstein undertook during his preparation of the 1913 paper in the "Zurich Notebook." For an analysis of what it reveals, see John D. Norton, "How Einstein found his Field Equations: 1912–1915," *Historical Studies in the Physical Sciences*, 14 (1984), 253–316; reprinted in Don Howard and John Stachel (eds.), *Einstein and the History of General Relativity: Einstein Studies*, Volume 1 (Boston: Birkhäuser, 1989), 101–159, and Jürgen Renn *et al.*, *The Genesis of General Relativity*, Volume 1, *Einstein's Zurich Notebook, Introduction and Source;* Volume 2, *Einstein's Zurich Notebook: Commentary and Essays* (Dordrecht: Springer, 2007).

that omission to be no failing, for he would prove that such equations are physically uninteresting. What resulted was his "hole argument."[27] It was published in four versions in 1913 and 1914. One was in an addendum to his joint paper with Grossmann.[28] The most complete version was in a 1914 review article.[29]

General covariance gives the theorist the power to represent the one physical reality in many different coordinate systems. What Einstein found, however, was that it also permitted a reverse capacity. One could fix the coordinate system and induce many apparently distinct physical realities in it and this could be done in a way that seemed to compromise determinism.

The essential idea can be conveyed in an analogy to different map projections. One sheet of paper can host many projections of the world. One of the oldest and best known is the Mercator projection of 1569 of Figure 4.1.

Many more projections are possible. Another is the Lambert projection of 1772 of Figure 4.2.

The continents in the two projections look rather different. Antarctica in the Mercator projection looks enormous in comparison to Antarctica in the Lambert projection. We know, of course, that this difference is purely an artifact of the different projections and represents nothing real.

Imagine, however, that one did not realize that the differences were artifactual. One would then imagine that these are maps of two different worlds: the Mercator world, with its enormous Antarctica, and the factually distinct Lambert world, in which Antarctica is a mere sliver. Worse, we can construct further factually distinct hybrid worlds. The Southern hemisphere may be extracted from the Mercator projection and the Northern from the Lambert projection. That would be yet another world depicted in Figure 4.3.

One would then be faced with an odd problem if one is trying to determine which of the Mercator or Lambert worlds is our world. We might check the

27 The hole argument and its resolution have been revived in more recent philosophy of space and time as presenting an insurmountable dilemma for certain versions of spacetime substantivalism. For a review, see John Earman and John D. Norton, "What Price Spacetime Substantivalism? The Hole Story," *British Journal for the Philosophy of Science*, 38 (1987), 515–525, and John D. Norton, "The Hole Argument," *The Stanford Encyclopedia of Philosophy* (Winter 2008 Edition), ed. Edward N. Zalta, URL = http://plato.stanford.edu/archives/win2008/entries/spacetime-holearg/.

28 Albert Einstein and Marcel Grossmann, *Entwurf einer verallgemeinerten Relativitätstheorie und einer Theorie der Gravitation.*

29 Albert Einstein, "Die formale Grundlage der allgemeinen Relativitätstheorie," *Königlich Preussische Akademie der Wissenschaften (Berlin), Sitzungsberichte* (1914), 1030–1085. For an account of these four versions and related issues, see John D. Norton, "Einstein, the Hole Argument and the Reality of Space," in J. Forge (ed.), *Measurement, Realism and Objectivity* (Dordrecht: Reidel, 1987), 153–188.

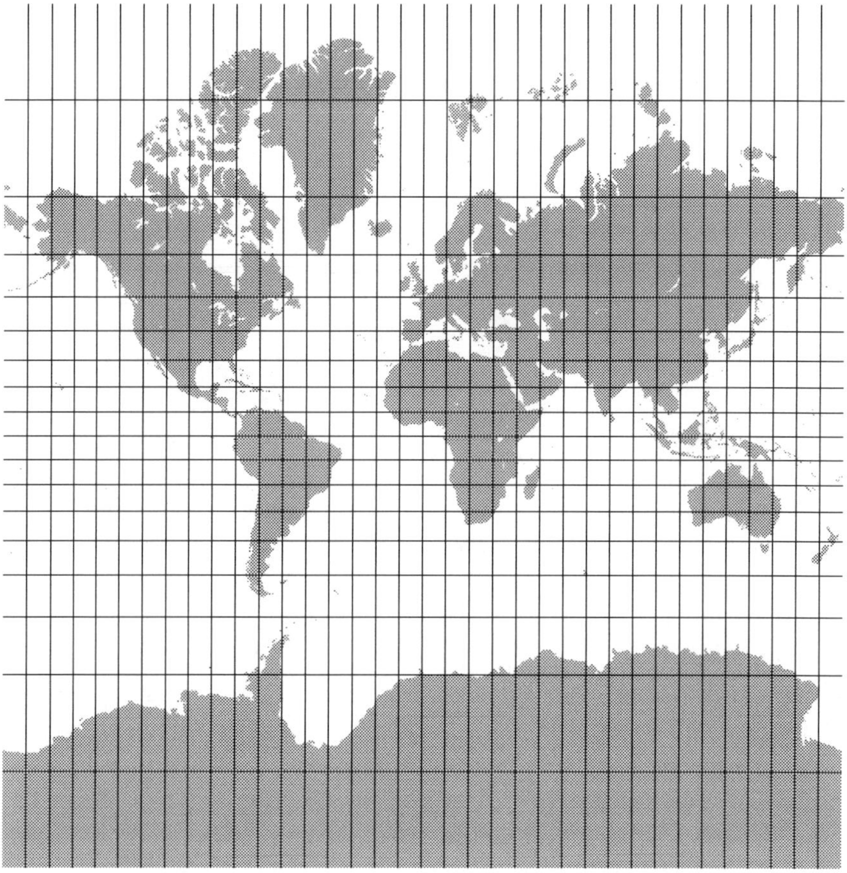

FIGURE 4.1 Mercator projection
THE TWO MAP PROJECTIONS ARE ADAPTED FROM THE US GEOLOGICAL SURVEYWEBSITES, HTTP://MCMCWEB.ER.USGS.GOV/DSS/IMGHTML/MERCATOR.HTML AND HTTP://MCMCWEB.ER.USGS.GOV/DSS/IMGHTML/LAMBERT0DEG.HTML

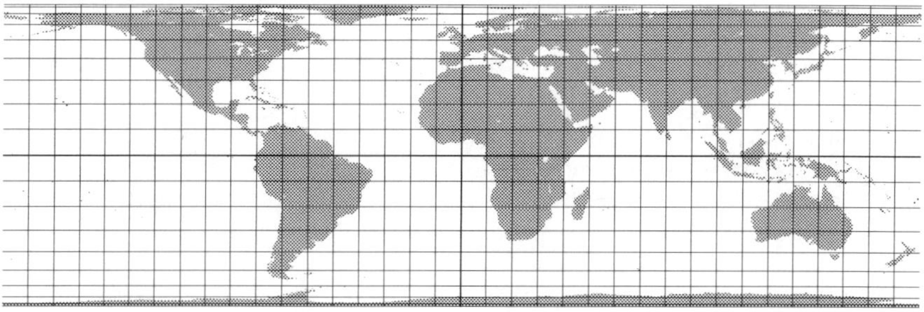

FIGURE 4.2 Lambert projection

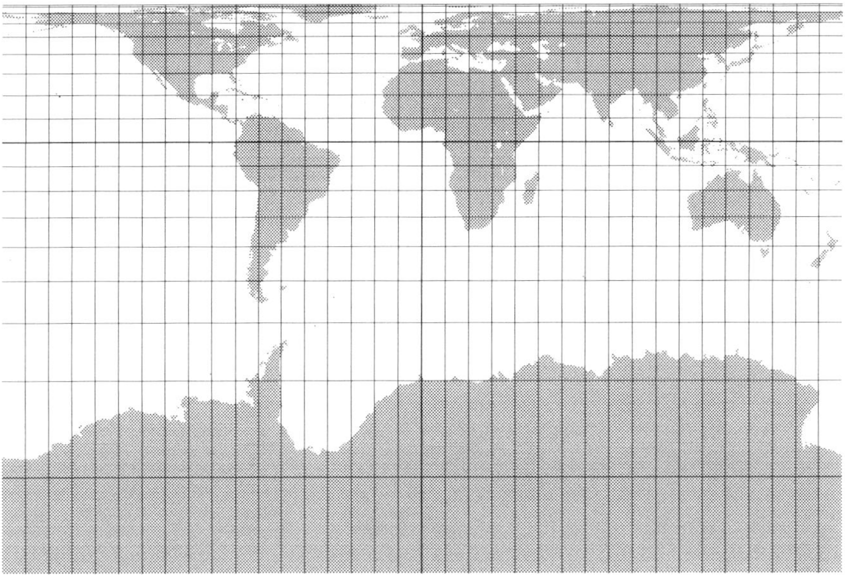

FIGURE 4.3 Hybrid mercator-lambert projection

world everywhere in the Southern hemisphere and find that everywhere in the Southern hemisphere our world conforms precisely with the Mercator world. We still could not know whether our world *in toto* is the Mercator world, for it could extend into the hybrid world of Figure 4.3 in which the Northern hemisphere is Lambertian. Perhaps that is our world. Even though we are dealing with a limited set of atlases, the geography of the Southern hemisphere does not fix the altas that applies to the Northern hemisphere.

This structure is essentially the one Einstein implemented in his hole argument. The sheet of paper that carries the different map projections corresponds to Einstein's spacetime coordinate system. The figures printed on the paper correspond to the fields of Einstein's theory. General covariance was the license that permitted him to spread his fields over the one spacetime coordinate system with the same freedom as we have in moving between map projections. In 1913, Einstein believed that the different spreadings of the fields corresponded to factually distinct realities. If we imagine that the Southern hemisphere of our map corresponds to the past of Einstein's spacetime and the Northern hemisphere to its future, the problem becomes a failure of determinism. Fixing the past of the spacetime does not fix its future.

What of the "hole?" Einstein realized that there was a sharpened version that was even more troubling. In the map analogy, instead of grafting hemispheres of different projections together, one could perform the grafting in

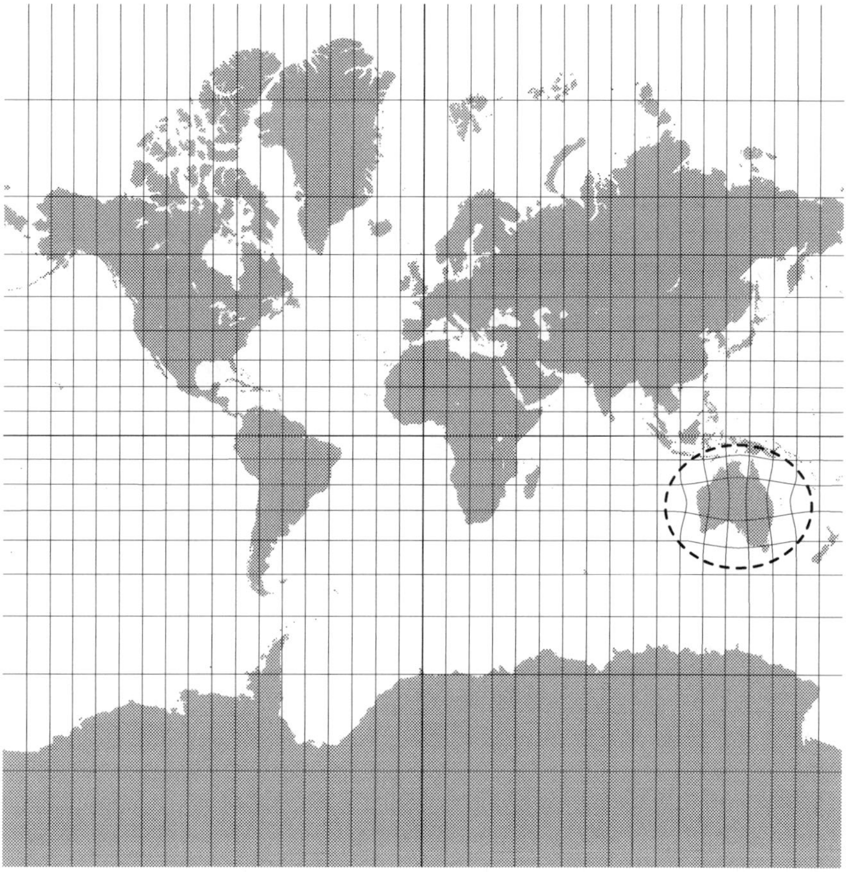

FIGURE 4.4 Mercator projection with a distorted hole

just a small portion of the page. We might have a projection that is everywhere Mercator, except in the region holding Australia, into which we graft some other projection, being sure to smooth the join neatly, as shown in Figure 4.4.

Allowing this possibility means that knowing the geography everywhere except Australia fails to fix the geography of Australia.

In Einstein's theory, the corresponding construction realizes a more severe failure of determinism. The hole corresponds to a small region of space that persists for a short time. It might just be a volume of space the size of a basketball that lasts for a second. Fixing everything in spacetime in its past, in its future and in all other parts of space throughout the universe still fails to fix what happens inside the hole. In 1913 and 1914, this result satisfied Einstein. It assured him that the generally covariant equations he could not find would not be worth finding. They would visit a radical form of indeterminism on his theory.

Matters had changed by the end of 1915. Then Einstein had returned to general covariance and had proposed the generally covariant gravitation field equations for which he became famous, the "Einstein field equations." Clearly some repair work was needed. Einstein now saw that, if he paid careful attention to what is real and what is not, the hole argument established neither indeterminism nor the physical inadmissibility of a generally covariant theory.

The map analogy lets us see how this comes about. While we have the freedom to spread our pictures of the continents in many ways over the one sheet of paper, there are no factual differences in geography to be found in the different spreadings.

The town of Alice Springs lies in the heart of Australia. Its exact placement on the sheet of paper will differ in the different map projections of the hole. However none of these differences will translate into verifiable geographical facts. Alice Springs lies roughly north of Adelaide on the south coast of Australia. Someone who leaves Adelaide and drives about 1500 km northward along the highways, mostly the Stuart Highway, will arrive there. The drive will intersect that of someone who leaves Darwin on the north coast of Australia and drives roughly 1500 km southward, again much of it on the Stuart Highway. The two projections will agree that the two travellers will meet at Alice Springs, as they will agree upon any other matter of geography that one can check. What is outside the hole fails to determine some aspects of what is within. But any aspects that remain undetermined are purely artifacts of the different projections; no geographic fact that can be checked is left uncertain.

Einstein's resolution was essentially identical to this analysis in geography, but with the claims carried over into spacetime theory. He wrote in his 1916 review article:[30]

> All our space-time verifications invariably amount to a determination of space-time coincidences. If, for example, events consisted merely in the motion of material points, then ultimately nothing would be observable but the meetings of two or more of these points. Moreover, the results of our measurings are nothing but verifications of such meetings of the material points of our measuring instruments with other material points, coincidences between the hands of a clock and points on the clock-dial, and observed point-events happening at the same place at the same time.

30 Albert Einstein, "Die Grundlage der Allgemeine Relativitätstheorie."

This is the "point-coincidence" argument.[31] Its immediate purpose is to establish the conclusion that the factual content of a physical theory is exhausted by the catalog of spacetime coincidences that it licenses. The different spreadings of the fields over the one coordinate system in the hole argument agreed in all spacetime coincidences. It now followed that any differences between the spreadings were purely artifactual; they correspond to no factual differences. The threat of the hole argument is averted.

The possibility of the multiple spreadings of fields had been inferred by Einstein from the requirement of general covariance, that is, the requirement that a physical theory can use any spacetime coordinate system. The hole argument had been deployed to demonstrate the untenability of general covariance. Einstein's position was now reversed. He was advocating general covariance strongly. The point-coincidence argument was called to his aid and it was extended into an argument for general covariance. The extension depended upon the fact that the two spreadings of the fields differ only in the spacetime coordinates at which the various spacetime coincidences may be found. Hence the assigning of spacetime coordinates is purely artifactual and has no physical content.[32] "The introduction of a system of reference [coordinate system] serves no other purpose than to facilitate the description of the totality of such coincidences," Einstein wrote immediately after the quote above. It follows that we should be free to use any coordinate system we like.

In his review article, Einstein did not make clear that this point-coincidence argument was his answer to the hole argument. It was introduced as an argument for general covariance. We can understand that Einstein would be reluctant to call any further attention to an argument that he had published four times but now deemed erroneous. That the point-coincidence argument was explicitly intended to resolve the hole argument was made clear when Einstein explained to correspondents why he felt authorized to restore the demand of general covariance. For our purposes, these explanations are interesting

31 Einstein's resolution is more severe than the one just sketched concerning map projections. He urges that all factual content reduces to space-time coincidences. In the map projection analogy, we took as a geographical fact that the driving distance from Adelaide to Alice Springs is 1500 km. We can adhere to the strictures of Einstein's argument by reducing facts about driving distances to facts about coincidences. If a car has tires with 2m circumference, then the fact of driving 1 km=1000m along the road reduces to the fact of 501 coincidences of a spot on the tire surface with different spots on the road.

32 In the map analogy, different map projections would place Alice Springs at different locations on the paper. Hence the locations on the paper by themselves have no geographical meaning.

because they contain strong statements about the division between what is real and what is not.

On January 3, 1916, he wrote to his friend, collaborator and confidant, Michele Besso:[33]

> Nothing is physically *real* but the totality of space-time point coincidences. If, for example, all physical happenings were to be built up from the motions of material points alone, then the meetings of these points, i.e. the intersection of their world lines, would be the only real things, i.e. observable in principle. (Einstein's emphasis)

A letter from Einstein to Paul Ehrenfest of December 26, 1915, contains similarly strong assertions concerning what is real:[34]

> The following considerations should replace [the 1914 presentation of the hole argument]. The physically real in what happens in the world (as opposed to what depends on the choice of the reference system) consists of *spatio-temporal coincidences*. [Einstein's footnote here: "and in nothing else!"] For example, the points of intersection of two world lines, or the assertion that they *do not* intersect, are real. Such assertions referring to the physically real are thus not lost because of any (single-valued) coordinate transformations.

This correspondence is also helpful in that it affirms that the map analogy captures Einstein's understanding. The analogy is close to one Einstein used in explaining the point-coincidence argument to Ehrenfest in another letter to him from late December 1915. Einstein took Ehrenfest's example of light from a star passing through an aperture to strike a photographic plate and then considered how the fields representing the system might be spread differently over a spacetime coordinate system. Einstein used a homely construction to illustrate how the resulting solutions of the field equations—Ehrenfest's A and B—are related. Ehrenfest was to trace a drawing of the system onto deformable paper. The different spreadings were then produced

33 Translation from John Stachel, "Einstein's Search for General Covariance, 1912–1915," 63–100, in Don Howard and John Stachel, eds., *Einstein and the History of General Relativity* (Boston: Birkhäuser, 1989), 86.

34 Translation from John Stachel, "Einstein's Search for General Covariance, 1912–1915," 63–100, in Don Howard and John Stachel, eds., *Einstein and the History of General Relativity* (Boston: Birkhäuser, 1989), 86.

merely by deforming the paper. Is there any factual difference between the two resulting figures? Einstein continued:[35]

> What is essential is this: As long as the drawing paper, i.e. "space," has no reality, the two figures do not differ at all. It is only a matter of "coincidences," e.g., whether or not the point on the plate is struck by light. Thus, the difference between your solutions A and B becomes a mere difference of representation, with physical agreement.

In all this, what is intriguing to a philosopher is to see that strong pronouncements concerning the real figure prominently in a major scientific discovery. Taken in isolation, they are strong, even programmatic, announcements of a fundamental principle: "Nothing is physically *real* but the totality of spacetime point coincidences," and "The physically real consists of *spatio-temporal coincidences* (and in nothing else!)." Even more striking is the strongly verificationist tone that underpins them. The real and the observable appear to be identified ("the only real things, i.e. observable in principle.") As a result, it has proven tempting to philosophers to regard these claims by Einstein as the anticipation of a grander verificationist view of science and one that, perhaps, should be regarded with suspicion for its extremism.[36]

We should approach that portrait of Einstein with caution. The Einstein who wrote these words was not the armchair philosopher grappling with the problem of reality and appearance at the most abstract level. The Einstein who wrote these words was jubilant at his great success with a scientific theory, general relativity, but physically and emotionally exhausted. He had struggled for nearly three years with his extraordinary new theory of gravity, in imperfect and misshapen form. Its defects were now finally identified and eliminated.

35 Translation from Don Howard and John D. Norton, "Out of the Labyrinth: Einstein, Hertz and Göttingen's Answer to the Hole Argument," in J. Earman, M. Janssen and J. Norton, eds., *The Attraction of Gravitation: New Studies in the History of General Relativity* (Boston: Birkhäuser, 1993), 30–62; 48.

36 For discussion of how Einstein's pronouncements have been received and interpreted philosophically, see John D. Norton, "Einstein, the Hole Argument and the Reality of Space," 154–155, and Don Howard, "Point Coincidences and Pointer Coincidences: Einstein on Invariant Structure in Spacetime Theories," *History of General Relativity IV: The Expanding Worlds of General Relativity*, H. Goenner, J. Renn, J. Ritter, and T. Sauer, eds. (Boston: Birkhäuser, 1999), 463–500. From the modern perspective, Einstein did not need such strong pronouncements for his purpose of defeating the hole argument. A weaker assertion, such as I have used in the exposition, would suffice. He need only assert that the factual content of the solution of his field equations is fixed by its catalog of spacetime coincidences, not that the coincidences themselves are all that is factual.

The struggle was over. All that remained was for Einstein to correct the errors of these three past years. One senses his weariness when he introduces his explanation of the error of the hole argument to Ehrenfest in his letter of December 26 1915.[37] "It is comfortable for Einstein. Each year he retracts what he wrote the previous year; now my duty is the extremely sad business of justifying my most recent retraction."

Einstein's proclamations about reality are not forward-looking, the anticipation of a new tradition of verificationism in philosophy. They are backward-looking, a convenient device that brings closure to an episode that Einstein now finds painful. In these circumstances, it is not surprising that Einstein would fail to give his pronouncements the pedantic precision that characterizes careful analysis by a professional philosopher.

Don Howard[38] has identified a significant and pertinent conflation. Are Einstein's coincidences "point-coincidences" or "pointer-coincidences?" The former are mathematical abstractions akin to idealized Euclidean points in geometry. They may be designated as the real in the sense that they are invariant, which means that they remain the same in all spacetime coordinate systems. The latter are the coincidences of macroscopic objects, such as pointer needles and scale marks. They are observable and hence real. Einstein's writing runs the two together. However the first, point-coincidences, is what Einstein needed to deflect his hole argument and restore general covariance. The second, pointer-coincidences, is what the ensuing tradition in verificationism needed to read in Einstein's writing if it was to claim him as their patron.

4.2 *The Incompleteness of Quantum Theory*

There is another, better known criterion of reality associated with Einstein. It too arose in the context of a significant problem in physical theory: Einstein's critique of quantum theory.

The Einstein of 1916 was the discoverer of general relativity. He was the theorist who stood in the vanguard of new work in physics. This was the same Einstein who had, in 1905, proposed the revolutionary concept of the light quantum. Now, in 1916, he continued his contributions to quantum theory with his "A and B coefficient" quantum analysis of heat radiation, laying the grounding for the modern theory of lasers.

37 Translation from John D. Norton, "Einstein, the Hole Argument and the Reality of Space," 169.

38 Don Howard, "Point Coincidences and Pointer Coincidences: Einstein on Invariant Structure in Spacetime Theories."

The Einstein of 1926, a decade later, was drifting into a different role. The energy of physics had been drawn into the emergence of the so-called "new quantum theory." In Schrödinger's famous formulation, each particle of the new theory was associated with a wave. Since a wave is spread out in space, one could associate no definite position with the particle, even though, on position measurement, the particle would always manifest in a definite position. Correspondingly, in general, a particle has no definite momentum, but it will always manifest a definite momentum upon momentum measurement. The best the new theory could provide was the probability that a particle would manifest in this position or with that momentum on the corresponding measurement.

Einstein joined his colleagues in recognizing that this new quantum theory was a worthy achievement that resolved accumulating difficulties of the "old quantum theory." However Einstein resisted one aspect of it resolutely. Does the quantum wave associated with a particle provide a complete description of the particle? Or are there further facts about a particular particle that are not expressed in the wave? The mainstream adopted the first view. Einstein urged the second view—incompleteness. A full accounting of why Einstein found himself a critic of the mainstream view of completeness would require a discussion of his reluctance to admit the arcane possibilities of non-locality and non-separability.[39]

However, in seeking the grounding of Einstein's discomfort, one cannot overlook his much repeated quip that God does not play dice.[40] If the quantum wave provides a complete description of the state of a particle, then quantum theory is indeterministic. Fixing the full state of the present does not fix the future. If we fix the quantum state of a particle now, the best we can recover for the future are merely probabilities for the particle being measured in this or that position. We have seen that Einstein recoiled from indeterminism when it was threatened in the hole argument. Einstein then described this failure of determinism as a failure of the "law of causality."[41] That reveals a decidedly nineteenth century aspect to Einstein's thinking, for in the nineteenth century, causality was purged of all its embellishments and reduced to

39 See, for example, Don Howard, "Einstein on Locality and Separability," *Studies in History and Philosophy of Science*, 16 (1985), 171–201.

40 For example, writing to Max Born on December 4, 1926, he remarked: "Quantum mechanics is very worthy of regard. But an inner voice tells me that this not yet the right track. The theory yields much, but it hardly brings us closer to the Old One's secrets. I, in any case, am convinced that He does not play dice" (Max Born, *The Born-Einstein Letters.* New York: Walker & Co. 1971, 91).

41 Albert Einstein, "Die formale Grundlage der allgemeinen Relativitätstheorie," 1066.

the simple notion of determinism. Hence the failure of determinism would be viewed as a failure of causality itself. This orientation is reflected in the general description in the earlier part of the twentieth century of the indeterminism of quantum theory as a violation of causality, a failure that many find inherently troubling. Einstein, it would appear, was sufficiently nineteenth century in his thinking to find the indeterminism of quantum theory an unacceptable violation of causality.[42]

While the origins of Einstein's discomfort with quantum theory may be diffuse, there was a single argument that Einstein favored as the way to establish the theory's incompleteness. The best-known presentation of the argument was given in his co-authored paper with Boris Podolsky and Nathan Rosen, the celebrated "EPR" paper.[43]

Establishing that the quantum wave provides an incomplete description of a particle is not straightforward. If one measures the position or momentum of a particle, under the standard account, its wave collapses to a new state with a definite position or momentum, according to the measurement undertaken. The old state is destroyed and one cannot preclude the possibility that the definite measurement outcome was created by the measurement process itself. Direct measurement no longer necessarily reveals the properties of particles possessed prior to measurement.

What the EPR paper recognized was that indirect measurement should succeed in revealing the properties really possessed by a particle, where direct measurement may fail. In classical physics, if two qualitatively identical particles are flung symmetrically from some central explosion, they will carry duplicate properties. When the first particle has moved some distance—say 100 m to the left—the other particle will have moved the same distance—100 m—to the right. Therefore measuring the position of one particle will reveal

42 We have now modified our notions of causality probabilistically so that quantum theory is no longer regarded as a mortal threat to causality. That, in my view, is no victory for causal metaphysics. Rather it provides one of many illustrations of the elasticity of causal notions that in turn reveals that a requirement of causality has no independent factual content. See John D. Norton, "Causation as Folk Science," *Philosophers' Imprint*, Vol. 3 (2003), No. 4; reprinted in H. Price and R. Corry, eds., *Causation, Physics, and the Constitution of Reality* (Oxford: Clarendon, 2007), 11–44.

43 See Albert Einstein, Boris Podolsky, Nathan Rosen, (1935), "Can Quantum-Mechanical Description of Physical Reality Be Considered Complete?," *Physical Review*, 47, 777–780; the same argument is developed in Albert Einstein, "Physics and Reality," *Journal of the Franklin Institute*, 221 (1936), 349–382; Albert Einstein, "Quanten-Mechanik und Wirklichkeit," *Dialectica*, 2 (1948), 320–24, and Albert Einstein, "Autobiographical Notes," 82–87.

the position of the other. Similarly, since momentum is conserved, the momentum of one particle is simply the negation of the momentum of the other. So measuring the momentum of one particle will tell us the momentum of the other. It turns out that quantum theory allows similarly correlated particles in what is known as an entangled state. So the same indirect measurements are possible for quantum particles.

Once that fact is recognized, the remainder of the analysis is straightforward. We prepare two entangled particles as above. We know that, were we to perform a position measurement on one particle, we could recover a definite position for it. Hence we would know that the other remote particle would reveal the corresponding, definite position on measurement. Since our measurement here cannot affect the remote particle some great distance away, we know that position property is possessed by the remote particle and not created by the measurement operation. That is already enough to establish the sought after incompleteness, for, in general, quantum theory does not allow a single definite position for the particle. The argument can then also be repeated for the momentum of the particle.

This simple and beautiful argument depends upon some apparently innocuous assumptions. One must assume that a measurement here cannot instantly affect a remote particle there; this is an assumption of locality. One must also assume that the two particles, once separated spatially, have independent existences with their own definite properties; this is an assumption of separability. The two assumptions must be made if we are to infer from measurements on local particles to the real properties possessed by distant particles entangled with them. The EPR authors recognized that assumptions along these lines were being made. While harboring no evident doubts, they asserted their version of the assumptions clearly in italic text at the start of the EPR paper:

> If, without in any way disturbing a system, we can predict with certainty (i.e. with probability equal to unity) the value of a physical quantity, then there exists an element of physical reality corresponding to this physical quantity.

This is the famous criterion of reality from the EPR paper of 1935. It is the foundation of the argument mounted by EPR and Einstein writing as a single author for the incompleteness of quantum theory.

All this is just the beginning of a long saga. The EPR paper failed to move the mainstream of physics. Einstein's antagonist, Niels Bohr, wrote a rebuttal that, sadly, was as obscure as it was influential. The EPR analysis did not receive the response it deserved until the work of John S. Bell. He demonstrated that no theory that agrees empirically with quantum theory could preserve locality and separability. Since these last two notions were, one way or another, necessary for the EPR reality criterion, Bell's work forced a choice between abandoning the empirical adequacy of quantum theory or abandoning the assumptions needed for the EPR criterion of reality. The mainstream of physics has chosen to abandon the latter.[44]

5 The Power of Platonism

Most of the examples of the philosophy we have seen so far in Einstein's work have empiricist, positivist or verificationist underpinnings. Hence we might conclude that Einstein's commitments are to empiricist and verificationist principles. To some extent, that was true. Yet, however strong these commitments may have been, they were subordinate to a deeper commitment. It was simple commitment to whatever ideology best led him to new theories. Hence, we should expect Einstein's commitments to empiricism and related approaches to be negotiable and even dispensable. And they were. Platonism is anathema to an empiricist. Yet, in the episode to be recounted, Einstein quite gladly adopted a mathematical Platonism when he sensed that it might be a more fertile aid to him in generating new theories.

During the research that led up to his discovery of the general theory of relativity, Einstein reflected explicitly on the methods he was using.[45] On the one side, he identified a physical approach. It took as the guide to new theories physical principles that were usually closely grounded in experience. They included the principle of relativity of his special theory and the principles of conservation of energy and momentum. The physical approach also gave special weight to limiting cases whose content is assured by physical reasoning. For weak gravitational fields, for example, his new theory had to replicate Newtonian theory. The purest embodiment of the physical approach came in

44 For a recent account of the incompatibility that requires only the most elementary notions from quantum theory, see John D. Norton, "Little Boxes: A Simple Implementation of the Greenberger, Horne, and Zeilinger Result for Spatial Degrees of Freedom," *American Journal of Physics*, 79:2 (2011), 182–188.

45 This conception of these two approaches was elaborated by a research group working in Berlin whose conclusions are given in Renn *et al.*, *The Genesis of General Relativity.* See also John D. Norton, "'Nature in the Realization of the Simplest Conceivable Mathematical Ideas': Einstein and the Canon of Mathematical Simplicity," *Studies in the History and Philosophy of Modern Physics*, 31 (2000), 135–170.

the thought experiments for which Einstein was famous. In them, our physical sensibilities would direct us inexorably towards a particular outcome.

This approach was contrasted with a formal or mathematical approach. According to it, we are guided to new theories by the mathematical properties of the structures involved. That transformations must form a group in the mathematical sense can be a powerful restriction. So also is the requirement of covariance, that the equations of the theory preserve their form in transformations among some stipulated range of spacetime coordinate systems. In this approach, mathematical theorems can reduce the viable theories to a very small selection. The purest embodiment of the approach is the use of formal naturalness and mathematical simplicity as a guide in theory selection.

Einstein's early inclinations had been strongly towards the physical approach, along with a discomfort and even distaste for the formal approach. This distaste is still evident in 1912 when he was turning to devote his attentions more fully to the developing general theory of relativity. One episode reveals it clearly. A gravitation theory that competed with Einstein's was formulated by Max Abraham by the simple expedient of transporting Newton's theory of gravity in special relativity in the mathematically simplest way. Einstein immediately proclaimed his displeasure with the theory to his correspondents, denouncing it as "totally untenable," "incorrect in every respect," and more. Most revealing, however, was that he was prepared to condemn the theory precisely because it had followed the formal approach. "… the thing is probably wrong," he wrote to Heinrich Zangger on 27 January 1912. "This is what happens when one operates formally, without thinking physically."[46]

After his move to Zurich in August of 1912, Einstein became embroiled in the new mathematical methods needed by his developing theory. With the assistance of a friend, the mathematician Marcel Grossmann, he began to learn the absolute differential calculus of G.C. Ricci and T. Levi-Civita (now called "tensor calculus"). As he developed his theory, Einstein continued to use his well-worked physical approach. However he could not ignore how the mathematical tools supplied by Grossmann offered certain natural structures from which to build his theory. Most importantly, it provided the Riemann curvature tensor as the structure that described the curvature of spacetime and from which the gravitational field equations should be built.

We are fortunate to have an intimate window onto Einstein's deliberations at this crucial time in the development of general relativity. His research notes

46 For the sources of these quotes and more similar ones, see John D. Norton, " 'Nature in the Realization of the Simplest Conceivable Mathematical Ideas': Einstein and the Canon of Mathematical Simplicity," §3.

have survived in the form of the "Zurich Notebook."[47] In it we can trace his earliest efforts to relate gravity and spacetime curvature and their ultimate fate. Einstein's expectation was that he could apply both approaches, physical and formal, and that they would agree. However, as his investigations proceeded, Einstein failed again and again to secure the agreement. Eventually he was forced to a choice: should he accept the results of the physical approach or those of the formal approach? Fatefully, Einstein chose in favor of the physical approach, abandoning general covariance. He struggled for nearly three more years with the resulting misshapen theory.[48]

Einstein's return to general covariance became a matter of public record in November 1915. Then he published four papers, one each week, with a series of proposals for gravitational field equations, based on the Riemann curvature tensor. It was a difficulty month, made all the more tense by the knowledge that David Hilbert in Göttingen, perhaps the greatest mathematician of the age, was in that same month working and publishing on the gravitational field equations of Einstein's theory. The month closed with the greatest achievement of Einstein's career. He completed the general theory of relativity and, in the process, received a welcome affirmation of its correctness. A jubilant Einstein found that his perfected theory could now account precisely for the anomalous motion of Mercury.

As that fateful month began, Einstein made no secret that his advances derived from a reversion to the mathematical or formal approach. In the first of the four papers, he reported:[49]

> I completely lost trust in the field equations I had chosen and looked for a way to restrict the possibilities in a natural manner. Thus I went back to the requirement of a more general covariance of the field equations, which I had left only with a heavy heart when I worked together with my friend Grossmann. In fact we had then already come quite close to the solution of the problem given in the following.

---

47 For an early account see John D. Norton, "How Einstein found his Field Equations: 1912–1915"; for a fuller account, see Jürgen Renn *et al.*, *The Genesis of General Relativity*, Volume 1, *Einstein's Zurich Notebook, Introduction and Source;* Volume 2, *Einstein's Zurich Notebook: Commentary and Essays*.

48 For an introductory synopsis of my view of what led Einstein astray in 1913, see John D. Norton, "A Conjecture on Einstein, the Independent Reality of Spacetime Coordinate Systems and the Disaster of 1913," in A.J. Kox and J. Einsenstaedt, eds., *The Universe of General Relativity, Einstein Studies,* Volume 11 (Boston: Birkhäuser, 2005), 67–102.

49 Albert Einstein, "Zur allgemeinen Relativitätstheorie," *Königlich Preussische Akademie der Wissenschaften* (Berlin), *Sitzungsberichte* (1915), 778–786; 778.

Einstein makes a similar report to Arnold Sommerfeld in correspondence at the end that month:[50]

> Once all trust in the results and methods of the earlier theory had gone, I saw clearly that a satisfactory solution could only be found through a connection to the general theory of covariants, i.e. to Riemann's covariant.

For our purposes, what is striking is just how willing a formerly scornful Einstein was to heap praise upon the fertility of the mathematical approach. He wrote in the first of the papers of November 1915:[51]

> Hardly anyone who has truly understood it can resist the charm of this theory; it signifies a real triumph of the method of the general differential calculus, founded by Gauss, Riemann, Christoffel, Ricci and Levi-Civita.

Einstein drew an important moral from this experience. Had he only taken the formal approach more seriously at the start, he would have spared himself much suffering. This moral entered into Einstein's methods. His subsequent search for a unified field theory depended essentially on seeking the mathematically simplest equations.

By the time of Einstein's 1933 Herbert Spenser lecture "On the Methods of Theoretical Physics," Einstein's advocacy of mathematical Platonism is explicit and powerful. There he wrote the following:[52]

> Our experience hitherto justifies us in believing that nature is the realization of the simplest conceivable mathematical ideas. I am convinced that we can discover by means of purely mathematical constructions the concepts and the laws connecting them with each other, which furnish the key to the understanding of natural phenomena. Experience may suggest the appropriate mathematical concepts, but they most certainly cannot be deduced from it. Experience remains, of course, the sole criterion of the physical utility of a mathematical construction. But the creative

50 Translation from John D. Norton, " 'Nature in the Realization of the Simplest Conceivable Mathematical Ideas': Einstein and the Canon of Mathematical Simplicity," 151.

51 Albert Einstein, "Zur allgemeinen Relativitätstheorie," 779.

52 Albert Einstein, "On the Methods of Theoretical Physics" (1933); in *Ideas and Opinions.* (New York: Bonanza, 1954), 270–276.

> principle resides in mathematics. In a certain sense, therefore, I hold it true that pure thought can grasp reality, as the ancients dreamed.

One should not mistake these words for the abstract musings of an armchair philosopher. They are the mature reflections of a philosophically sophisticated Einstein, reporting the methods that had worked in his earlier researches and that he hoped would lead him to his unified field theory.

6 Conclusion

Albert Einstein took philosophy seriously. He read it, he wrote it, and he engaged in exchanges with the leading philosophers of his time. Most importantly, philosophical analysis was incorporated directly into his theorizing in physics. He was a physicist who could use the word "epistemological" in a physics paper. However Einstein was not a philosopher. His concern was physics and his allegiance was to whatever instrument would advance his theorizing. Hence, taken in isolation, Einstein's philosophical commitments may appear capricious, changing at whim. But that is a short-sighted appraisal. It merely reflects that the Einstein who wrote philosophy was not a dogmatic philosopher who would defend his system come what may. Rather, it reflects an Einstein who used philosophy pragmatically for other purposes.

One might find this assessment of Einstein's philosophical commitments slighting. However Einstein was quite self-aware and it is, I believe, his own assessment. Responding to critics later in life in early 1949, he wrote:[53]

> … no sooner has the epistemologist, who is seeking a clear system, fought his way through to such a system, than he is inclined to interpret the thought-content of science in the sense of his system and to reject whatever does not fit into his system. The scientist, however, cannot afford to carry his striving for epistemological systematic that far. He accepts gratefully the epistemological conceptual analysis; but the external conditions, which are set for him by the facts of experience, do not permit him to let himself be too much restricted in the construction of his conceptual world by the adherence to an epistemological system. He therefore must appear to the systematic epistemologist as a type of unscrupulous

53 Albert Einstein, "Remarks Concerning the Essays Brought Together in this Co-operative Volume," (1949), in P.A. Schilpp, ed., *Albert Einstein: Philosopher-Scientist*, 665–688.

opportunist: he appears as *realist* insofar as he seeks to describe a world independent of the acts of perception; as *idealist* insofar as he looks upon the concepts and theories as free inventions of the human spirit (not logically derivable from what is empirically given); as *positivist* insofar as he considers his concepts and theories justified *only* to the extent to which they furnish a logical representation of relations among sensory experiences. He may even appear as *Platonist* or *Pythagorean* insofar as he considers the viewpoint of logical simplicity as an indispensable and effective tool of his research.

References

Born, Max, *The Born-Einstein Letters.* New York: Walker & Co. 1971.

Earman, John, and Norton, John D., "What Price Spacetime Substantivalism? The Hole Story," *British Journal for the Philosophy of Science*, 38 (1987), 515–525.

Earman, John, Janssen, M., and Norton, John D., eds., *The Attraction of Gravitation: New Studies in the History of General Relativity*, Boston: Birkhäuser, 1993.

Einstein, Albert and Grossmann, Marcel, *Entwurf einer verallgemeinerten Relativitätstheorie und einer Theorie der Gravitation*, Leipzig: B.G.Teubner, 1913, (separatum); with addendum by Einstein in *Zeitschrift für Mathematik und Physik*, 63 (1914), 225–261.

Einstein, Albert, "Autobiographical Notes," in P. A. Schilpp, ed., *Albert Einstein: Philosopher-Scientist*, Evanston, IL: Library of Living Philosophers (1949), 2–95.

Einstein, Albert, "Die formale Grundlage der allgemeinen Relativitätstheorie," *Königlich Preussische Akademie der Wissenschaften (Berlin), Sitzungsberichte*, 1914, 1030–1085.

Einstein, Albert, "Die Grundlage der allgemeinen Relativitätstheorie," *Annalen der Physik*, 49 (1916), 769–822; translated as "The Foundation of the General Theory of Relativity," in Albert Einstein et al., *The Principle of Relativity*, trans. W. Perrett and G.B. Jeffrey, New York: Dover, 1952, 111–164.

Einstein, Albert, "Ernst Mach," *Physikalische Zeitschrift*, 17 (1916), 101–104.

Einstein, Albert, "Gibt es eine Gravitationswirkung, die der elektrodynamischen Induktionswirkung analog ist?," *Vierteljahrsschrift für gerichtliche Medizin und öffentliches Sanitätswesen*, 44 (1912), 37–40.

Einstein, Albert, "On the Methods of Theoretical Physics," 1933; in *Ideas and Opinions.* New York: Bonanza, 1954, 270–276.

Einstein, Albert, "Physics and Reality," *Journal of the Franklin Institute*, 221 (1936), 349–382.

Einstein, Albert, Podolsky, Boris, and Rosen, Nathan, "Can Quantum-Mechanical Description of Physical Reality Be Considered Complete?," *Physical Review*, 47 (1935), 777–780.

Einstein, Albert, "Quanten-Mechanik und Wirklichkeit," *Dialectica*, 2 (1948), 320–24.

Einstein, Albert "Remarks Concerning the Essays Brought Together in this Co-operative Volume," 1949; in P.A. Schilpp, ed., *Albert Einstein: Philosopher-Scientist*, 665–688.

Einstein, Albert, *The Meaning of Relativity*, (1922); 5th Expanded Edition, Princeton: Princeton University Press, 1956.

Einstein, Albert, *Über die spezielle and die allgemeine Relativitätstheorie* (*Gemeinverständlich*), Braunschweig: Friedr. Vieweg & Sohn, 1917; 15th expanded edition translated by R.W. Lawson as *Relativity: the Special and the General Theory*, London: Metheun, 1954.

Einstein, Albert, "Zum gegenwärtigen Stande des Gravitationsproblems," *Physikalische Zeitschrift*, 14 (1913), 1249–1262.

Einstein, Albert, "Zur allgemeinen Relativitätstheorie," *Königlich Preussische Akademie der Wissenschaften* (Berlin), *Sitzungsberichte* (1915), 778–786.

Einstein, Albert "Zur Elektrodynamik bewegter Körper," *Annalen der Physik*, 17 (1905), 891–921; translated as "On the Electrodynamics of Moving Bodies," in Albert Einstein *et al.*, *The Principle of Relativity*, trans. W. Perrett and G. B Jeffrey, New York: Dover, 1952, 37–65.

Herneck, F., "Zwei Tondokumente Einsteins zur Relativitätstheorie," *Forschungen und Fortschritte* 40 (1966), 133–135; translated in John Stachel *et al.*, *The Collected Papers of Albert Einstein*, Volume 2, *The Swiss Years: Writings, 1900–1909*, Princeton: Princeton University Press, 1989.

Howard, Don, "Albert Einstein as a Philosopher of Science," *Physics Today*, 58, No. 12 (Dec. 2005), 34–40.

Howard, Don and Stachel, John, eds., *Einstein and the History of General Relativity*, Boston: Birkhäuser, 1989.

Howard, Don, "A Peek behind the Veil of Maya: Einstein, Schopenhauer, and the Historical Background of the Conception of Space as a Ground for the Individuation of Physical Systems," in *The Cosmos of Science: Essays of Exploration*, John Earman and John D. Norton, eds., *Pittsburgh-Konstanz Series in the Philosophy and History of Science*, vol. 6., Pittsburgh: University of Pittsburgh Press; Konstanz: Universitätsverlag, 1997, 87–150.

Howard, Don, "Einstein on Locality and Separability," *Studies in History and Philosophy of Science*, 16 (1985), 171–201.

Howard, Don, "Einstein's Philosophy of Science," *The Stanford Encyclopedia of Philosophy*, Edward N. Zalta (ed.), URL = https://plato.stanford.edu/entries/einstein-philscience/.

Howard, Don, "Point Coincidences and Pointer Coincidences: Einstein on Invariant Structure in Spacetime Theories," *History of General Relativity IV: The Expanding Worlds of General Relativity*, H. Goenner, J. Renn, J. Ritter, and T. Sauer, eds., Boston: Birkhäuser, 1999, 463–500.

Hume, David, *A Treatise of Human Nature*, ed. P.H. Nidditch, 2nd ed., Oxford: Clarendon Press, 1978.

Mach, Ernst, *The Science of Mechanics: A Critical and Historical Account of Its Development*, 6th ed., trans. T. J. McCormach, LaSalle, Illinois: Open Court, 1960.

Newton, Isaac, *Mathematical Principles of Natural Philosophy and his System of the World*, 1729; tr. Andew Motte, revised Florian Cajori, Berkeley: University of California Press, 1934.

Norton, John D., "A Conjecture on Einstein, the Independent Reality of Spacetime Coordinate Systems and the Disaster of 1913," in A. J. Kox and J. Einsenstaedt, eds., *The Universe of General Relativity, Einstein Studies,* Volume 11, Boston: Birkhäuser, 2005, 67–102.

Norton, John D., "Causation as Folk Science," *Philosophers' Imprint,* Vol. 3 (2003), No. 4; reprinted in H. Price and R. Corry, eds., *Causation, Physics, and the Constitution of Reality*, Oxford: Clarendon, 2007, 11–44.

Norton, John D., "Einstein's Investigations of Galilean Covariant Electrodynamics prior to 1905," *Archive for History of Exact Sciences* 59 (2004), 45–105.

Norton, John D., "Einstein, the Hole Argument and the Reality of Space," in J. Forge, ed., *Measurement, Realism and Objectivity*, Dordrecht: Reidel, 1987, 153–188.

Norton, John D., "How Einstein found his Field Equations: 1912–1915," *Historical Studies in the Physical Sciences*, 14 (1984), 253–316; reprinted in Don Howard and John Stachel, eds., *Einstein and the History of General Relativity: Einstein Studies*, Volume 1, Boston: Birkhäuser, 1989, 101–159.

Norton, John D., "How Hume and Mach Helped Einstein Find Special Relativity," in M. Dickson and M. Domski, eds., *Discourse on a New Method: Reinvigorating the Marriage of History and Philosophy of Science*, Chicago: Open Court, 2010, 359–386.

Norton, John D., "Little Boxes: A Simple Implementation of the Greenberger, Horne, and Zeilinger Result for Spatial Degrees of Freedom," *American Journal of Physics*, 79:2, 2011, 182–188.

Norton, John D. "Mach's Principle before Einstein," in J. Barbour and H. Pfister, eds., *Mach's Principle: From Newton's Bucket to Quantum Gravity: Einstein Studies*, Vol. 6, Boston: Birkhäuser, 1995, 9–57.

Norton, John D., " 'Nature in the Realization of the Simplest Conceivable Mathematical Ideas': Einstein and the Canon of Mathematical Simplicity," *Studies in the History and Philosophy of Modern Physics*, 31 (2000), 135–170.

Norton, John D., "The Hole Argument," *The Stanford Encyclopedia of Philosophy* (Winter 2008 Edition), ed. Edward N. Zalta, URL = http://plato.stanford.edu/archives/win2008/entries/spacetime-holearg/.

Renn, Jürgen, *et al., The Genesis of General Relativity,* Volume 1, *Einstein's Zurich Notebook, Introduction and Source;* Volume 2, *Einstein's Zurich Notebook: Commentary and Essays*, Dordrecht: Springer, *2007*.

Ryckman, Thomas, *Reign of Relativity: Philosophy in Physics 1915–1925*, Oxford: Oxford University Press, 2007.
Stachel, John, "Einstein's Search for General Covariance, 1912–1915," in Don Howard and John Stachel, eds., *Einstein and the History of General Relativity*, Boston: Birkhäuser, 1989, 63–100.
Stachel, John, *et al.*, eds., *The Collected Papers of Albert Einstein, Volume 2, The Swiss Years: Writings 1900–1909*, Princeton: Princeton University Press, 1989.

CHAPTER 5

Nature, God, and Scientific Method

Edward L. Schoen

Despite the long, historical interplay between religion and the sciences, a number of writers from the middle of the last century to the present have argued for a rigorous separation of contemporary scientific thought from anything distinctively religious. Often, this clean separation has been grounded in the conviction that science employs a special method with some distinctive set of epistemic practices that somehow imposes limits upon the kinds of things that can be studied scientifically. So, for example, Bertrand Russell claimed that science uses observations to discover particular facts about the world in order to formulate laws. Since Russell also believed that "... what science cannot discover, mankind cannot know,"[1] anything that could not be studied in this particular way fell utterly beyond the scope of human knowledge. This, of course, provided his basis for repudiating traditional Christian creeds, with their references to an essentially unobservable, incorporeal God. But even writers sympathetic to religious belief have embraced the conviction that distinctively scientific methods carry serious metaphysical implications. Langdon Gilkey claimed that he could identify three canons of scientific method:

> ... first, the *empirical* canon that no concept is permissible except one that grows out of and can be checked in sensible and so sharable experience; second, the *naturalistic* canon that no supernatural explanatory cause is permissible in scientific explanation—that only natural or human causes may be appealed to, and third, that scientific explanations are in terms of *universal* and so *necessary* relations and not purposes or intentions.[2]

---

1 Bertrand Russell, *Religion and Science* (Oxford: Oxford University Press, 1961), 243. See 8 for his characterization of the scientific method. I also would like to thank Professor Philip MacEwen for the opportunity to contribute this Chapter, and for his suggestions and encouragement along the way.

2 Langdon Gilkey, "Religion and Science in an Advanced Scientific Culture," *Zygon* 22, No. 2 (June 1987), 172. Emphasis in original. See also 173–174.

 | DOI:10.1163/9789004415270_007

Unlike Russell, Gilkey extended the range of human understanding beyond the limits of science. He thought that religion could investigate questions about the ultimate origins and grounds of natural processes. Religion also could delve into nonobjective, personal and historical matters as well as raise the question of God. Despite expanding the scope of human understanding, however, Gilkey shared Russell's fundamental conviction about the metaphysical limits of scientific method. Consequently, areas that Gilkey believed to be open to religious investigation remained forever closed off from science, due to restrictions imposed by its methodology.

More recently, David Ray Griffin defined scientific naturalism in the minimal sense as "… the doctrine that the universe involves an extremely complex web of cause-and-effect relations; that every event occurs within this web, having causal antecedents and causal consequences; and that every event exemplifies a common set of causal principles."[3] According to Griffin, the repudiation of this metaphysical doctrine is utterly unthinkable. It must be taken as cognitively normative, because it is so pervasively and deeply embedded in current scientific practice. Consequently, traditional Christian creeds that clash with minimal naturalism must be jettisoned or, at least, reformulated so as to render them compatible with this contemporary scientific constraint.

Despite the popularity of such lines of thinking, the burden of this discussion will be to challenge the contention that scientific methods impose some stringent range of metaphysical restrictions. To the contrary, it will be argued that the connections between scientific methods and specific metaphysical commitments are extremely loose. Indeed, there may not be any significant connections at all, certainly none that would point in the direction of one or another of the usual standard arrays of metaphysical doctrines, be they naturalist or supernaturalist, materialist, idealist or dualist. If the history of the development of scientific methods suggests anything at all, it counsels a tentative, rather open-ended pluralistic stance, something culminating in little more than metaphysical agnosticism.

1 The Unity of the Sciences

Many of those who try to extract metaphysical constraints from scientific practice are convinced that genuinely scientific investigation is characterized,

3 David Ray Griffin, *Two Great Truths: A New Synthesis of Scientific Naturalism and Christian Faith* (Louisville and London: Westminster John Knox Press, 2004), 2.

indeed unified, by some singular, distinctive method. Towards the middle of the twentieth century, A.J. Ayer stood as perhaps the most celebrated proponent of some such normative scientific method, a method he believed would exclude all sorts of things from the range of legitimate scientific investigation. Invoking a verifiability criterion of factual significance, he identified genuinely scientific utterances as factually meaningful. Then, he sharply distinguished such utterances from the unintelligible mutterings of metaphysicians and religious believers. When his verifiability criterion ran into technical difficulties, one of Ayer's critics, Karl Popper, substituted falsifiability as the mark of true scientific method. Sadly, Popper's suggestion fared no better, turning out to be susceptible to the same kinds of criticisms that had plagued the verificationists. But this did not deter other philosophers of the time. Dismissing such setbacks as temporary nuisances, Rudolf Carnap outlined an ambitious program of rational reconstruction for the sciences. In the second edition of his *Logical Structure of the World,* he envisioned a comprehensive unification of all scientific disciplines, including the social sciences. Carnap used formal logical analysis to erect his singular scientific edifice, grounding it all in a common fund of elementary sensory experiences.[4]

Thomas Kuhn, of course, rejected such constructs as pure, speculative fantasy. Scanning the history of science, he found no sweeping methodological consensus anywhere. Instead, there were only more localized, modest forms of scientific unity. According to Kuhn, scientists were bound together into research communities by their common commitment to specific paradigms. Despite their pervasiveness, the influence of such paradigms was so delimited that scientists from different research perspectives shared nothing in common, not even the same world. Under the pressure of criticism, however, Kuhn eventually backed away from this extreme position, substituting a list of five methodological universals in its place. Despite claiming that these five were common to all of the sciences, Kuhn quickly evaporated his universal core in a sequence of qualifications. Not only did he claim that scientists legitimately

4 See Alfred Jules Ayer, *Language, Truth and Logic*, 2nd ed. (New York: Dover Publications, Inc., 1952); Karl R. Popper, *The Logic of Scientific Discovery* (New York and Evanston: Harper and Row, Publishers, 1968), and Rudolf Carnap, *The Logical Structure of the World and Pseudoproblems in Philosophy*, trans. Rolf A. George (Berkeley and Los Angeles: University of California Press, 1969). For a general summary of the development of verifiability and falsifiability criteria as well as a helpful survey of the most important critical responses to assorted positivistic proposals, see Frederick Suppe, ed., *The Structure of Scientific Theories*, 2nd ed. (Urbana: University of Illinois Press, 1977). Supplemental material may be found in Richard Boyd, Philip Gasper and J.D. Trout, eds., *The Philosophy of Science* (Cambridge: The MIT Press, 1991), Chapter 1.

could modify the list of five with various additions and subtractions, but values "... like accuracy, consistency and scope may prove ambiguous in application, both individually and collectively; they may, that is, be an insufficient basis for a *shared* algorithm of choice."[5] Ultimately, then, except perhaps at the level of superficial terminology, Kuhn identified no genuinely universal scientific methodology at all.

Following Kuhn's lead, countless philosophers of science have pronounced the death of Logical Positivism. While they may have disputed the details of his analyses, most enthusiastically joined him in the repudiation of rigidly fixed, universal scientific methods. Imre Lakatos limited the scope of methodological agreement to individual research programs. Even there, the negative heuristic emerged as the only fixed, hard core. According to Lakatos, the positive heuristic of research programs could change and develop over time. Larry Laudan went farther down the diversity road. He rejected any long-term stasis in science, even within specific research traditions. For Laudan, scientists were bound together by a common commitment to three basic items: a group of factual claims, a set of methodological rules, and finally, an assemblage of goals, aims, and values. With respect to all three, scientists enjoyed a flexibility of options regarding how to proceed. At every juncture, no matter which of their theoretical, methodological or axiological commitments scientists might choose to protect, they could reasonably choose to stand firm elsewhere instead. Since each choice would affect the subsequent course of reasonable scientific development, research traditions maintained only a loose kind of evolutionary continuity. As with the evolutionary tree of biological species, each node and branch on the tree of reasonably developing science would grow as it did for intelligible reasons. Nevertheless, had fixed commitments been different, a tree of another, equally reasonable shape would have emerged. Over the long history of scientific development, Laudan required nothing to endure forever.[6]

Despite these and similar analyses, broad, global sketches of sweeping unities in the sciences continue to thrive. In *Concilience: The Unity of Knowledge,*

5 Thomas S. Kuhn, "Objectivity, Value Judgment, and Theory Choice," in Thomas S. Kuhn, *The Essential Tension: Selected Studies in Scientific Tradition and Change* (Chicago: University of Chicago Press, 1977), 331. See 322 for Kuhn's list of the five standard criteria. For his original view, see Thomas S. Kuhn, *The Structure of Scientific Revolutions,* 2nd ed., Enlarged (Chicago: University of Chicago Press, 1970), especially 10, 94, 103, 126ff, and 150–151.

6 See Imre Lakatos and Alan Musgrave, eds., *Criticism and the Growth of Knowledge* (London and New York: Cambridge University Press, 1970), 133–138, and Larry Laudan, *Science and Values: The Aims of Science and Their Role in Scientific Debate* (Berkeley: University of California Press, 1984), especially Chapter Two.

Edward O. Wilson paints a dazzling picture, where all of the sciences, and even the humanities, are depicted as knit together by relations of analysis and synthesis, one comprehensive network ultimately grounded in physics. Along rather different lines, Arthur Peacocke envisions a hierarchy of sciences, where the concept of emergence provides the key to linking various levels. But no matter how attractive or persistent such dreams may be, as with the work of Carnap, they dissipate into sheer speculation under the pressure of detailed scrutiny. Although Wilson enthusiastically endorsed the reduction of one field to another, he allowed hazy, hopeful promises of future reductions to stand in for actual, convincing examples. In defense of his own vision, Peacocke pointed to numerous instances of phenomena he believed to be irreducible to lower levels in his hierarchies. Skeptics, however, saw no such neat hierarchical relations among the sciences. Even when bolstered with ideas about supervenience and downward causation, the concept of emergence appeared more mystifying than illuminating—more like a label pasted over the problem of connecting the sciences than a promising solution to it.[7]

When actual cases are studied closely, even narrow stretches within scientific disciplines emerge as disconcertingly disparate. In physics, often considered a paradigm among the sciences, there are persistent signs of troubling disconnection. After decades of debate, many physicists still doubt whether wave and particle theories are genuinely compatible. While some take solace in mathematics, claiming that a unified, formal treatment of the data is all that is needed, others are not so sure. Perhaps mathematical formalisms only mask the problem, swapping the convenience of computational devices for any real, penetrating unity of theoretical understanding. Then there are the potent clashes between quantum theory and general relativity, where lasting inconsistencies fuel the pursuit of string theory. Whether any genuinely unified theory, string or otherwise, will eventually emerge is anyone's guess.[8]

Of course, there is no need to plumb the esoteric depths of elementary physics to uncover signs of disunity in actual scientific practice. Other, more accessible sources conspire, not merely to signal radical multiplicity in scientific

7 See Edward O. Wilson, *Consilience: The Unity of Knowledge* (New York: Vantage Press, 1998). For an early but clear defense of Peacocke's position, see A.R. Peacocke, *Creation and the World of Science* (Oxford: Clarendon Press, 1979). For critical reflections on emergence, downward causation, and related concepts, see Philip Clayton and Paul Davies, eds., *The Re-Emergence of Emergence: The Emergentist Hypothesis from Science to Religion* (Oxford: Oxford University Press, 2006).

8 See Brian Greene, *The Elegant Universe: Superstrings, Hidden Dimensions, and the Quest for the Ultimate Theory* (New York: W.W. Norton & Company, 1999).

practice, but to recommend it as cognitively virtuous. As they become increasingly technical, Patrick Suppes noted that the languages of various branches of science diverge, rather than converge. He concluded:

> ... it is especially the experimental methods of different branches of science that have radically different form. It is no exaggeration to say that the handbooks of experimental method for one discipline are generally unreadable by experts in another discipline ...[9]

Instead of distress, Suppes took heart from this ever increasing diversity and fragmentation. He declared:

> Personally, I applaud the divergence of language in science and find in it no grounds for skepticism or pessimism about the continued growth of science. The irreducible pluralism of languages of science is as desirable a feature as is the irreducible plurality of political views in a democracy.[10]

This cheerful attitude is reinforced by studies in cognitive linguistics. George Lakoff identified serious conceptual incompatibilities in sophisticated scientific contexts. Far from being cognitively detrimental, he viewed such clashes as highly profitable, often indispensable. For example:

> Many functioning scientists ... who understand electricity only as a fluid tend to make systematic errors in certain kinds of problems—those where the crowd metaphor works better. Students who understand electricity only as a crowd of electrons tend to make mistakes on a different set of problems—those where the fluid metaphor works better ... Knowing how to solve problems in electrical circuitry involves knowing which metaphor to use in which situation.[11]

Inconsistent construals are commonplace, sometimes occurring within the brief span of a single sentence. Consider the scientific remark: "We weren't able to use red objects in the experiment because there is no single wavelength that can be perceived as focal red."[12] Here, the switch is between two quite

9 Patrick Suppes, "The Plurality of Science," in Janet A. Kourany, ed., *Scientific Knowledge: Basic Issues in the Philosophy of Science* (Belmont: Wadsworth Publishing Company, 1987), 320.

10 Suppes, "The Plurality of Science," 319.

11 George Lakoff, *Women, Fire, and Dangerous Things: What Categories Reveal about the Mind* (Chicago and London: University of Chicago Press, 1987), 305.

12 Lakoff, *Women, Fire, and Dangerous Things: What Cateegories Reveal about the Mind*, 214.

different understandings of red, one whereby red is a sensuously perceptible quality of certain objects, and the other, whereby red is construed in terms of reflected wavelengths of light, wavelengths that may not sensuously present anything as red in the perceptual sense at all. This and similarly incoherent construals do not reduce scientific claims to unintelligibility. To the contrary, scientists easily switch among alternative conceptualizations, so easily that they rarely even notice the shifts involved.

Ian Hacking has located indispensable discontinuities in the activities of experimental research teams. He reported:

> Various properties are confidently ascribed to electrons, but most of the confident properties are expressed in numerous different theories or models about which an experimenter can be rather agnostic. Even people in a team, who work on different parts of the same large experiment, may hold different and mutually incompatible accounts of electrons. That is because different parts of the experiment will make different uses of electrons. Models good for calculations on one aspect of electrons will be poor for others. Occasionally, a team actually has to select a member with a quite different theoretical perspective simply to get someone who can solve those experimental problems. You may choose someone with a foreign training, and whose talk is well-nigh incommensurable with yours, just to get people who can produce the effects you want.
>
> But might there not be a common core of theory, the intersection of everybody in the group, which is the theory of the electron to which all the experimenters are realistically committed? I would say common lore, *not* common core. There are a lot of theories, models, approximations, pictures, formalisms, methods, and so forth involving electrons, but there is no reason to suppose that the intersection of these is a theory at all.[13]

As Hacking emphasized, teams of diverse experimenters may be committed to all sorts of differing descriptions, theories, analyses, methods, goals, and values, yet still join together in a collective exploration of the mysteries of electrons. Even beliefs about the very existence of electrons need not coincide. Some team members may accept the reality of electrons, while others, more inclined to antirealism, might deny their existence. In fact, shared cognitive

13 Ian Hacking, "Experimentation and Scientific Realism," in Kourany, ed., *Scientific Knowledge: Basic Issues in the Philosophy of Science*, 389–390.

commitments are so extraneous to team membership that some individuals might actually refuse to embrace any distinctively scientific orthodoxy at all.

This is particularly evident, when it comes to the operation of scientific apparatus. In every laboratory, machines must be tuned, calibrated and properly operated, if they are to produce useful, reliable results. Sometimes, only certain individuals possess the uncanny, often incomprehensible, capacity to tweak the best performance from finicky pieces of essential equipment. Those people are welcomed eagerly into the very heart of scientific research projects. No matter how they manage to coax machines into high levels of performance, if technicians can get necessary equipment to hum just right, no one else will care how bizarre their beliefs, goals, methods or values might be. Outside the laboratory, such people might not even count as scientists. Their very sanity could be open to question. But none of that matters. Like professional plumbers with a feel for just the right amount of pressure to apply to compression nuts or a sense of just how much joint compound is enough, but not too much, such technicians offer vital skills, abilities and understanding. Without them, countless ongoing research projects would grind to a halt.

2 The Methodological Constraints of Science

Clearly, it is not necessary to insist upon some all-encompassing vision of scientific unity to press the case for metaphysical limitations. It is enough to claim that certain methodological practices characterize science typically and persistently, if not universally. Among such localized scientific practices, two are especially prominent. First is the drive to uncover and identify scientific laws. To someone like Griffin, the presumption of a thoroughgoing, causal regularity in nature seems to foreclose any possibility of supernatural agency or divine intervention. Second is the empirical nature of science. For someone like Russell, the pivotal need for sensory evidence seems to preclude anything essentially unobservable, like an incorporeal God, from falling within the scope of scientific investigation. Since I have discussed the impact of these two constraints fairly extensively elsewhere, they will receive only brief attention here. That will permit more space for a third concern, one raised by Ludwig Wittgenstein and subsequently pressed by Peter Winch and D.Z. Phillips.

2.1 *Regularity in Nature*

Rigid regularity is surprisingly hard to come by. Kids practice their tennis strokes by bouncing balls off garage doors, never expecting the balls to rebound in exactly the same way. Ptolemaic astronomers looked for perfect circles in the skies, but mostly found frustration. Kepler searched endlessly for regular celestial patterns, trying all sorts of shapes for planetary orbits. Even when he settled on elliptical ones, he never could plot exactly where the planets would be, but only their general vicinity. Armed with the latest astrophysical theories and dazzling banks of supercomputers, contemporary astrophysicists still tolerate slop—undoubtedly less slop than Kepler, but still slop.

Superficially, the historical drift of theory toward predictive precision may inspire hope for the future. After all, the predictive slop of today is drastically reduced from the time of Kepler, not to mention Ptolemy. But less predictive slop is a far cry from no slop. Even under highly controlled laboratory conditions, using equipment of extreme precision, exact replication of experimental results is rare. And laboratory conditions are extremely artificial environments, where almost all of the vicissitudes and flux of normal life are excluded. Moreover, many of the most accurately predicted regularities are not about real things at all. In plotting planetary orbits, astronomers never engage the full complexities of the real world. That would require calculating the gravitational attractions among absolutely all the specks of dust in the universe. For that matter, not only do astronomers ignore almost everything else in the universe, they do not even try to calculate real attractions among actual planets. If they did, they would have to recalculate every time a grain of sand shifted on the beach. Instead, they allow point masses, things that do not exist at all, as substitutes for the real planets. Without such substitutions, even the calculation of the simplest orbits would remain utterly beyond the capacities of the finest scientific minds.

Even in those rarest of cases, under highly simplified, artificial conditions, where exact replications of data actually do occur, scientists rarely celebrate. If precise replications happen too frequently, suspicions grow. Laboratory technicians begin to wonder whether the results have been faked. After all, even in the most deterministic settings, genuine data sets never plot precisely onto smooth curves. And the growing evidence of indeterminism in nature has only increased the scientific tolerance for slop. Statistical regularities may be the most that scientists can hope for in those settings.

Reflections like these have inspired some theorists to make room for divine action in areas of scientific slop.[14] Unfortunately, while there may be significant promise in this strategy, anyone with Leibnizian metaphysical proclivities

14 For a recent collection of essays exploring this sort of strategy, see Nicholas Saunders, ed., *Divine Action and Modern Science* (Cambridge: Cambridge University Press, 2002). For skepticism regarding the binding regularity of scientific laws, see Keith Ward, *God, Chance and Necessity* (Oxford: Oneworld Publications, 1996).

would not find it especially attractive. Leibniz imagined the Creator as a supremely efficient architect. Consequently, God's designs embodied the kind of impeccable elegance that would gladden the hearts of mechanical engineers. Indeed, the clockwork machinery of nature was so finely crafted and smoothly lubricated that, according to Leibniz, "*it is not even possible to conceive of events which are not regular.*"[15]

Charles Darwin, of course, trashed such clockwork imagery with his detailed accounts of the excesses, causal redundancies, and inefficiencies of the biological world. For the purposes of this discussion, however, it is not necessary to mobilize any such Darwinian themes. Contrary to first impressions, there is plenty of room for divine action, even in the most severely regimented of natural worlds. Even if this world were regulated by the most rigidly universal of scientific laws, there would still be ample space for divine intervention by supernatural agents. All that is needed is the recognition of a multiplicity of scientific goals.

Contrary to monolithic depictions of the scientific enterprise, not all scientists are trying to do the same thing. Field biologists wish to record accurate, detailed accounts of animal behavior. Taxonomists want to devise comprehensive classification schemes. Ecologists identify interdependencies among organisms in ecosystems. Physiologists elucidate the intricacies of blood chemistry, while geneticists chart patterns of inheritance. Medical technicians run trials to determine the efficacy of drugs, documenting side-effects. Particle physicists search for new particles. Astrophysicists estimate the ages of galaxies and plot the evolution of stars. These and countless other activities go on simultaneously, sometimes in loosely associated or interconnected ways, but often in blissful isolation.

Out of this multifarious array, only two scientific goals are of special relevance here. The first is the familiar quest to find patterns of regularity in nature. When scientists uncover such patterns, they often try to codify their findings in statements of scientific law. Having identified patterns in nature, scientists frequently move on to a second task, attempting to decipher whatever is responsible for those regularities. When successful in this latter quest, researchers typically encapsulate their findings in theories about the causal activities of various forces or entities. These two goals, though obviously related, should not be conflated. Studying peas, Gregor Mendel identified various patterns of inheritance, regularities now embodied in Mendel's Laws. But as to

15 Gottfried Leibniz, *Discourse on Metaphysics*, in Philip P. Wiener, ed., *Leibniz Selections* (New York: Charles Scribner's Sons, 1951), 296.

how those patterns were produced, he remained baffled. The best he could do was label the agents responsible for patterns of inheritance as "factors." Only now, a century after Mendel's death, is any clear understanding emerging concerning just how the patterns of inheritance, codified in Mendel's Laws, are naturally generated. Thanks to decades of biological research, Mendel's vague "factors" are slowly being replaced by a sophisticated genetic theory, replete with genes, chromosomes, the helical structure of DNA, and all the rest.

Armed with this contrast between lawful patterns of regularity and the forces or entities responsible for producing those patterns, it is possible to envision ways that divine agents might enter into the natural course of events without disrupting scientific laws. Suppose, for the sake of simplicity, that the patterns of the orbiting planets were to conform precisely to Kepler's Laws. In the natural course of events, this would be due to the operation of a natural force, the force of gravity. Imagine, however, as theologians often do, that there is some powerful divine agent, God, who can directly do anything, including creating or destroying the universe, in whole or in part. Now suppose, perhaps due to the divine penchant for acting in mysterious ways, that God grew weary of the force of gravity and chose to destroy it. Normally, this loss of gravitational attraction among the planets would become immediately evident. The planets would fly off in all directions. But imagine that God chose to intervene, to keep the planets spinning in their Keplerian orbits, not by recreating gravitational attraction, but by direct, divine action. In that case, thanks to supernatural agency, the planets would move across the heavens. Yet, there would be no interruption, disruption, or violation of Kepler's Laws.

No doubt, to many contemporary ears, this scenario sounds bizarre. And it does not take much imagination to make matters worse. Instead of eliminating gravity once and for all, suppose that God chose to act intermittently, maybe spinning the planets supernaturally on Tuesdays and Thursdays, but letting gravity do the job for the rest of the week. If God were to spin the planets exactly the way that gravity does, how could anyone tell the difference? And if no one could detect any difference, how could anyone know which was happening now? For that matter, without any detectable difference, would there really be any genuine difference at all?[16]

16 The conceptual coherence of this kind of example has been defended, and its implications explored, in Edward L. Schoen, *Religious Explanations: A Model from the Sciences* (Durham: Duke University Press, 1985), Edward L. Schoen, "Religious Explanations: A Reply to Professor McMullin," in *Faith and Philosophy* 5, No. 3 (July 1988), and Edward L. Schoen, "David Hume and the Mysterious Shroud of Turin," in *Religious Studies* 27, No. 2 (June 1991).

This kind of thought experiment, with its attendant conceptual vertigo, will be addressed more systematically in the last section of this paper, where it will be connected to early modern reflections on the Clock of Strasbourg. Here, it is only necessary to mention two points in passing. First, it must be conceded that there is a strong, dominant tradition that condemns as theologically onerous any such direct, divine dabbling, especially anything as seemingly capricious as Tuesday-Thursday dabbling. Xenophanes thought it unfitting for God to be so engaged. Many medieval Christian thinkers distinguished God's primary causal activity as creator and sustainer from the everyday operation of secondary causes. Deists were willing to let God create natural laws, but thought it was better for those laws to govern the world. More recently, D.Z. Phillips repudiated crudely interventionist conceptions of God as primitive superstitions, while David Ray Griffin worried that such conceptions of God would render the problem of evil insoluble.[17]

Fortunately, for the purposes of this discussion, any such distinctively theological concerns may be left aside, though it might be useful to note that this tradition of distancing God from the course of ordinary events is by no means the only one. To mention just one theological alternative, Samuel Clarke advocated the biblical model of God as king. On that model, God's absence from the everyday world would amount to dereliction of duty. As Clarke explained:

> … 'tis not a *Diminution*, but the true *Glory* of his Workmanship, that *nothing* is done without his *continual Government* and *Inspection* … If a *King* had a *Kingdom*, wherein all Things would continually go on *without* his Government or Interposition, or *without* his attending to and ordering what is done therein: It would be to *him*, merely a *Nominal* Kingdom; nor would he in reality deserve at all the Title of King or Governor.[18]

The second point has to do with apologetics. Historically, much philosophical debate has centered about proofs and various sorts of evidence for or against the existence of God. Generally, discussions about divine intervention, particularly about the possibility of miracles, have followed suit. John Locke famously

---

17 See D.Z. Phillips, *Faith and Philosophical Enquiry* (New York: Schocken Books, 1971), 39–40, 101–105, 127–132, and D.Z. Phillips, *The Concept of Prayer* (New York: The Seabury Press, 1981), especially Chapter 6. For Griffin's concerns, both scientific and theological, see Griffin, *Two Great Truths: A New Synthesis of Scientific Naturalism and Christian Faith*, 2–3, 11–22, and 55–57.

18 Samuel Clarke, *Dr. Clarke's First Reply*, in Samuel Clarke, *The Works* (New York: Garland Publishing, Inc., 1978), Vol. IV:590–591.

worried over the way that miracles might be used as signs to certify various prophets as genuine or their teachings as true. In this connection, divine interventions that disrupted the natural course of events, especially if they did so in striking ways, would seem particularly germane. By contrast, the kind of divine intervention depicted here, where planets continue to circle in precisely the way they would naturally, appears utterly pointless, at least so far as matters of apologetics or evidence are concerned.

For the purposes of this discussion, this second point may be conceded. For apologetic or evidential purposes, this sort of divine intervention might carry little or no weight. It probably would not offer much basis for mounting defenses of the faith. Nor would it tend to encourage the skeptical to revise their views about the existence of God. But none of this implies that such interventions are impossible. Nor does it follow that such happenings would be devoid of religious meaning or point. After all, there is much more to religious life than proofs and apologetics. Indeed, for many actively religious people, proofs and apologetic manoeuvres are little more than idle, intellectual curiosities. Of vastly more significance, at least for many Christians, is the vivid conviction that God is lovingly, directly, and actively engaged in their everyday lives. While such people may not care one way or the other what is keeping the planets in orbit, they do care deeply that God is working in their lives, not merely thoughtfully watching them from a comfortable seat in the balcony. Special occasions, like surprising recoveries from illness, are never enough to satisfy those who thirst most for God's presence. They want to feel the active involvement of God in their day-to-day endeavours, when everything is normal and all is ticking along with mundane, even dreary, regularity.

This is certainly not the place to assess the validity of such religious wants or feelings. For those who seek diligently, maybe God can be found in the daily grind—or maybe not. As many scholars seem to believe, maybe such people are just deluded. Perhaps their ideas of a God so intimately engaged or actively loving are primitively superstitious or religiously otiose. Be all that as it may, the point here is only to recognize that there is nothing about the tight regularity of nature that excludes the possibility of a God acting directly in the world.

2.2 *Observing the Unobservable*

Turning to the empirical character of the sciences, there can be no doubt that human perceptual capacities are limited. Human eyes are insensitive to infrared and ultraviolet light. Some people cannot see colors, while others cannot see at all. Biologists claim that bats and dogs can hear highly pitched sounds, far in excess of the range of human ears. Beyond such obvious physiological limits, however, some have argued that there are other, more metaphysical, constraints. Thomas Aquinas said:

> It is impossible to see God by the power of sight or by any other sense or sensitive power. Any power of this kind is … the proper activity of some corporeal organ. Such activity must belong to the same order as that of which it is the activity, hence no such power could extend beyond corporeal things. God, however, is not corporeal, … hence he cannot be seen by sense or imagination but only by the mind.[19]

More recently, Robert Oakes wrote:

> … it seems unproblematic that *being incorporeal* is … a property that God has essentially, and that the function '*X* is incorporeal' entails the function '*X* is unobservable'. Hence, since 'If God exists, he is incorporeal' expresses a necessary truth, it should be intuitively clear that the proposition expressed by 'If God exists, he is unobservable' is *also* a necessary truth.[20]

Despite the appearance of logical rigour here, it is important to proceed cautiously. After all, perceived items need not appear as they actually are. Even granting the controversial assumption that things have essences, essential natures, or essential characteristics, it does not follow that such features must be displayed sensuously. Suppose it were possible to manufacture a perfect cube out of pure gold. Presumably, the edges of that cube would be absolutely straight and all of its corners would form right angles, perhaps essentially so. Chemists might insist that the atomic number of that gold must be 79 and the radius of its atoms 1.79 angstroms. Those familiar with reflective properties and optical laws might argue that any gold of such purity must be yellow in color. Even if all of this were a matter of necessity, be it logical, mathematical, or physical, none of that would prohibit the golden cube from appearing in all sorts of ways. Beneath the rippling surface of a stream, its edges might appear crooked. From various perspectives, its right angles might look narrowly acute or broadly obtuse. No matter how essential its atomic number or the radius of its atoms might be, those features need not be sensuously presented at all.

19 *Thomas Aquinas, Summa Theologicae,* Vol. 1, *The Existence of God,* Part One: Questions 1–13, Thomas Gilby, O.P., General Editor (Garden City: Image Books, 1969).

20 Robert Oakes, "Religious Experience, Sense-Perception, and God's Essential Unobservability," *Religious Studies* 17, No. 3 (September 1981).

Even if pure gold should turn out to be essentially yellow, that golden cube could easily appear to be some other color, given the right lighting conditions. If spray painted, it might appear bright purple, even under optimal lighting conditions.

Anything that can be sensuously discriminated from its surrounding perceptual backdrop may serve as an object of perception, no matter what kind of thing it may be and no matter how peculiar its appearance. Imagine an ordinary case, where the primary focus of visual attention is on a piece of paper. Normally, that piece of paper is not the only item given in the visual field. It is sensuously apprehended within a wider perceptual background, of which the observer may be conscious, but in a less prominent way. Moreover, the paper is not displayed fully, completely, or as it really is in itself. Although the object of visual attention is the entire piece of paper, it is presented perspectivally. Only this side of the paper is visually displayed. Under various circumstances, the sharp edges may appear blurry, the right angles of the corners may seem acute, or the white surface may look reddish in the rosy light of dawn.

Of course, the paper serves as the object of this perception only so long as it remains the primary focus of visual attention. If the percipient's attention were to shift, the primary object of perception would shift accordingly. Staring fixedly in the direction of the piece of paper, attention might shift from the paper to the adjacent pencil or inkwell. For that matter, it might shift to the spatial relation between the paper and the inkwell, to the color or shape of the inkwell, or even to the intensity of the color or the symmetry of the shape of the inkwell. With suitable shifts of attention, not only may things, parts, phases, organizations, and processes become the primary objects of perception, but, among other items, various relations, qualities, or even qualities of qualities could be perceived as well. No matter what its metaphysical kind, whatever can be singled out for isolated sensuous attention may serve as an object of perception.[21]

The limits of human perceptual capacities are set, not by standard observational techniques, but by the human capacity to sensuously discriminate items from their perceptual backgrounds. Normally, fingerprints are invisible because they blend into their perceptual backgrounds so seamlessly. But this is hardly an insuperable obstacle to detectives, who regularly render invisible fingerprints visible just by dusting them. Even if there were something in the very nature of fingerprints that made them invisible to normal percipients under

21 See Edmund Husserl, *Ideas: General Introduction to Pure Phenomenology*, W.R. Boyce Gibson, trans. (New York: Collier books, 1962).

normal conditions—that is, even if they were essentially invisible, dusting would immediately solve the problem. Dusted fingerprints stand out dramatically from their surrounding perceptual environs.

The same goes for observations in scientific contexts. Microbes are too small for anyone to see with the naked eye. If they have a nature or essence, perhaps a genetic one, then it is reasonable to suppose that the very nature or essence of microbes determines many of their features, possibly even their size. If so, then it would seem that microbes are essentially as small as they are. If human visual capacities also were a matter of essence so that, by their very nature, unaided human eyes could not resolve microbes as perceptual objects, none of this would have the slightest bearing on whether or not humans could see microbes. Thanks to the invention of microscopes, bright lights and, in some cases, special staining techniques, observers now regularly isolate microbes visually from their surrounding environments. With the recognition that anything that can be sensuously isolated from its surrounding background can be perceived comes the realization that all sorts of imperceptible items, even essentially imperceptible ones, may be observed. Not only do optical microscopes make it possible to see microorganisms too small to be resolved by unaided human eyes but electron microscopes allow the luxury of seeing things that are too small even to reflect visible light. With the aid of night vision goggles, hunters can watch the movements of deer, even when it is too dark out to see. With the aid of television and video recorders, people can watch events on the other side of the world, whether those episodes are happening now or transpired years ago. All of this is possible, often with the assistance of technology, only because human perception does not demand that things appear as they actually are. All that is required is the ability to isolate items sensuously from their surrounding perceptual environments.

For those without much of a religious life, it may be difficult to imagine where divinities, particularly essentially incorporeal ones, might show up in the course of everyday experience. Within the context of established religious traditions, however, such encounters seem common. According to Homer, Athena once disguised herself as a sailor and appeared to Telemachus.[22] Moving to the Christian tradition, the Creed of Nicea claims that Jesus was "… God from God, Light from Light, true God from true God, begotten not created, of the same essence as the Father …"[23] If this important Christian affirmation

22 See Robert Fitzgerald, trans., *Homer: The Odyssey* (Garden City: Anchor Books, 1963).

23 John H. Leith, ed., *Creeds of the Churches: A Reader in Christian Doctrine from the Bible to the Present*, 3rd ed. (Atlanta: John Knox Press, 1982).

should turn out to be correct, then whoever saw Jesus, saw God, someone of the same essence as the Father.

Skeptics, of course, might doubt the veracity of claims like these, perhaps with good reason. But whatever legitimate reasons for doubt there may be, none stems from any distinctively metaphysical limitations of human perception. So long as that sailor or Jesus could be visually isolated from their surroundings, they could be seen. If that sailor were, in fact, the goddess Athena in disguise, then in seeing the sailor, Telemachus would have seen an Olympian goddess. Similarly, if God the Father were essentially incorporeal and eternal, and if Jesus were of this same essence as the Father, then anyone who saw Jesus in his bodily form would have seen an essentially eternal and incorporeal God. Whatever, in fact, may have been seen on those occasions would depend upon the identities of the items seen, not upon the ways in which they showed up. Since items need not appear as they actually are, Olympians, even essentially female ones, might appear as male sailors, just as an eternal, incorporeal god might show up as a preacher in Galilee. Of course, there may be all sorts of reasons why divinities would not bother to show up in such ways. Xenophanes thought it would be undignified to do so. Perhaps, as with Medusa, perceptual encounters with gods would be lethal. Maybe there are no such divinities. But whatever obstacles there may be, essential unobservability is not one of them.[24]

2.3 *The Methodological Constitution of Kinds*

According to D.Z. Phillips, the impact of scientific methodological practice is much more penetrating than anything considered to this point. Methodological constraints, be they regularity, observability or anything else, do not merely delimit the range of what may be studied but actually constitute the kinds of realities that can be encountered. Phillips was convinced that God's reality is the reality of a kind, rather than that of an individual falling under some kind. As such, God's reality is constituted by the practices embedded in the linguistic behavior of believing communities, just as the reality of physical objects is constituted by the methodological practices of scientific communities. As he explained:

24 This line of argument has been elaborated and defended much more extensively in Edward L. Schoen, "Perceiving an Imperceptible God," in *Religious Studies* 34, No. 4 (1998), and Edward L. Schoen, "The Sensory Presentation of Divine Infinity," in *Faith and Philosophy* 7, No. 1 (January 1990).

> When God's existence is construed as a matter of fact, it is taken for granted that the concept of God is at home within the conceptual framework of the reality of the physical world ... But to ask a question about the reality of God is to ask a question *about a kind of reality*, not about the reality of *this* or *that*, in much the same way as asking a question about the reality of physical objects is not to ask about the reality of this or that physical object.[25]

For Phillips, the implications of this point were dramatic. Because the reality of kinds is determined by methodological practices, to switch practices is to transform the kinds of reality studied. So, if the reality of God were approached through the methods of science, God's reality would be transformed into the kind of reality enjoyed by physical objects. At the level of individual religious practice, this would have disastrous consequences. It would transform genuine religion into mere superstition.[26]

Despite the attractions of so neat a correlation of practices with kinds of reality, there is no historical evidence that methodological practices actually do determine the reality of kinds. To the contrary, kinds of reality display so much methodological recalcitrance that they often demand sweeping methodological change. Confident that their mathematical methods were sufficient for investigating everything from music to justice, the early Pythagoreans developed rigorous techniques for constructing various geometrical shapes and for proving assorted truths about them. In sympathy with Phillips, it might be conceded that the mathematical nature of Pythagorean geometry was constituted by its ancient methodological practices of mathematical definition,

---

25 Phillips, *Faith and Philosophical Enquiry*, 3; see also 70. For the background to this line of argument, see Ludwig Wittgenstein, *Lectures and Conversations on Aesthetics, Psychology and Religious Belief*, compiled from notes taken by Y. Smythies, R. Rhees, and J. Taylor; Cyril Barrett, ed. (Berkeley and Los Angeles: University of California Press, 1967), Peter Winch, *The Idea of a Social Science and Its Relation to Philosophy* (London: Routledge & Kegan Paul, 1958), and Peter Winch, *Ethics and Action* (London: Routledge & Kegan Paul, 1972). For a sampling of the critical literature, see Michael J. Coughlan, "Zande Witchcraft," in *Sophia* 24, No. 3 (October 1985), Colwyn Williamson, "Witchcraft and Winchcraft," in *Philosophy of the Social Sciences* 19, No. 4 (December 1989), and Colwyn Williamson, "Following a Rule," in *Philosophy* 64, No. 250 (October 1989). For objections to Winch's concept of the reality of God, see Peter Byrne, "Arguing About the Reality of God," *Sophia* 19, No. 3 (October 1980). For parallels between scientific and religious epistemic practices, see Schoen, *Religious Explanations: A Model from the Sciences* and Edward L. Schoen, "The Roles of Predictions in Science and Religion" in *International Journal for Philosophy of Religion* 29, No.1 (February 1991).

26 See D.Z. Phillips, *Faith and Philosophical Enquiry*, 103–104.

construction, and demonstration. Unfortunately, none of this was particularly helpful in determining the mathematical relation between the length of one of the sides of an ideally constructed isosceles right triangle and its hypotenuse. No matter how hard they tried, the early Pythagoreans were unable to provide any exact numerical value for it. When they finally realized that there was no such value, Pythagorean celebrations of mathematical prowess dissolved into a crisis of incommensurables. To the utter shock of the ancient world, isosceles right triangles proved to be scandalously independent methodologically. Only later, with the development of radically new mathematical procedures, did it become possible to express the elusive proportion as 1 to $\sqrt{2}$. At least with respect to this one aspect of isosceles right triangles, instead of ancient methodological practice determining the nature of triangular reality, the nature of triangular reality itself demanded sweeping methodological change. Rather than Pythagorean methodological practices specifying criteria for the intelligibility of isosceles right triangles, the true nature of those triangles became intelligible only after methods changed.[27]

Aristotle's method of scientific demonstration dictated that heavenly bodies must be a very special kind of reality, utterly different from anything found on earth. His scientific method of demonstrating truths from intuitively known first principles demanded that heavenly bodies be composed of a special, divine substance that always travels in perfect circles. Being simple, heavenly bodies could never change or decompose. Consequently, Aristotle said that "… the nature of the stars is eternal just because it is a certain kind of substance …"[28]

Notwithstanding the power of Aristotelian deductive methods, the kinds of reality found in the heavens eventually proved methodologically recalcitrant. While they conformed to some of Aristotle's expectations, at least closely enough to allow Ptolemaic systems to flourish for many centuries, Galileo argued that superlunary items were not the kind of things that Aristotle's methods required them to be. Using his own observational techniques, aided by telescopic magnification, he concluded that sunspots could not be formations close to the earth, because they revealed no perceptible parallax. The observed

27 See John Mansley Robinson, *An Introduction to Early Greek Philosophy: The Chief Fragments and Ancient Testimony with Connecting Commentary* (Boston: Houghton Mifflin Company, 1968), Chapter 4, and Philip Wheelwright, ed., *The Presocratics* (Indianapolis: The Odyssey Press, 1966), 205–207.

28 Aristotle, *Metaphysics*, W.D. Ross, trans., in Richard McKeon, ed., *The Basic Works of Aristotle* (New York: Random House, 1941), 1073a, 34–35. Aristotle's main line of argument regarding these points can be found in *Metaphysics*, Book XII. For the full details of his demonstrations, see Aristotle, *On the Heavens*, 269a, 31–32. Also see *On the Heavens*, Book I, Chapter 2.

shapes of sunspots as well as various changes they underwent indicated both that they were on the surface of the sun and that they were constantly being generated and destroyed.[29]

Superficially, this clash between Galileo's understanding of the kinds of things that could be found in the sky and that of the Aristotelians may seem to confirm the idea that kinds of reality are constituted by methodological procedures. When the Aristotelians used their methods for studying the heavens, they found only items of a divine, unchanging kind. But when Galileo looked to the skies, his methods disclosed a different kind of reality, one that included changeable kinds of things. But this does not do justice to the actual debate. Despite their methodological differences, Galileo and the Aristotelians agreed that, whatever else they were, sunspots were fluctuating, irregularly shaped, dark blobs. Neither side was sure about the true nature of the blobs. Galileo entertained a number of different theories, including the idea that sunspots might be something like clouds. Many Aristotelians argued that the blobs were disturbances in the upper atmosphere. The more cynical among them thought the blobs were optical aberrations produced by imperfections in Galileo's telescopes. For both sides, location was critical. Confident that sunspots were on or near the surface of the sun, Galileo backed his view with countless observations and arguments, attacking real as well as imagined Aristotelian objections with every ounce of critical skill he could muster. The Aristotelians, on the other hand, insisted with equal intensity that sunspots were neither on the sun nor anywhere else in the heavens. They were located somewhere on or near the surface of the earth.[30]

The two sides fought with such fury over location, not out of any passion for cartography, but for methodological reasons. Both Galileo and the Aristotelians easily conceded that they knew almost nothing about the exact nature of sunspots. What little they did know, however, was enough to pose a methodological threat of revolutionary proportions. Because those dark blobs

29 See Stillman Drake, trans., *Discoveries and Opinions of Galileo* (Garden City: Doubleday Anchor Books, 1957), 90–92 and Stillman Drake, trans., *Galileo Galilei: Dialogue Concerning the Two Chief World Systems, Ptolemaic & Copernican,* 2nd ed. (Berkeley, Los Angeles and London: University of California Press, 1967), 54–55. For the genuine strengths of Ptolemaic systems as well as the pressures that led eventually to the success of Copernican alternatives, see Thomas S. Kuhn, *The Copernican Revolution: Planetary Astronomy in the Development of Western Thought* (New York: Vintage Books, 1957).

30 For some of the Aristotelian analyses of sunspots, see Drake, trans., *Discoveries and Opinions of Galileo*, 90–91; 98. Galileo discussed the fuller range of possibilities on both sides of the debate and defended his own position in his *Letters on Sunspots*, which may be found in this same volume.

fluctuated so irregularly, whatever else they were, sunspots were not the kind of things that could be located in an Aristotelian sky. If those changeable, dark blobs really were in the heavens, then Aristotelian methods for studying the skies would have to be replaced by more adequate investigative techniques, perhaps by the ones newly introduced by Galileo. The Aristotelians could resist this pressure for methodological reform only by insisting that sunspots were somewhere below the orbit of the moon. While they were able to stave off the inevitable for a time, the Aristotelians eventually were forced to abandon their methods due to the methodological recalcitrance of the kinds of things that Galileo found in the sky.

Some things are so recalcitrant as to resist every method of investigation. As a committed Copernican, Johannes Kepler wondered why the number of known planets circling the sun was six rather than some other number. Coupling the ancient Greek discovery of five regular polyhedra with a cosmology inspired by Plato's *Timaeus*, he devised a method for constructing a possible answer. If the structure of the solar system were conceived in terms of just the right nesting of the five regular polyhedra, the locations of the six planets could be correlated with the various surfaces of that imaginary geometrical assembly. Unfortunately, the astronomical data collected by Tycho Brahe did not mesh closely enough with the surfaces of Kepler's polyhedra. So he tried another method. No matter what method he tried, however, Kepler never could find a satisfactory explanation for the number of planets circling the sun. Since that time, other astronomers have tried everything from Kepler's original geometrical approach to the latest, most exotic methods of contemporary evolutionary cosmology. While new planets have been discovered, the actual number of solar planets still resists understanding by any method yet devised.

When it came to orbits, Kepler enjoyed more success. Following the methodological tradition of his day, he tried to find an epicyclic description of the path of Mars. Sadly, this method failed and so Kepler used his data about the positions of Mars to devise a new method. His calculations revealed not only that the velocity of Mars varied, but that it varied in a mathematically regular way. The velocity of Mars was inversely related to its distance from the sun. Kepler's mathematical studies also revealed that the orbit of Mars was an elongated kind of shape, one that could not be duplicated by any plausibly simple compounding of circles upon circles. Seizing upon a physical analogy, he guessed the orbit to be shaped like an egg. Ultimately, more calculations refuted this initial guess, pointing him toward an ellipse instead.[31]

31 For a preliminary account of Kepler's methodological struggles, see Stephen Toulmin and June Goodfield, *The Fabric of the Heavens: The Development of Astronomy and Dynamics*

In his *Astronomia Nova*, Kepler painstakingly chronicled the steps that led him to his revolutionary rejection of Aristotelian astronomical methods. Step by step, he documented not only how each of his hypotheses fared with Brahe's data, but precisely how various aspects of those data drove him through a succession of methodological innovations. Like Galileo, Kepler discovered that instead of Aristotelian methods constituting the kind of reality to be found in the skies, the kind of reality found there determined which methods were most effective in his quest for scientific understanding.

Repeatedly, scientists who have clung rigidly to methodological constraints have been stymied by the kinds of reality they have encountered. Robert Boyle, like so many other early modern scientists, was self-consciously committed to corpuscularianism. But while Isaac Newton struggled heroically to understand gravity in corpuscularian terms, he eventually gave up in despair.[32] The subsequent failure to find any such corpuscularian account of gravity motivated later scientists to abandon the methodological constraints of purely mechanistic explanations, not only for gravity, but for electromagnetism as well. When Charles Darwin spotted temporally developmental patterns in the biological world, he felt the need to substitute a new, evolutionary method of explanation for the traditionally static, Creationist technique. More recently, the strangely paradoxical, discontinuous, and statistical nature of quantum phenomena has exerted unprecedented pressures for methodological change. Indeed, the pressures have been so great that some theorists have even argued for relaxing the constraints of classical logic.[33] Whether such an extreme conclusion is warranted by the data may be questionable. What is unquestionable, however, is the intensity of the pressure for methodological change that may be brought to bear upon scientists by the kinds of reality they encounter. Like their predecessors, contemporary scientists confront kinds of reality that are so resistant to current methods as to demand methodological reform—in some instances, dramatic methodological reform.

(New York: Harper and Row, Publishers, 1961), 198–208. For a more detailed treatment, see E.J. Dijksterhuis, *The Mechanization of the World Picture: Pythagoras to Newton*, C. Dikshoorn, trans. (Princeton: Princeton University Press, 1986), 303–323.

32 See Florian Cajori, trans., *Sir Isaac Newton's Mathematical Principles of Natural Philosophy and His System of the World* (Berkeley: University of California Press, 1934), 547.

33 See Albert Einstein and Leopold Infeld, *The Evolution of Physics: from Early Concepts to Relativity and Quanta* (New York: Simon and Schuster, 1938), 291–292 and Hilary Putnam, "The Logic of Quantum Mechanics," in Hilary Putnam, *Mathematics, Matter and Method: Philosophical Papers I* (Cambridge: Cambridge University Press, 1975), 184.

3 Clockworks and the Underdetermination of Theories

To this point, several reasons for skepticism regarding any tight connections between scientific methodological practices and specific metaphysical commitments have been sketched. Monolithic visions of science as a single, unified edifice changelessly devoted to the investigation of some stable range of metaphysical kinds founders on the recognition of so vast a jumble of methods, activities and goals, both within and across scientific fields. Even where localized pockets of unity may be found, as in common assumptions of lawful regularity or the devotion to empirical method, there is considerable flexibility regarding the kinds of reality that may be encountered in the course of scientific investigation. The range stretches from the material to the immaterial, the observable to the unobservable, and the natural to the supernatural. Historically, of course, there have been many kinds of things that have resisted scientific understanding. When confronted with methodologically recalcitrant kinds of things, however, scientists have responded by changing methods. Instead of scientific methods rigidly dictating what kinds of things could be studied, the kinds of things studied often dictated revisions—sometimes radical revisions—in scientific method.

To appreciate just how loosely the methods of science are tied to specific metaphysical claims, it is helpful to consider an early modern thought experiment. Like many writers today, Robert Boyle was captivated by a sweeping vision of natural order. Unlike many today, he also was deeply religious. Consequently, Boyle envisioned the creator of the universe as a kind of cosmic clockmaker. At the time of creation, God originally produced matter and dictated the laws of nature. With all the manipulative skills of a crafty engineer, God gave matter its initial push, but in such a way as to achieve specific, anticipated results. In nature,

> As in the … Clock of *Strasburg*, (sic) the several Pieces making up that curious Engine, are so fram'd and adapted, and are put into such a motion, that though the numerous Wheels, and other parts of it, move several ways, and that without any thing either of Knowledge or Design; yet each performs its part in order to the various Ends for which it was contriv'd, as regularly and uniformly as if it knew and were concern'd to do its Duty.[34]

---

34 Robert Boyle, *Some Considerations Touching the Usefulness of Experimental Natural Philosophy: The First Part*, in *The Works of Robert Boyle*, ed. Michael Hunter and Edward B. Davis (London: Pickering and Chatto Publishers Ltd., 1999), Vol. 3:248.

Boyle realized, however, that a devastating, skeptical threat lurked at the very heart of this clockwork imagery. He conceded:

> … as confidently as many atomists and other naturalists presume to know the true and genuine causes of the things they attempt to explicate, yet very often the utmost they can attain to in their explications is that the explicated phenomena *may* be produced after such a manner as they deliver, but not that they really *are* so. For, as an artificer can set all the wheels of a clock a-going as well with springs as with weights, … so the same effects may be produced by divers causes different from one another …[35]

Here, Boyle joined Descartes, who had come to the same skeptical conclusion years earlier. As Descartes had argued:

> … just as there may be two clocks made by the same workman, which though they indicate the time equally well and are externally in all respects similar, yet in no wise resemble one another in the composition of their wheels, so doubtless there is an infinity of different ways in which all things that we see could be formed by the great Artificer …[36]

In their reflections, both Descartes and Boyle assume the same distinction mentioned earlier in this discussion: the contrast between the search for patterns of regularity and the quest to understand whatever is behind those patterns and is responsible for producing them. In their reasoning about clocks, the movements across the faces of clocks represent the recurring patterns of regularity in nature, while the inner clockworks signify the mechanisms of nature that stand behind those patterns, producing them. Sometimes, the mechanisms responsible for producing patterns may be inspected directly, as when clock cases are opened to look inside. In many scientific settings, however, natural mechanisms remain hidden. In those cases, scientists often must rely upon analogies. Typically, they begin with an initial analogy, often crude

---

35 Robert Boyle, "An Essay, Containing a Requisite Digression, concerning Those that would Exclude the Deity from Intermeddling with Matter," in M.A. Stewart, ed., *Selected Philosophical Papers of Robert Boyle* (Indianapolis: Hackett Publishing Company, 1991), 169.

36 René Descartes, *The Principles of Philosophy*, Principle CCIV, Discussion, in Elizabeth S. Haldane and G.R.T. Ross, trans., *The Philosophical Works of Descartes,* Vol. I (Cambridge: Cambridge University Press, 1969), 300.

and obviously inadequate. Then, through a variety of theoretical techniques, including the skillful use of predictions, they refine their analogies, sometimes to high levels of sophistication.[37]

Unfortunately, as clockworks so vividly illustrate, any number of quite diverse mechanisms can produce exactly the same patterns. The hands of a clock might be driven by anything from gears and springs to ropes, weights, and pulleys. In the same way, patterns in nature might be produced by any number of diverse, natural mechanisms. Since both Descartes and Boyle believed that God could move matter directly, this opened yet another possibility. Instead of creating natural mechanisms to do the job, God could have chosen to govern the natural world straight away, without the assistance of intermediary natural mechanisms. If that were the case, there would be no natural mechanisms at all behind the observable regularities of nature.

Here, then, is an early modern account of the underdetermination of theories by evidence. Without having recourse to the direct inspection of natural mechanisms, there seems to be no theoretical way to discriminate the correct description of the inner workings of nature from any number of strikingly different alternatives. To use the modern clockwork analogy, no matter how carefully a posit of wheels and springs might be refined, no matter how predictively accurate, simple, elegant, or explanatorily successful it might become, the hands on the face of nature might, in fact, be driven by something as radically different from gears and springs as a system of weights and pulleys. Indeed, if God is thrown into the mix, no matter how sophisticated or adequate a theory about gears and springs may be, the actual agent responsible for patterns in nature might not even be material. It might be a supernatural, incorporeal divinity.

The last few centuries of scientific discovery and philosophical development have only expanded the range of alternative possibilities. Boyle restricted his list of natural mechanisms to corpuscularian ones. Today, the natural options seem endless. All sorts of combinations of electronic devices, computer chips, vibrating quartz crystals, electric motors, or gasoline engines can replace wheels and springs or ropes and pulleys. Many of these not only look different from mechanical push-pull systems but they operate on entirely different principles and laws. Due to his conception of God as supreme architect, Boyle narrowed the range of natural mechanisms to exclude inefficient, makeshift, redundant, or backup designs. For contemporary scientists, who view nature

---

37 See Rom Harré, *The Principles of Scientific Thinking* (Chicago: University of Chicago Press, 1970), 33–62.

through more Darwinian eyes, there is plenty of room for extravagance, waste, and jerry-rigging. Unlike Boyle, many contemporary physicists also leave open the possibility of brute, uncaused regularities, cases where no mechanisms at all, natural or otherwise, are responsible for patterns in nature.

Confronted with so many theoretical alternatives, Descartes sought refuge in religious faith. So long as humans refrained from pridefully overstepping the natural bounds of reason, an always beneficent God would not lead them astray. Especially in cases where successful explanations covered wide ranges of phenomena, Descartes believed that high levels of probability were justified, even though no one could be absolutely certain that the best scientific theories were true.[38]

While few have embraced the Cartesian strategy for escaping doubt by way of proving the existence of a trustworthy God, many, including Boyle, have been impressed by the capacity to account for wide ranges of phenomena. As Boyle argued:

> … the more numerous and the more various the Particulars are, whereof some are *explicable* by the assign'd Hypothesis, and some are *agreeable* to it, or at least are not dissonant from it, the more valuable is the Hypothesis, and the more likely to be true. For 'tis much more difficult, to finde (sic) an *Hypothesis* that is not true which will suit with *many* Phaenomena, especially if they be of various kinds, than but with a *few*.[39]

There can be no question of the explanatory power of single mechanisms accounting for wide ranges of diverse phenomena. Indeed, it is difficult to imagine that such broad explanatory success could be purely coincidental. Surely, the mechanisms responsible for such diverse ranges of phenomena must actually be as they are depicted in successful theories. Nevertheless, the clock at Strasbourg vividly illustrates why Boyle ultimately resisted the temptation to move with any assurance from explanatory success to truth. The internal mechanism of that spectacular clock was responsible for all sorts of diverse phenomena, from the turning of the hands across its face to the plotting of assorted planetary positions, equinoxes, phases of the moon, and eclipses. The

38 See Descartes, *The Principles of Philosophy,* 301–302 and René Descartes, *Discourse on Method,* in *Descartes: Discourse on Method and Other Writings,* trans. Arthur Wollaston (Baltimore: Penguin Books, 1969), 95–97.

39 Robert Boyle, *Experiments Notes, &c., about the Mechanical Origin of Qualities,* in Michael Hunter and Edward B. Davis, eds., *The Works of Robert Boyle,* Vol. 8 (London: Pickering and Chatto Publishers, Limited, 2000), 325.

Strasbourg mechanism regularly struck the hours and quarter hours. It also animated various figurines, including a mechanical cockerel. Yet, without looking inside, it was impossible to guess the nature of the clockworks involved, whether they were gears and springs or ropes and pulleys. For that matter, there was no telling how many mechanisms nestled behind the face. If there were more than one, there was no telling how the various tasks might be divided among them. Countless alternative designs could easily produce exactly the same results.

Consequently, Boyle reluctantly admitted:

> ... if God be allowed to be, as indeed he is, the Author of the Universe, how will it appear that He, whose Knowledge infinitely transcends ours, and who may be suppos'd to operate according to the Dictates of his own immense Wisdom, should, in his Creating of things, have respect to the measure and ease of Humane Understandings, and not rather, if of any, of Angelical Intellects, so that whether it be to God, or to Chance, that we ascribe the Production of things, that way may often be fittest or likelyest for Nature to work by, which is not easiest for us to understand.[40]

It may be enticing to think that the skeptical concerns generated by this thought experiment spring from the limitations of analogical reasoning. After all, analogies are notoriously imprecise and regularly misleading. While it is easy to tolerate analogies as useful heuristic devices, perhaps especially in the initial stages of theory formulation, many scientists hope to purge them entirely from sophisticated scientific theories. Particularly in physics, many believe that rigorously mathematical formalisms should replace analogical reasoning. Ideally, the very best scientific theories should be structured as formal, axiomatic systems.

Whether analogies are so easily dispensable, even within the narrow confines of formal mathematics, is a matter of considerable controversy.[41] But even if they are, it would be a mistake to blame analogical reasoning for theoretical

---

40 Boyle, *Some Considerations Touching the Usefulness of Experimental Natural Philosophy: The First Part*, 257. For a much fuller discussion of modern thought experiments about the clock at Strasbourg, see Edward L. Schoen, "Clocks, God and Scientific Realism," in *Zygon* 37, No. 3 (September 2002).

41 For connections of metaphors to scientific theories, see Schoen, "Clocks, God and Scientific Realism." For accounts of the role of metaphors in mathematics, see Philip Kitcher, *The Nature of Mathematical Knowledge* (New York and Oxford: Oxford University Press, 1984) and George Lakoff and Rafael E. Núñez, *Where Mathematics Comes From: How the Embodied Mind Brings Mathematics into Being* (New York: Basic Books, 2000).

underdetermination. The same kinds of ontological variability show up in even the most rigorously formal, axiomatic systems. Consider the recent history of elementary number theory. In the last century, various formal, axiomatic treatments of elementary number theory were devised. Each of these theories was as fully adequate as the others, and all were theories about numbers. But they did not all refer to the same things. Both Zermelo and von Neumann identified the natural numbers with classes ingeniously constructed out of the empty class, but their constructions differed. Consequently, Zermelo's theory referred to items that diverged from the referents of von Neumann's theory. Nor did their choices exhaust the possibilities. As W.V.O. Quine noted:

> We are free to take 0 as anything we like, and construe S [successor] as any function we like, so long merely as the function is one that, when applied in iteration to 0, yields something different on every further application.[42]

In short, elementary number theory underdetermines its referents. So far as formal, axiomatic treatments of elementary number theory are concerned, all sorts of different sequences of things may serve equally well as numbers.

Over the last hundred years or so, investigations into the foundations of logic and mathematics have revealed that the ontological plasticity displayed by elementary number theory is by no means unique. To the contrary, any formalized, axiomatic system constructed along classical lines displays the same kind of tolerance for alternative ontologies. When specifying formal systems, logicians typically provide a vocabulary of primitive symbols, together with rules for constructing well-formed formulae. Then, a list of well-formed formulae is offered to serve as the axioms of the system, together with rules for inferring additional well-formed formulae. The resulting system, officially at least, is nothing more than a collection of meaningless strings of symbols from which other meaningless strings of symbols can be generated. For these meaningless strings of formulae to become meaningful bearers of truth, a semantics, or interpretation, of the symbolism is required. Typically, such an interpretation consists of a value assignment associating various symbols with a specified range of items, usually called a universe of discourse. Any mapping onto a universe of discourse that makes the formalized theory true is called a model of the theory. For those who rhapsodize about Theories of Everything, such an axiomatized, formal theory would be a dream come true.

---

42 W.V.O. Quine, *Set Theory and Its Logic*, Revised Edition (Cambridge: The Belknap Press, 1969), 81.

According to the strong form of the Löwenheim-Skolem Theorem, for any true theory with an indenumerable universe of discourse, that is, a universe with uncountably many items, there is a denumerable model for it. This means, ontologically speaking, that for purposes of truth, all those extra entities are unnecessary—excess baggage. So far as the truth of the theory is concerned, all that is needed is the denumerable subset, not the entire, indenumerable universe. But that is not the end of the matter. Proxy functions are formal, technical devices for mapping one denumerable universe onto another. So, with the assistance of proxy functions, all of the items from one denumerable universe of discourse may be swapped out for substitutes, without transforming any of the true sentences of a formalized theory into falsehoods. Again, ontologically speaking, so far as truth is concerned, it simply does not matter whether the referents of a formalized theory are the items in the original, perhaps indenumerable, universe of discourse, a denumerable subset, or some entirely different collection of swapped out proxies. As with elementary number theory, all sorts of alternatives serve equally well. Consequently, as Hilary Putnam remarked:

> ... even if we have constraints of whatever nature which determine the truth-value of every sentence in a language *in every possible world,* still the reference of individual terms remains indeterminate.[43]

To reinforce this conclusion, Putnam sketched the way that a theory, ostensibly about cats, could be interpreted as referring to cherries instead.[44]

Of course, the natural reaction to such formalized pyrotechnics is to protest that anyone who bothered to look could easily tell the difference between cats and cherries. In scientific contexts, while the ontological status of hidden mechanisms may remain open to question, few, if any, of the mechanisms of nature remain hidden today. Thanks to recent technological advances, all sorts of ways are available to look behind the face of nature to inspect its clockworks. Electron microscopes, Geiger counters, mass spectrometers and particle accelerators, not to mention a mountain of other devices, now

43 Hilary Putnam, *Reason, Truth and History* (Cambridge: Cambridge University Press, 1981), 33. For the ontological impact of the Löwenheim-Skolem Theorem and proxy functions, see W.V.O. Quine, *Ontological Relativity and other Essays* (New York and London: Columbia University Press, 1969). For the constraints upon domains of discourse for elementary number theory, see Quine, *Set Theory and Its Logic,* particularly Chapter 4.

44 See Putnam, *Reason, Truth and History,* 33–35, 217–218.

provide scientists with observational access to even the most secret recesses of nature.

Unfortunately, this line of protest gains leverage only by assuming that things in nature must ultimately appear to be as they actually are. But no matter how attractive this assumption may be, it is false. Observed items need not ever appear as they actually are. Even if perceived items have essential natures, those essences need not be displayed sensuously. So far as human observational capacities go, there are no metaphysical guarantees. An essentially female goddess may look like an old sailor, just as something too small to be visible may look as big as a house. Something that appears to be as furry as a cat may, in fact, turn out to be a cherry. In formalized contexts, even the most precipitous of metaphysical chasms can be bridged without difficulty. Cats may be mapped onto cherries, cherries onto events, events onto spatio-temporal locations, spatio-temporal locations onto sequences of number, and sequences of numbers onto cats.

The upshot of all this is not to claim that ancient mariners are young goddesses or that cats are cherries, but only to underscore how loosely scientific practices are connected to kinds of reality. Traditionally, metaphysics has been concerned to mark out the most basic kinds of things. Historically, for all sorts of reasons, all manner of fundamental, categorial contrasts have been drawn. Aristotle thought that sublunar items were radically different from the kinds of things that circled the heavens. Others have marked off the temporal from the eternal, corporeal from incorporeal, finite from infinite, abstract from concrete, observable from unobservable, material from ideal, and physical from spiritual.

This is not the place to indulge in such metaphysical debates nor to decipher the true ontological boundaries or divisions of the real. The concern here is only to challenge the general policy, popular since the time of Pythagoras, of trying to extract metaphysical constraints from scientific practice. Invariably, such Procrustean strategies fail. Even the most successful scientific techniques do not dictate what sorts of things can be investigated or what kinds of things must exist. The investigations of Galileo and Kepler flouted Aristotelian prohibitions about what could be found in the sky. Notwithstanding Boyle's corpuscularian strictures, scientists today are perfectly at ease with gravity and electromagnetism. Relativistic conceptions of spacetime now replace Newtonian absolutes. Not only do contemporary physicists ignore Kant's insistence that space be Euclidean, string theorists cannot even agree about how many dimensions to ascribe to the natural world. If there is any lesson at all to be learned from the history of science, it is that there is no telling what the future may hold, what kinds of things scientists may repudiate or advocate, what metaphysical boundaries may be respected or violated.

Despite this embarrassing history of Procrustean failure, many writers today remain every bit as ontologically confident as Aristotle and Boyle were in the past. Edward O. Wilson still believes that he can extract metaphysical clues from contemporary scientific practice. Like Russell, Gilkey, Griffin, and so many others, he believes that a sophisticated, contemporary scientific mindset excludes the possibility of a dynamically intervening, supernatural God. In Wilson's grand, unified scheme of reduction, all of reality boils down to the entities of physics. Unlike Russell, Wilson does not repudiate the existence of God altogether, though he certainly marginalizes the divine. He says: "I am an empiricist. On religion I lean toward deism but consider its proof largely a problem in astrophysics."[45] So far as the traditional blessings of religion go, Wilson seeks them in science, saying: "Perhaps science is a continuation on new and better-tested ground to attain the same end. If so, then in that sense science is religion liberated and writ large."[46]

No doubt, many of the world's religious traditions are comfortable with distant gods. Some, like Tibetan Buddhism, may be content to repudiate gods altogether. But these traditions hardly represent the full spectrum of human religious sensibility. For many people, the vibrant heart of religion is an intimate, often testy, ongoing entanglement with God. The vicissitudes of this intense, frequently personal relationship are recorded in the sacred texts of many traditions, including the Hebrew Psalms. Sometimes, there is the secure sense of God's protective presence, as when David declared: "Even though I walk through the valley of the shadow of death, I fear no evil, for Thou art with me …"[47] Other times, there is a sense of deep betrayal, a burning rage at divine abandonment. Psalm 44 bitterly demands: "… why dost Thou sleep, O Lord? Awake, do not reject us forever. Why dost Thou hide Thy face, And forget our affliction and our oppression?"[48] For anyone on this spectrum, relations with God could hardly be construed as some esoteric problem for astrophysics.

Even today, in the most scientifically advanced cultures, reports of messy, everyday intimacies with God are as prevalent as they were among the ancient desert psalmists. The purpose here is neither to endorse nor challenge such reports, but only to claim that nothing about contemporary scientific practice precludes their veracity. Suppose, as Wilson seems to believe, that

45 Wilson, *Consilience: The Unity of Knowledge*, 263.

46 Wilson, *Consilience: The Unity of Knowledge*, 7.

47 Psalm 23:4. New American Standard Bible translation.

48 Psalm 44:23–24. New American Standard Bible translation.

consciousness really could be reduced to brain chemistry, brain chemistry to physics, and on down to whatever physicists happen to believe is rock bottom, be it strings, superstrings, or who knows what else. Indeed, suppose something vastly more precise were known, that a specific rush of chemicals in the brain were sufficient to provide a scientific explanation of the peaceful feelings enjoyed by David in Psalm 23. None of this would demonstrate anything of metaphysical consequence.

It is perfectly possible, of course, that things might actually be just as Wilson and his colleagues take them to be. Cats might just be cats. However, the clock at Strasbourg suggests plenty of alternatives. Making room for deism, perhaps God began by creating the general principles of chemistry and biology, then let those divinely ordained principles do the rest of the work. This might have pleased Robert Boyle, but it would not have satisfied David, who felt God's active presence much closer than that. Perhaps, then, God chose to behave in the more kingly fashion envisioned by Samuel Clarke. Maybe God suspended the normal operation of natural mechanisms in order to produce David's peaceful feelings by direct, divine action. Alternatively, God may not have bothered to suspend natural causal processes at all, yet still managed to act directly. After all, there is no reason why God and natural processes cannot act simultaneously. Causal redundancies are commonplace. Perhaps, then, God joined in with brain chemistry to produce the feelings that brain chemicals alone would have produced anyway.

Clearly, of course, from a scientific standpoint, such divine behavior seems entirely pointless. From the standpoint of causal efficacy, it is utterly superfluous. Still, as with a clock driven alternately by natural and divine mechanisms, or simultaneously by both mechanisms, the fact that God's involvement might be capricious, superfluous, or causally redundant does not rule out the possibility that God might so act anyway.[49] From the theoretical perspective of the sciences, such divine behavior might appear senseless. To Leibniz, who prized the values of an architect or clockmaker, it might seem inelegant, inefficient, or even wasteful. As Clarke noted, however, kings need not share the sensibilities of mechanical engineers. Or, switching to another Biblical metaphor, like a parent, God's reasons for acting need not be scientific ones.

When reading a book with her child, a mother may help turn the pages. Enfolding the child's hand in her own, she may push each leaf over in a way that, from a causal point of view, is utterly pointless. The page may turn no

49 For an extended defense of this line of reasoning, see Schoen, *Religious Explanations: A Model from the Sciences*, particularly chapters 4, 5 and 6, and Schoen, "Religious Explanations: A Reply to Professor McMullin."

differently than if the child had acted alone. To the child, however, there is all the difference in the world. By her causally superfluous act, the mother enters intimately into the personal world of her child, expressing her devotion and deepening their mutual ties. Indeed, the intimacy may be deepest when the child realizes not only that the mother is helping turn the page, but is helping in a way that is not really needed to get the page turned at all. Just as a mother often has reasons for directly helping to turn a page, besides the obvious one of getting the page turned, so God may have reasons for directly producing peaceful feelings, besides the obvious one of getting those feelings produced. The bond of intimacy only intensifies when God's involvement is causally superfluous, an act of grace, pure love. To explain the sheer occurrence of peaceful feelings scientifically, it may be enough to cite the rush of chemicals. To grasp the significance of Psalm 23, it may be necessary to see God's hand in the midst of it all.

The same goes for the opposite end of the spectrum, represented in the words of Psalm 44. Even those devoid of any personal religious life can readily appreciate the absence felt by Eli Wiesel at Auschwitz or Buchenwald. God failed to interfere in human affairs, failed to disrupt the natural order of things. But there is another, more subtle sense of abandonment, one that surfaces when life chugs along just as it should. This is a longing for God's presence in the routine course of everyday life. It does not wish for disruptions in natural causal processes. Like a child turning the page in the absence of the mother, the problem is not with the turning of the page, which the child can easily do alone, but with the missing hand and the love it signifies.

For all their epistemic power, the methods of science cannot determine if, when, or where God may offer an enfolding hand. For those whose eyes of faith have seen God's hand, science offers no reason for doubt. For those who have sought God's hand, but not found it, Wilson's Gilead of science offers no balm.

References

Aquinas, Thomas, *Summa Theologicae*, Vol. 1, *The Existence of God*, Part 1: Questions 1–13, Thomas Gilby, O.P., General Editor, Garden City: Image Books, 1969.

Ayer, Alfred Jules, *Language, Truth and Logic*, 2nd ed., New York: Dover Publications, Inc., 1952.

Boyd, Richard, Gasper, Philip, and Trout, J.D., eds., *The Philosophy of Science*, Cambridge: The MIT Press, 1991.

Byrne, Peter, "Arguing About the Reality of God," *Sophia* 19, No. 3 (October 1980), 3–11.

Cajori, Florian, trans., *Sir Isaac Newton's Mathematical Principles of Natural Philosophy and His System of the World*, Berkeley: University of California Press, 1934.

Carnap, Rudolph, *The Logical Structure of the World and Pseudoproblems in Philosophy*, trans. Rolf A. George, Berkeley and Los Angeles: University of California Press, 1969.

Clarke, Samuel, *Samuel Clarke: The Works*, New York: Garland Publishing, Inc., 1978.

Clayton, Philip and Davies, Paul, eds., *The Re-Emergence of Emergence: The Emergentist Hypothesis from Science to Religion*, Oxford: Oxford University Press, 2006.

Coughlan, Michael J., "Zande Witchcraft," in *Sophia* 24, No. 3 (October 1985), 4–15.

Dijksterhuis, E.J., *The Mechanization of the World Picture: Pythagoras to Newton*, C. Dikshoorn, trans., Princeton: Princeton University Press, 1986.

Drake, Stillman, trans., *Discoveries and Opinions of Galileo*, Garden City: Doubleday Anchor Books, 1957.

Einstein, Albert and Infeld, Leopold, *The Evolution of Physics: from Early Concepts to Relativity and Quanta*, New York: Simon and Schuster, 1938.

Galilei, Galileo, *Dialogue Concerning the Two Chief World Systems, Ptolemaic and Copernican*, trans. Stillman Drake, 2nd ed., Berkeley, Los Angeles and London: University of California Press, 1967.

Gilkey, Langdon, "Religion and Science in an Advanced Scientific Culture," *Zygon* 22, No. 2 (June 1987), 165–178.

Greene, Brian, *The Elegant Universe: Superstrings, Hidden Dimensions, and the Quest for the Ultimate Theory*, New York: W.W. Norton and Company, 1999.

Griffin, David Ray, *Two Great Truths: A New Synthesis of Scientific Naturalism and Christian Faith*, Louisville and London: John Knox Press, 2004.

Hacking, Ian, "Experimentation and Scientific Realism," in *Scientific Knowledge: Basic Issues in the Philosophy of Science*, ed. Janet A. Kournay, Belmont: Wadsworth Publishing Company, 1987.

Haldane, Elizabeth S. and Ross, G.R.T., trans., *The Philosophical Works of Descartes*, Vol. 1, Cambridge: Cambridge University Press, 1969.

Harré, Rom, *The Principles of Scientific Thinking*, Chicago: University of Chicago Press, 1970.

Hunter, Michael and Davis, Edward B., eds., *The Works of Robert Boyle*, London: Pickering and Chatto Publishers Ltd., Vol. 3, 1999; Vol. 8, 2000.

Husserl, Edmund, *Ideas: General Introduction to Pure Phenomenology*, W.R. Boyce Gibson, trans.. New York: Collier Books, 1962.

Kitcher, Philip, *The Nature of Mathematical Knowledge*, New York and Oxford: Oxford University Press, 1984.

Kuhn, Thomas S., "Objectivity, Value Judgment, and Theory Choice," in Thomas S. Kuhn, *The Essential Tension: Selected Studies in Scientific Tradition and Change*, Chicago: University of Chicago Press, 1977.

Kuhn, Thomas S. , *The Copernican Revolution: Planetary Astronomy in the Development of Western Thought*, New York: Vintage Books, 1957.

Kuhn, Thomas S., *The Structure of Scientific Revolutions,* 2nd ed., Enlarged, Chicago: University of Chicago Press, 1970.

Lakatos, Imre, and Musgrave, Alan, eds., *Criticism and the Growth of Knowledge,* London and New York: Cambridge University Press, 1970.

Lakoff, George, *Women, Fire, and Dangerous Things: What Categories Reveal about the Mind*, Chicago and London: University of Chicago Press, 1987.

Lakoff, George and Núñez, Rafael E., *Where Mathematics Comes From: How the Embodied Mind Brings Mathematics into Being,* New York: Basic Books, 2000.

Laudan, Larry, *Science and Values: The Aims of Science and Their Role in Scientific Debate*, Berkeley: University of California Press, 1984.

Leibniz, Gottfried, *Discourse on Metaphysics*, in Philip P. Wiener, ed., *Leibniz Selections,* New York: Charles Scribner's Sons, 1951.

Leith, John H., ed., *Creeds of the Churches: A Reader in Christian Doctrine from the Bible to the Present*, 3rd ed., Atlanta: John Knox Press, 1982.

McKeon, Richard, *The Basic Works of Aristotle*, New York: Random House, 1941.

Oakes, Robert, "Religious Experience, Sense-Perception and God's Essential Unobservability," *Religious Studies* 17, No. 3 (September 1981), 357.

Peacocke, A.R., *Creation and the World of Science,* Oxford: Clarendon Press, 1979.

Phillips, D.Z., *Faith and Philosophical Enquiry*, New York: Schocken Books, 1971.

Phillips, D.Z., *The Concept of Prayer*, New York: The Seabury Press, 1981.

Popper, Karl R., *The Logic of Scientific Discovery*, New York and Evanston: Harper and Row, Publishers, 1968.

Putnam, Hilary, *Reason, Truth and History,* Cambridge: Cambridge University Press, 1981.

Putnam, Hilary, "The Logic of Quantum Mechanics," in Hilary Putnam, *Mathematics, Matter and Method: Philosophical Papers I,* Cambridge: Cambridge University Press, 1975.

Quine, W.V.O. , *Ontological Relativity and other Essays,* New York and London: Columbia University Press, 1969.

Quine, W.V.O. , *Set Theory and Its Logic*, Revised Edition, Cambridge: The Belknap Press, 1969.

Robinson, John Mansley, *An Introduction to Early Greek Philosophy: The Chief Fragments and Ancient Testimony with Connecting Commentary*, Boston: Houghton Mifflin Company, 1968.

Russell Bertrand, , *Religion and Science,* Oxford: Oxford University Press, 1961.

Saunders, Nicholas, ed., *Divine Action and Modern Science,* Cambridge: Cambridge University Press, 2002.

Schoen, Edward L, "Clocks, God and Scientific Realism," in *Zygon* 37, No. 3 (September 2002), 555–580.

Schoen, Edward L., "David Hume and the Mysterious Shroud of Turin," in *Religious Studies* 27, No. 2 (June 1991), 209–222.

Schoen, Edward L., "Perceiving an Imperceptible God," in *Religious Studies* 34, No. 4 (1998), 433–455.

Schoen, Edward L., *Religious Explanations: A Model from the Sciences*, Durham: Duke University Press, 1985.

Schoen, Edward L., "Religious Explanations: A Reply to Professor McMullin," in *Faith and Philosophy* 5, No. 3 (July 1988), 310–314.

Schoen, Edward L., "The Roles of Predictions in Science and Religion" in *International Journal for Philosophy of Religion* 29, No. 1 (February 1991), 1–31.

Schoen, Edward L., "The Sensory Presentation of Divine Infinity," in *Faith and Philosophy* 7, No. 1 (January 1990), 3–18.

Stewart, M.A., ed., *Selected Philosophical Papers of Robert Boyle*, Indianapolis: Hackett Publishing Company, 1991.

Suppe, Frederick, ed., *The Structure of Scientific Theories*, 2nd ed., Urbana: University of Illinois Press, 1977.

Suppes, Patrick, "The Plurality of Science," in Janet A. Kourany, ed., *Scientific Knowledge: Basic Issues in the Philosophy of Science*, Belmont: Wadsworth Publishing Company, 1987.

Toulmin. Stephen and Goodfield, June, *The Fabric of the Heavens: The Development of Astronomy and Dynamics*, New York: Harper and Row, Publishers, 1961.

Ward, Keith, *God, Chance and Necessity*, Oxford: Oneworld Publications, 1996.

Wheelwright, Philip, ed., *The Presocratics*, Indianapolis: The Odyssey Press, 1966.

Williamson, Colywn, "Following a Rule," in *Philosophy* 64, No. 250 (October 1989), 487–504.

Williamson, Colwyn, *"Witchcraft and Winchcraft,"* in *Philosophy of the Social Sciences* 19, No. 4 (December 1989), 445–460.

Wilson, Edward O., *Consilience: The Unity of Knowledge*, New York: Vantage Press, 1998.

Winch, Peter, *Ethics and Action*, London: Routledge & Kegan Paul, 1972.

Winch, Peter, *The Idea of a Social Science and Its Relation to Philosophy*, London: Routledge & Kegan Paul, 1958.

Wittgenstein, Ludwig, *Lectures and Conversations on Aesthetics, Psychology and Religious Belief*, compiled from notes taken by Y. Smythies, R. Rhees, and J. Taylor; Cyril Barrett, ed., Berkeley and Los Angeles: University of California Press, 1967.

Wollaston, Arthur, trans., *Descartes: Discourse on Method and Other Writings*, Baltimore: Penguin Books, 1969.

CHAPTER 6

Charles De Koninck, John Leslie, and the Conceptual Parameters of Science

Elizabeth Trott

1 Introduction: Science and Philosophy—Two Sides of the Same Coin*

Science has seldom left the philosopher's domain. Scientific requirements of repeatability and predictability mean nothing without consciousness to observe and record, criticize and work with scientific results. Science also purports to be in the business of truth, and that fact ensnares it with philosophical precedents, for truth is both relative (to time, space, and causes that lead us to adapt to change) and yet aspires to the universal, assuming that facts and laws turn out to be not contingent but determinants of future existence. If idealism places the mind in a logically necessary relation (neither material nor contingent) to all experiences without regards to their categorization (mental, physical, imaginative, religious, speculative, aesthetic, etc.), then it is no surprise that idealist philosophers have more readily embraced the changes and the revolutions in the scientific world than almost any other members of the scholarly academy.[1]

---

* I am grateful to Luke Bowman, MA, for his comments and questions on the last section of the paper ("The Good World") and for his careful reading of an earlier draft of the whole paper.

1 Some thinkers, suspicious of idealism, who tend to support a materialist view of the world, have gone to great lengths to illustrate how science supports a God-less universe. Richard Dawkins, author of such books as *The Selfish Gene* (Oxford: Oxford University Press, 1976), *The Blind Watchmaker* (London: Penguin Books, 1986), and *The God Delusion* (Boston: Houghton Mifflin, 2006), has waged a life-long campaign against the inference that the involvement of the conscious mind in the universe gives grounds for there being a grand mind (God) which stands as a first cause or accounts for any design properties that we discern. Adhering to Darwin and a theory of slow gradualism (bit-by-bit changes let by natural selection), Dawkins has pressed his case in favour of empirical data established by years of excruciatingly careful research. Nothing in his position precludes the idealist commitment, however, quite likely because arguments for or against the existence of God tend towards much more sophistication than blind support of Darwinian conclusions and also, as we shall discover with John Leslie, design does not necessarily entail a single designer.

 | DOI:10.1163/9789004415270_008

Idealist philosophy has never been rejected by the scientific community. It may have been ignored, but for scientists who have put pen to paper, the role of consciousness has not been discarded. Fred Hoyle in his book, *Encounter with the Future*,[2] began to expand on his assuredness with steady state theories of the universe. Steady-state physics was based on the theory that there must be a perpetual creation of matter to counter the weakening of physical states with the perpetual expansion of the universe, for such expansion would soon threaten the idea of the universe's infinite existence. What will remain constant or universal about physics, Hoyle suggested, will be the mathematical constructs that framed the applications of scientific laws. In short, mathematics and laws of physics will become the directing forces behind new discoveries, not experiments, material ontologies, and telescopes.[3] The laws of change themselves might not be dependent on predictable results; they might produce different predictions in different environmental universes. For Hoyle, the search for the constants of nature will eventually give way to the abstract laws of physics.

Such a pronouncement that aspires to truth about future scientific possibilities still requires human consciousness. Laws are only understood as "laws" because of the power of mind which creates their formulations and works with them. The appeal of the possibility that future knowledge will be developed through the application of such laws, Hoyle states, lies in their simplicity and "elegance."[4] Yet, these terms are rooted in the human mind as the terms of aesthetics and value. Hoyle does not escape the need for the constructs of consciousness, no matter how far reaching his vision of possible universes was and has become.

The legacy of the role of human consciousness in scientific inquiry, an involvement that makes the free floating truth of science logically impossible, stems back much further than Fred Hoyle. Certainly the shift from classical mechanics to relativity theory shook the scientific world with its commitment to an independent measurable material universe. The materialist metaphysics which many 19th century scientists had promoted in their desire to rid themselves of the trammels of holding God to be the first cause was threatened by theories that suggested truth was relative to human theoretical constructs. The resistance of the scientific academy to Einstein and Eddington was understandable after a 300 year struggle. Some philosophers had already tackled the dualism upon which scientific materialism rested, however. As soon as the

2 Fred Hoyle, *Encounter with the Future*, New York: Simon and Schuster, 1965.

3 *Encounter with the Future*, 102.

4 *Encounter with the Future*, 107–108.

Herbert Spencers of the world began to celebrate the liberation of the facts of science, a process enlivened by the evolutionary theory of Darwin, others began to remind them of the incoherent world views that would ensue.

In Canada, John Clark Murray, a philosopher at McGill University in Montreal, published the "The Dualistic Conception of Nature" in *The Monist* in 1896. Murray strongly suggested that, in order for theories of science to have legitimacy, they must be transferrable to a variety of fields. Nature is a unified presence. There could not just be one set of laws for physical phenomena and another for mental phenomena. The laws of nature apply to all the components of nature, including people. Murray argues for the supremacy of reason, the source of the laws of mental phenomena, as the source of scientific theory as well. "Science therefore must give to the reality of this power a prominence equal to the reality claimed for the force of non-moral causation …"[5] Murray specifically targeted John Tyndall (1820–1893), a promoter of the materialist theory of the universe and author of the celebrated "Belfast Address" (1874).[6]

Philosophical positivists who leaped at the chance to rid the world of mind soon found themselves bogged down by the fact that there could never be absolute certainty in science because there could be no real application of a theory to all cases in all places, spaces, and times. No theory could ever be tested according to those criteria. Jonathan Powers in his book *Philosophy and the New Physics* traces the reaction of the positivists to the new discoveries and theories of quantum physics.[7] Others came to accept the reuniting of philosophical consciousness and the scientific mind. "In England some commentators … concluded that the dependence of the measures of mass, length and time upon frame of reference, meant they could no longer be regarded as constituents of an independent real world."[8]

Powers references Sir James Jeans,[9] the notable British physicist of the first part of the 20th century who openly acknowledged the unavoidable link

5 John Clark Murray, "The Dualistic Conception of Nature," *The Monist*, Vol. 6: No. 3 (April), 1896, 382–395; 395.

6 For a further discussion of Murray's concerns about dualism in science and the Belfast Address, see Elizabeth Trott, "John Clark Murray: Religion, Science, and the Unity of Nature," *Religion and Science in Early Canada*, ed. Douglas Rabb, Kingston, Ontario: Ronald P. Frye and Co., 1988, 43–74.

7 See Jonathan Powers, "The Absolute World," for a discussion of the failed efforts of the positivists to separate physics from the observer's conscious mind (in *Philosophy and the New Physics*, Metheun and Co. Ltd., 1982, 89–93).

8 "The Absolute World," 93.

9 "The Absolute World," 93; Sir James Jeans, *Physics and Philosophy*, Mineola, NY: Dover Publications, 1981; first published Cambridge: The University Press, 1942.

between the conscious mind and physics, and between physics and philosophy, in his book, *Physics and Philosophy*. Jeans had previously published an article in *The New York Times* declaring the new world of physics to be "more like a Thought than a Machine."[10] *Physics and Philosophy* established the close link between the conscious mind and its determination of world constructs, a connection he documented from Plato to Einstein.

Thomas Kuhn in his ground-breaking book, *The Structure of Scientific Revolutions*, revealed not only the links between mind and constructs we create that refer to or *mean* matter, but he also documented how these constructs *determine* the kind of research those purporting to call their results "scientific" have adopted, knowingly or unaware. Kuhn writes with compelling force:

> The need to change the meaning of established and familiar concepts is central to the revolutionary impact of Einstein's theory ... Just because it did not involve the introduction of additional objects or concepts, the transition from Newtonian to Einsteinian mechanics illustrates with particular clarity the scientific revolution as a displacement of the conceptual network through which scientists view the world.[11]

Nothing more clearly establishes the role of concepts of the mind in the progression and alteration of the "world" that science returns to us through its theoretical constructs. Yet in recognizing the inseparability between mind and science, nothing commits us to God, a first cause or designer. Developing philosophical arguments that are designed to support religious propositions derived from scientific observations is not a defining mission for idealist philosophers of science.[12]

The two Canadian philosophers to be discussed in support of the above claim are separated by three decades. Charles De Koninck (1906–1965) was a student at the Louvain in Belgium when the scientific revolutions involving quantum mechanics, evolutionary theory, and the push towards a secular

10 Sir James Jeans, "A mystic universe," *The New York Times*, January 28, 1928, 14.

11 Thomas Kuhn, *The Structure of Scientific Revolutions*, Chicago: University of Chicago Press, 1962, 102.

12 In his book *The God Delusion*, clearly written for popular consumption, Richard Dawkins relates his story of being invited to an Intelligent Design conference sponsored by the Charles Templeton Foundation as the token atheist (151–159). The present writer, a philosophical idealist, can verify the nature of the story, having been the invited token atheist (and, as a result, quite ostracized) at a conference sponsored by the same organization held at the University of New Brunswick in Fredericton NB., September 16, 2000. The invitation to publish with that organization was declined.

academy were underway. The great physicists of the 20th century were rethinking the world in terms of Einstein's work which was becoming widely accepted by 1917–20. De Koninck's writings reflected the need to accommodate the scientific revolutions on his doorstep. John Leslie (1940-) arrived at the University of Guelph in Canada having received his M. Litt. from Oxford. His book, *Value and Existence,*[13] initiated his lifelong interest in the extent to which principles of rationality delimit the conceptual parameters of science. His reflections on the anthropic principle and on immortality stand as conceptual limits to the truth claims of science. We begin with Charles De Koninck.

2 Charles De Koninck

Charles De Koninck came to Laval University in Quebec City from Louvain University in Belgium in 1934, having written a dissertation in the philosophy of science. Shortly after he was hired, he was promoted to a full professorship and later assumed the position of Dean of the Faculty of Philosophy. Throughout his career, De Koninck's concerns with science and its relation to nature "… remained always at the centre of his thought."[14] De Koninck was a participant in the Distinguished Professors Program at the University of Notre Dame and gave the Whidden Lectures in English at McMaster University (1959), published a year later as *The Hollow Universe.*[15] This book was his only major publication in English until the fortunate release of *The Writings of Charles De Koninck.*[16]

During the quiet and not-so-quiet revolutions in Quebec of the nineteen fifties and sixties, the philosophical community was anxious to throw off the shadow of the Church and eschew past associations with Thomistic doctrines. Religious orthodoxy as the measure of philosophical wisdom was rejected by

13 John Leslie, *Value and Existence,* Oxford: Blackwell Press, 1979.

14 Leslie Armour, Introduction to *The Writings of Charles De Koninck,* ed. and trans. Ralph McInerny, Notre Dame: University of Notre Dame Press, 2008, Vol. 1: 2.

15 Charles De Koninck, *The Hollow Universe,* London: Oxford University Press, 1960 (hereafter HU).

16 De Koninck served as an advisor on religious commissions for the Church of Rome, and he served on a number of government commissions in Canada, in particular the Tremblay Commission which was given the mandate of making explicit Quebec's vision of Canada. The Tremblay Commission published a five-volume report in 1956. De Koninck was invited to speak and present his ideas throughout the United States, Mexico, Argentina, Venezuela, and Europe, including extended speaking tours in Louvain, Paris, and Rome. He died in Italy in 1965. At Université Laval, the Pavillon Charles-De Koninck is named in his honour. De Koninck spent his early childhood in Detroit. Later, he was instrumental in establishing exchange programs for students from the United States to study at Laval.

many academics in philosophy. Charles De Koninck, an unapologetic Catholic philosopher, continued to write, undeterred by these rebellions. Nevertheless, in exploring his thoughts on the nature and limits of science and biology, we shall discover that he shares the goals of many later idealist philosophers.

3 The Idealist Mission

In a review of an anthology of writings about idealism,[17] David Crossley writes:

> Idealism offers a distinctive perspective on the methods of philosophical inquiry and on the nature of reality ... [including] a strong emphasis on the search for unity, for the connections between and among diverse aspects of experience, and the many spheres of human interest, such as natural science, psychology, ethics, and politics. While each distinct science or area of study tries to explain and to provide a coherent account of the phenomenon falling within its spheres, such studies are limited in that they must begin from various conceptions or 'working hypotheses' ... Since there is an emphasis on the unity of things, on the systematic coherence of reality, many of the Idealists ... felt there could not be any gap between theoretical and practical concerns."[18]

The point of doing science (and generating scientific theories) raises the troublesome question of what gives truth claims their validity. Is empiricism as a basis of knowledge claims sufficient to justify beliefs and explain actions of human experience? Or do empirical claims rest on the existence of some conception of nature (of what there is) that itself requires some further conception in order to produce comprehensive truths which explain all of our experiences, not just the statistically measurable ones? This question provides a context in which De Koninck takes his place.[19]

De Koninck shared the idealist's desire to unify knowledge. In particular, in HU he sought to unite the claims of scientific theory with the actual affairs of

17 William Sweet, ed., *Idealism, Metaphysics, and Community*, Burlington, Vermont: Ashgate Publishing, 2001.

18 *Dialogue*, Vol. XLII, No. 3 (Summer 2003), 607.

19 R.G. Collingwood has admirably traced the various answers to this question in his book, *The Idea of Nature* (Oxford: Clarendon Press, 1945). Whether science is independent of mind or mind-dependent is, according to Collingwood, contingent on one's view of history.

human experience. He argued that unification could only be achieved through a discussion of values which are essential to our understanding of the complexities of the universe. Writing before De Koninck, philosopher Alfred North Whitehead[20] posited God as an infinite 'lure' towards which actions and events were directed and he wanted to include not just human actions but all events which revealed teleological creative acts.[21] Writing after De Koninck, John Leslie has insisted in much of his work that "The lives of separate individuals are logically incapable of existing each in isolation from the others. …" Creation's goodness need not "… be crammed into a single such world."[22] Leslie argues that value concepts are logically necessary to give reason any focus towards good ends. Writing at the same time as De Koninck, J. Bronowski[23] argues that developments in the arts and science since the Renaissance have given rise to modern values.

The search for a theory of nature that unites theory and practical human affairs in science has not been shared by many. In *The Hollow Universe,* De Koninck references the ancient Greek philosophers Democritus, Plato, and Aristotle. Hegel and Kant are given brief acknowledgment and Wittgenstein also surfaces. De Koninck's main figures of debate are scientists and mathematicians. Philosophers with different conceptions of God, such as William James, A.N. Whitehead, the 20th century idealists, or fellow Canadians, receive no attention at all. Perhaps the lack of communication between Catholic philosophers and their non-Catholic colleagues, accounts for the silence, or perhaps De Koninck believed that to unify scientific claims with philosophical ones, one had to do battle with scientists and mathematicians. HU refers to scholars and researchers such as Eddington, Einstein, Reichenbach, Poincaré, and Russell. We turn now to De Koninck's view of nature.

4 De Koninck on Nature and Scientific Abstractions

Nature, for De Koninck, means whatever physicists, chemists, and biologists study—whatever is the source of truths that mathematicians purport to represent to us. However, he cautions, we must be careful 1) to recognize the limits of scientific abstractions in explaining nature and the world, 2) to

20 A.N. Whitehead, *Process and Reality: An Essay in Cosmology,* London: Macmillan, 1929.
21 See Collingwood, *The Idea of Nature*, 169–170.
22 John Leslie, *Value and Existence,* 168.
23 J. Bronowski, *Science and Human Values*, New York: Harper and Row, 1956.

acknowledge the inadequacies of biological explanations, and 3) to realize the dangers of the uses to which scientific determinism may be put.

Let us begin with the limits of scientific abstractions. In the preface to HU, De Koninck writes: "If we generalize this new scientific outlook, as it is commonly understood, if we accept it as the one true way of thinking about nature, the only wonder left to us will be wonder at the hollowness of the universe, both of nature and of thought"[24] The hollowness that he refers to is the seemingly emptiness of the world that results as science abstracts more and more from our experience. In his chapter on "Mental Construction and Test of Experience," he gives us an example of the hollowness that surrounds scientific claims:

> Sheer mathematical constructs, endowed with that hollowness which renders them so efficient as tools, must make over into their own likeness the natural world to which they are applied. The science which uses these symbolic devices, mathematical physics, thus finds itself able to reveal the world of nature only to the extent in which it can be caught in what has been termed the selective screen of symbolic construction.[25]

Here De Koninck shares metaphysical ground with objective idealism. There is something out there; nature is not just an invention of subjective consciousness. But our knowledge of it will be shaped by the conceptual structures through which experience is ordered. These orders we will discover to be increasingly complex and interdependent as our ability to use them and develop theories further advances.

Science functions in terms of theory and theory reduces the world to symbols and equations. Are we to consider the world we experience as a source of truth to be the real world or is the world of the physicist the true representation of reality? In the "Introduction" to Volume 1 of De Koninck's writings, Leslie Armour offers an example to illustrate De Koninck's concerns about the increasing abstractions in the world of science. The example deals with a glass of water. If the symbols and equations of the scientist document the real world, then how do we accommodate the claims that science offers as knowledge about the glass of water? For the physicist, the water is made of hydrogen and oxygen, which are composed of atoms and molecules. There is nothing in the water in addition to the atoms and molecules.[26]

24 HU, xii.

25 HU, 51.

26 *The Writings of Charles De Koninck*, Vol. 1:20–21.

However, the atoms are not wet, nor even a liquid. The glass of water is, for De Koninck, a different object from the one captured by the concepts of physics. Yet here we must sort out what an object is, for there are not two objects, the glass of water for the physicist and the glass of water for the thirsty, such as you and me. There are two different conceptualizations of an object in the real world. The 'object' the physicist studies is a theoretical abstraction of the 'object' you and I experience. Science, with its powerful abstractions and capacities to reconstruct and anticipate events in the world, cannot render all the truths there are about the world. De Koninck did not want to suggest, however, that the conceptualization of the glass of water that you and I experience is *only that*, a set of ideas. He clearly defended Aristotle's world of material objects sustained through causal relations. His point is that scientists cannot capture in equations the feel of cool water in our throats on a hot day; they cannot tell us all there is to know about a glass of water. Hot and cold as experiential properties are predicated of living beings, not scientific measurements. If we revere science as the only source of reliable knowledge, our understanding of the world will be seriously impoverished.

De Koninck notes that no equation in science carries with it a mandate of care for those who are exposed to the powers of nature and yet who are unprepared to cope with such powers.[27] Experience alone fills the emptiness of the symbolic universe. Ultimately, for De Koninck, nature, in its vast interrelatedness, is sustained by God. God provides intrinsic unity to the world of objects we experience.

Nature itself, De Koninck suggests, is a great work of art, inexhaustible in its depth of understanding and certainly not fully explained by science. "The aim of natural science, even when employing mathematical tools, then, can only be to learn everything possible about the Art that fashioned natures."[28] For De Koninck, the grand artwork before us, Nature, has a divine intelligence behind it, God.[29]

---

27 For example, positive experiences we have with nature incline us to create wilderness parks for recreation and spiritual renewal. Negative experiences we have with nature may incline us to create shelters for the homeless, post high-water warnings along river banks, and use mosquito lotion.

28 HU, 76.

29 De Koninck explains the role of experience in understanding nature with an unexpected appeal to the concept of art. Art is the process whereby the essence or purpose of things is expressed. Artists, as we know them, reveal to us a world bigger than appearance first suggests. They combine and contrast many meanings in their artworks that we may not expect to find in the objects of the world. The artist brings forth meanings in ways not even imagined by the scientist who is restricted by his or her universal abstractions. Science reveals what is true of like things, such as atoms, unfettered by individual experiences.

5 Nature and Science: United by Common Values

Although De Koninck is writing within a Thomistic framework and invokes God as a grand artist, his intent is not *to* rescue the world from science, nor deepen the chasm between the world of experience and the world of mathematical abstractions, but to demonstrate the close affinity between science, experience and religion. Leslie Armour explains De Koninck's mission:

> De Koninck noted ... that since science necessarily abstracts selectively from a given reality which is invariably richer than the scientific descriptions which result, there is always a case to be made for the proposition that concrete reality has a natural priority over scientific description.[30]

De Koninck's intent was to demonstrate that religious or experiential orderings of the world can be understood as not conflicting with the scientific orderings.

Armour suggests that De Koninck left sufficient evidence in his writings to indicate that his ultimate intent was to demonstrate that science and religion imply each other. Each does so by having a similar aim—wisdom. Wisdom is not just a vast storehouse of knowledge. Wisdom is knowing what to do with what one knows, being able to identify the good toward which one's experiments must be directed, and acknowledging that such a good is a feature of the universe.[31] We know, for example, that the abstractions of science resulting solely from a scientific desire to discover new knowledge are capable of generating powers that can destroy the planet. We struggle now not to do so. Yet what reason could science alone possibly have for the planet and the human race to continue to exist, other than that their continued existence provides the materials for scientific endeavours? The planet and the human race do

However, science cannot reveal to us all of the multiple meanings that individual objects have or can inspire in the mind of an artist. The only reason that we sustain objects which have long lost their usefulness (for example, by displaying them in museums), or promote the protection of species in the living world, is our continuing interest in them. Our interest in some objects may have been inspired by the exposure to equations which capture their historical situations by carbon dating, and in species by documenting their genetic codes. What is more likely true for the majority of people, this interest is sustained by the narratives we associate with objects that we preserve and, in the case of species, by the properties that delight and amaze us when we encounter different species in zoos, films, travel experiences (natural settings), and the arts.

30 Armour, Introduction to *The Writings of Charles De Koninck*, Vol. 1:25.

31 See further explanation of this point in Armour, Introduction to *The Writings of Charles de Koninck*, Vol. 1:31–35.

provide an agenda for scientific research. But the justification for doing scientific research without the concept of good seems circular, however, or, to use De Koninck's word, "hollow." The justification would spell out as follows: scientists strive to keep the planet in existence by doing research to understand the mechanisms of its existence so that they can continue to do research to understand the mechanisms of its existence. Without some concept of good, the scientific community would be the most self-centered profession imaginable.

De Koninck does not put the question, "Why should the world continue to exist?," so bluntly, as the reason not to destroy the planet is self-evident to him. The wisdom given to us by the Divine Intelligence—wisdom that is available to us even though we may not actively court it—is diffused throughout the scientific community. Science proceeds in a context of seeking truth. What scientists discover or inherit from century-old traditions is that scientific research progresses more successfully towards truth when working in a context of honesty, cooperation, and accountability. Such a moral context for scientific work seems to match the context within which we learn about and experience the objective world. A learning context which promotes deception and lies soon becomes ineffective.

As an example of the need for a moral context for scientific truths, consider the sick babies in China, reported in a Toronto newspaper. The sickness of the babies can be explained scientifically; there was a toxic substance in the milk powder. The scientists of China can establish that fact. The more difficult question is the cause of the babies receiving the milk at all, for they had to be given the milk to become ill. Can we abstract from the concrete world of babies a scientific explanation that will provide that piece of the puzzle? The explanation that was reported in the *Toronto Star* of why 16,000 babies became so ill came from a Chinese journalist whose blog appeared on a website, one that was quickly shut down thereafter by Chinese authorities: "Systemic corruption, bureaucratic inefficiencies and the lack of a strong civil society of non-governmental organizations and a free press."[32] Total control of the media prevented the scientific explanation from being shared by hospitals in need of the information. The capacity of the Chinese government to shut down the flow of information is a technological capacity. Blocking information through technological channels, such as shutting down websites, emails, and newspapers, requires considerable scientific expertise. To have total control of the people in China is a scientific aim, in the sense that science aims at control and power over the contents of the world. Without the diffused wisdom of the

32 Bill Schiller, "A crisis rooted in two Chinas," *Toronto Star*, September 21, 2008, A4.

good as a goal for all individuals all of the time, science can become as much a source of harm as of good.

De Koninck, no doubt, would have regarded the tragic situation there as a failure of the leaders to grasp the intrinsic relations that tie science to the world of individuals and their concrete realities. Reason as the spiritual energy for the human soul fuels both the scientific and the moral engines of our relations with the world around us. Reason can grasp purposes and goodness in the world. Religion embodies those truths. Sick babies are not just statistics. De Koninck's spiritualism is not of the kind that suggests we should pray more to get good results. His faith is in human reason as a force which is neither solely calculative nor directed by other worldly aspirations, but as a spiritual capacity that is multifaceted.[33] Good is served when all of reason's capacities are harmoniously employed. Reason is pre-emptive of science; it is not produced nor directed by scientific discoveries alone.[34]

6 De Koninck and the Inadequacies of Biological Explanations

Sociobiology, De Koninck's target, sprang to life after the discovery of the DNA molecule (1952) and the rapid increase in research on human and non human genes. Scientific claims about the evolution of the human being began to turn anthropology from a study which documented empirical observations (ones constructed, of course, by the conceptual orders of the viewer). A new kind of measurable evidence was now available to lend credence to Darwin's evolutionary theory. The "systematic study of the biological basis of human behaviour" reached its pinnacle with the publication of E.O. Wilson's tome, *Sociobiology,* in 1975, an abridged version of which was printed in 1980.[35] In due time, Wilson maintained, we would be able to predict, document, and therefore have to accept the way nature has laid out the human social world because science says so. The forces of genetic adaptations to a competitive world were unstoppable.

---

33 De Koninck regards the multiplicity in nature as evidence of the eternal and infinite character of God. God's goodness is not collected in the sum of nature's parts but is revealed as nature becomes known to us through an intersection of the eternal and the internality of diverse living forms.

34 This focus on reason puts De Koninck on similar ground with many early Canadian philosophers. For those interested in this historical connection, see Leslie Armour and Elizabeth Trott, *The Faces of Reason*, Waterloo: Wilfred Laurier University Press, 1982.

35 Edward O. Wilson, *Sociobiology*, abridged edition, Cambridge: Harvard University Press, 1980, 4.

This highly influential volume was published after De Koninck's death in 1965, but 697- page books are not written overnight and the 64 pages of bibliographical references reveal a vast percentage published between the early 1930's and the early 1960's. De Koninck, a man of practical affairs, was abreast of the latest developments in science and the direction the research was taking horrified him. Where was the capacity for ethical choice that our status as free agents permits? It was seemingly being obliterated by the Harvard researchers. Even the title of Chapter 1 of Wilson's book, "The Morality of the Gene," mocks the history of ethical thought. De Koninck sensed what was afoot acutely.

De Koninck titles the third and final chapter of HU, "The Lifeless World of Biology." He begins, "The title of this lecture may seem a contradiction, and perhaps also a piece of impertinence."[36] De Koninck challenges one increasingly popular thesis during his time, that biological objects can be studied without much concern for their status as living or dead. Granting the fact that 'living and dead' as biological distinctions face increasing obstacles as the world of physics probes the biological, De Koninck launches a reverse attack on standard scientific methods. "Adhering to the general scientific principle that we explain the complex in terms of what is at least less complex"[37] results in the claim that we should be more familiar with the features of a dead amoeba than a dead horse. De Koninck recognizes that there is something rather counterintuitive with this starting position and he suggests working in reverse. Study what is obvious and known to us, not the obscurity of a world we cannot experience.[38] Begin with what we can know through consciousness and its evaluative capacities of our particular experiences.

De Koninck's common sense approach to the questions of biology determinedly reflected his vision of a universe that can be conceived holistically. The knowledge that some biological organisms not only experience sensations but can reflect on them and know them is critical for De Koninck to understanding how we prioritize complex life forms. Unless we hold this capacity to know and reflect in some esteem, why should we value human life any more

36 HU, 79.

37 HU, 82.

38 De Koninck's advice was not obvious to those who followed, probably because they were not familiar with De Koninck and the nature of his reasoning. In *The Blind Watchmaker*, Richard Dawkins writes as follows: "If we wish to understand how a machine or living body works, we look to its component parts and ask how they interact with each other. If there is a complex thing that we do not yet understand we can come to understand it in terms of its simpler parts that we do already understand" (11). De Koninck's fear that we would approach the human species with all of its interactive complexity in the same way that we approach the analysis of machines seems to have had solid grounds.

than amoebic life? Biologists, he thought, were drifting away from acknowledging this reflective capacity of complex organisms and, in doing so, seemed to be ignoring the increasing uncertainties of knowledge claims that physicists were beginning to recognize. He expresses his frustrations with the biologists of his day and their proclivity to make sweeping claims about truth as follows:

> Is it credible that some biologists continue to state that their science is working with physical phenomena entirely accessible to our understanding, by which they mean life is explained in terms of physics and chemistry?. … of the science of nature, physics is the most exact, but I have yet to meet a modern physicist who speaks with any such confidence about what he knows.[39]

De Koninck quotes Eddington, about whose work he wrote his doctoral dissertation: "As he digs to the foundations of the physical world, the scientist finds himself '*treating a universe which is the theatre of unknowable actions and operations*'."[40]

The biologist of his day is much too confident for De Koninck's liking. De Koninck's desire to unify the world of knowledge is being thwarted by an obvious separation between the scientific disciplines. The biologist also reduces evolution to mere superficial phenomena, "accidental variations of the same thing."[41] This attitude results in a dismissal of the various concepts of purpose, intent, need, choice, and decision that we employ to account for the activities of biological entities, both complex and simple. These concepts, which help us to understand the metaphysical categories of things in social relations, would be unintelligible for the biologist. De Koninck continues: "If an organism has no purpose, it is no organism at all: it is not a body equipped with tools making possible the fulfillment of needs, but simply a mass in which appear a number of functionless appendages."[42]

De Koninck reminds us that purposes and their evaluation, as good or evil, long preceded scientific abstraction. Scientific discoveries have not lessened human activity, with its purposes and judgments, or changed the role such judgments play in our lives. Having all of our appendages rather than some of the normal supply is judged as a condition that is good or bad for our lives.

39 HU, 91. De Koninck's doctoral dissertation (Louvain, 1934) is entitled, "La philosophie de Sir Arthur Eddington."

40 HU, 91.

41 HU, 94.

42 HU, 97–98.

This state of completion, or lack thereof, may be simply recorded as a statistic in a medical notebook, but such information is immediately transformed into a value judgment by those who can reflect on its implications.

De Koninck would have us re-evaluate the distinction we often make between things that happen which can be explained, for example, by a causal relation between them, and things which happen by chance. Suppose we resist the process of evaluating the events of nature as good or bad. What purpose would the concept of a 'balance of nature' serve? Appealing to such an idea as a way of explaining actions would be unnecessary.[43] De Koninck reminds us, as would the logical scientist, that the concept of chance has no meaning unless we have the oppositional idea of order. Order is associated with outcomes, predictions, and expectations. In identifying orders that are and can be repeated, the scientist can plan and predict. Purpose as a concept captures the sense of outcome or expectation that experiments embody. Does it make sense to explain events (in a world so infused with the idea of purpose) with little regard for the concept of purpose? Trying to do so could hardly account for life as we know and experience it.[44]

Throughout the chapter on "The Lifeless World of Biology," De Koninck has one mission: that we recognize the affinity between nature and reason. In agriculture we learn how things in nature grow in order to benefit from that knowledge and improve upon the products of nature which we need to survive. We engage in fallowing, we replant trees, and we attack predators of nature's resources. Insofar as we are benefactors of natural resources, we are capable of contributing to the welfare of such resources. Insofar as we are consumers of natural resources, we are predators and, as such, are capable of making the conditions required for the survival of such resources much worse. Reason helps us to grasp the required cooperation with nature that is needed. Biology as De Koninck knew it, however, would have us disregard these narratives of human experience.

7 De Koninck and the Limits of Deterministic Scientific Theories

De Koninck warns us that we need to include in our scientific theorizing that which can be known about the experiences of being human, not just the molecular biological data associated with the human species. The information

43 HU, 102.

44 HU, 104–105.

that biology makes available to us must be fitted into a holistic theory which is related to other sciences and humanities and is structured by minds and their purposes about the world.[45] De Koninck's reluctance to abandon purpose as part of scientific explanations stems from his increasing awareness of the consequences that result from the increasing fragmentation of knowledge, where each discipline operates in isolation from others. Should the increasing differentiation and isolation of new specialties in scientific research fragment our understanding of the human being, the fallout will be a loss of our humanity: "The fatal consequences of abandoning all thought of the subject as a whole, to become absorbed and lost in independent investigation of single aspects of it, is illustrated everywhere."[46]

The brutality of a world toward which a mechanistic vision would propel us should be of no concern to those persons who accept the mechanistic view of the human species. If that is the kind of knowledge to be revered, then we will gradually evolve towards being the extensions of the abstractions we have heralded as total truths.[47] One consequence of which De Koninck writes is the reductionist tendency to regard everything as a kind of machine. De Koninck grants that computer-minded people may readily assign the capacity to think to machines, but that physicists are not so able to abandon the possibility of unanswered questions.[48] Unanswered questions are expected by idealist

45 E.O. Wilson would later explain that purposes were simply genetic adaptations to the need for territory and resources; see *Sociobiology* (1980), 273–275.

46 HU, 112–113.

47 See Epilogue, HU, 115–127.

48 De Koninck was not alone in recognizing that science could not tell us everything about the world. Sir James Jeans (1877–1946) was working in the fields of astronomy, thermodynamics, and hear radiation. He also published books that were about physics, philosophy, and the universe. *The Mysterious Universe* (London: Cambridge University Press, 1930) and *Physics and Philosophy* (Cambridge: University Press, 1942) are still in print. Jeans, although promoting the fact that mathematics was the key for unlocking secrets about the universe, also acknowledged that there was more uncertainty than certainty in a world where quantum mechanics substituted probabilities for mathematical certainties and calculations. "Nature seems very conversant with the rules of pure mathematics," he wrote, yet he also paraphrased Newton in saying that "The great ocean of truth rolls, unexplored, beyond our reach" (*Physics and Philosophy*, 217).

De Koninck also urged us to keep the great ocean of truth, the "whole," in sight as the target of understanding. This vision has received increasing attention since his death. Metallurgist Cyril Stanley Smith, writing in 1978, echoes De Koninck's concerns that we stay rooted in what we know and in what we experience, and that we not turn to abstractions as the sole echoes of truth. "Everything—meaning and value as well as appropriateness of individual human conduct or the energy state of an atom—depends upon the interaction of the thing itself and its environment" (C.S. Smith, "Structural Hierarchy

philosophers of science and they stand as no threat to the unity of accepted truths. This focus on unity is the ground of De Koninck's belief that the different branches of knowledge will complement and augment one another.

8 John Leslie and the Platonic Universe of Universes

John Leslie arrived at The University of Guelph from Oxford University in 1968. He retired as Professor Emeritus and Fellow of the Royal Society of Canada in 1996 and moved to Victoria, British Columbia, where he participates

---

in Science, Art and History," *On Aesthetics in Science*, ed. Judith Weschsler, Cambridge, MA.: MIT Press, 1978, 9–53; 41).

Biologist Richard Lewontin in *Biology as Ideology* (Concord, Ontario: Anansi, 1991) disagrees with De Koninck's commitment to a holistic approach to science, one that conceives of the world as a coordinated whole directed toward good. But he also does not think that "… at every level the world is made up of bits and pieces that can be isolated and that have properties that can be studied in isolation" (15).

It is worth quoting Lewontin as his words capture exactly the concerns about which De Koninck was sounding the alarm. In the introduction to his book, *Biology as Ideology,* Lewontin urges that neither of the above theories (the holistic or the bits and pieces determinism) will work. He cites the false paths that the 'bits and pieces determinism' sets up" "We will consider how biological determinism has been used to explain and justify inequalities within and between societies and to claim that those inequalities can never be changed. We will see how a theory of human nature has been developed using Darwin's theory of evolution by natural selection to claim that social organization is also unchangeable because it is natural … We will see how simple economic relationship masquerading as facts of nature can drive the entire direction of biological research and technology" (16).

John Dupré in *The Disorder of Things* (Cambridge, MA.: Harvard University Press, 1993) reinforces De Koninck's warnings about the fragmentation of knowledge. He argues that few scientists believe that the disciplines of science are clearly distinct. One of his goals is to refute that principle: "The disunity of science is not merely an unfortunate consequence of or limited computational or other cognitive capacities, but rather reflects accurately the underlying complexity of things" (7). We have not yet grasped the intricate relations between abstractions, things, and human values. "Determinism is contradicted by the majority of our most successful scientific theories" (189). Probabilistic theories "… explain apparently quite deterministic behaviour in terms of the interactions of very large numbers of constituents homogenous in certain respects" (189). Determinism is just "… one range of possible metaphysical positions of causal order" (193).

There is no shortage of current literature that has recognized the concerns that De Koninck was writing about in 1960. At the time he was challenging the accepted orthodoxy of science, foreseeing the unfortunate uses to which such approaches to science could be put. In his desire both to accommodate science and yet maintain the perspective that the good of humanity must be in the forefront of research, De Koninck was a man very far ahead of his times.

actively with the Philosophy Department at the University of Victoria. Leslie was made University Professor Emeritus by the University of Guelph in 1999. He continues to publish and speak at international conferences. Our exploration of Leslie's thought will refer to some but not all publications from his prolific career.

John Leslie's writings address issues in philosophical cosmology and the relation between philosophy and science.[49] Leslie's aim is to establish a unified metaphysics of existence. Rather than there being a world of human experience, a world of scientific inquiry, and a world of divine causal determinism, Leslie seeks to demonstrate that all three different 'worlds' are basically the same world. Leslie's major philosophical writing began with the publication of *Value and Existence*.[50] His most recent publication, *Immortality Defended*,[51] relies on principles he has developed over his career, notions of ethical necessity, a pantheistic view of creation, and the conjecture that if the universe is infinite and that claim is bolstered by the need for an infinite consciousness to make it legitimate, then the idea of immortality is not incoherent. Leslie offers three theories of immortality to make this claim, one of which is Einstein's four dimensionality principle that universes are not locked into the time sequences which we experience. Thus our lives may not be exhausted by the

---

49 A clear explanation of the relation between philosophy and science, particularly with respect to cosmological inquiry, can be found in Ernan McMullan, "Is Philosophy Relevant to Cosmology?", *Modern Cosmology and Philosophy,* ed. John Leslie (Amherst, NY: Prometheus Books, 1998), 36–56. McMullan points out that science, like philosophy, relies on certain theories based on probably conjectures. Models of the universe can be thought of as ideas as well as presumed accounts of matter. "How reliable an insight do the models of contemporary cosmology give us into the structures of the world on the largest scale? This raises all the issues made familiar by the realism-idealism as well as the realism-instrumentalism debates. The idealist would challenge the assumption of an extra-mental real to whose structures scientific theory has to conform. And the instrumentalist would argue that the success of a theory gives no warrant for believing in the existence of the entities postulated in the associated model" (37).

Cosmology can be a testing ground for philosophers' theories about science as well as physicists' theories about matter. If we think of the cosmos as everything in existence, which can include universes within universes, it becomes less surprising that science and philosophy soon merge onto the same speculative, theory-laden turf.

50 John Leslie's publications include *Value and Existence* (Oxford: Blackwell, 1979) (hereafter VE), *Universes* (New York: Routledge, 1989), *The End of the World* (New York: Routledge, 1996), *Modern Cosmology and Philosophy*, ed. John Leslie (Amherst, NY: Prometheus Books, 1998), and *Immortality Defended* (Oxford: Blackwell, 2007), plus many article in *Mind, The Philosophical Review*, and other prestigious journals.

51 Hereafter ID.

'time-restrained-being-alive-experience' we imagine, and use, in order to reflect on our conjectures about the not-being-alive alternative.[52]

9 Platonism and Existence

Leslie's idealism can first be identified with his neo-Platonic metaphysics. His commitment to Platonism is one which acknowledges the Platonic universal world of forms that has so challenged scholars. By Platonic forms Leslie means a commitment to a necessary and eternal realm of abstract truths/ facts/ realities. "Plato holds that innumerable facts are independent of all of the things of our world. Instead of being human inventions, countless realities (mathematical, ethical, etc.) are real necessarily and eternally."[53]

This is a world of necessary propositions, the necessity of which is more established by rational probabilities than by logic. By this I mean that many propositions are meaningful in that they do not purport to promote logical contradictions. Others stand as meaningful beyond their logical coherence. In order to provide refutation of their inclusion in any position put forth, they would require arguments to indicate the probability of their being meaningful or having reference as being so low that we would dismiss them. Leslie finds Plato's concept of eternal truths as highly probable truths. Platonic truths could not be discredited even if there were to be nothing in existence by which existence they could be established.

Thus hypothetical conditionals that are logically linked provide a unifying structure for "reality". Hypothetical conditionals, of course, are propositions that are intelligible to conscious beings. Consciousness grounds all claims about reality. Leslie's examples help us to understand his worldview:

> Suppose 4 groups of 3 roses came to exist in some garden, making 12 when folks counted them expertly. The mathematical reality met with here—that any 4 groups of 3 things, *were they ever to exist*, would always include exactly 12 things—would not have depended on the actual existence of any one expert at counting, or of any roses or other countable objects.[54]

---

52 ID, 60–61. In ID, Leslie does subscribe to Platonic realities (abstract and eternal truths, one of which is goodness (see Chapter 2, 16–35). The fact of evil can also be contemplated as a Platonic truth, one dialectically related to good. Leslie would seem to urge that our proclivity as humans is towards good, but neither empiricism nor logic can establish this claim with any assurance.

53 ID, 16.

54 ID, 16–17.

Following Plato's emphasis on the good as being the ground of all existing things, Leslie extends these eternal irrefutable hypothetical conditionals to include ethical truths.[55] For example, if there were a world that was full of misery, then such a world would be undesirable and the requirement that it not exist is a necessity in the same sense that 3x4=12 is a necessity. Leslie then wants to argue that, inasmuch as there can be ethical necessities of negative value, there can be ethical necessities of positive or intrinsic value, one being that the world exists.

Leslie introduces the possibility of a world existing out of ethical necessity in Chapter 3 of ID, though his consideration of this idea originates with his first published book.[56] Ethical necessity means that if there is to be good in any sense of the word, then it ought to exist. If there is reason (requirement) for a world of misery not to exist and if the claim that nothing ought to exist is somewhat vacuous, the proposition being contradictory in nature (in the way that round squares are contradictory) and therefore not qualifying as an abstract truth, then there is reason (a requirement) for good to exist, as all other alternatives are rationally undesirable.

10 Leslie on Consciousness

Leslie's idealism is found in his commitment to consciousness but it is a commitment with qualifications. Leslie does not subscribe to the Neo-Hegelian position that, as conscious beings, we participate in the consciousness which grounds the universe. Rather, he supports evolutionary theory and the fact that beings with brains evolved long after the origins of the universe, origins perhaps explained by the big bang theory. In no way does Leslie want to compromise the findings of science or current theories in physics. On the other hand, he is not content to let neutrinos and atoms comprise all there is, explaining why the universe exists, or why anything at all exists. One reason is that patterns, which enable predictions and hence further discovery in knowledge, seem hard to deny. Patterns and their recognition are peculiar to the mind of persons. Leslie induces from their existence a cosmic or divine mind, the thinking of which gives existence to those patterns. While our acquaintance with such a mind is not direct, neither was the insight of scientists about the existence of neutrinos and their energy-carrying capacities the result of any

55 ID, 17.

56 See VE, 56–65.

direct experience of such entities as neutrinos. There seems to be no other explanation for certain scientific events.[57] Leslie urges that there is no better explanation for the existence of patterns and, eventually, of observers who could recognize, observe, and document those patterns.[58]

The divine mind carries weight as a construct of explanation because such a mind makes comprehensible abstract and eternal Platonic truths, in particular the truth of good in the universe as a state undeniably more rationally convincing for the existence of anything at all rather than evil. Leslie will go on to give ethical requiredness creative force which makes possible self-reflection, recognition of intrinsic value, and our capacity to grasp the intricately systematic unity of the fraction of an infinite universe—possibly universes—which we are able to contemplate.[59]

11 God's Existence or Non-Existence and God's Relevance to Science

John Leslie is not engaged in a set of inquiries that will establish the existence of God. Whereas De Koninck simply presumed God's existence and then moved on to establish that science and value are inextricably linked, Leslie is unperturbed by the actual challenge of establishing God's existence, or the task of denying the idea of God to be meaningful. It is almost a linguistic choice should the reader want to employ theistic terminology to grasp the overview that Leslie is articulating.

Where the concept of God becomes optional is in the pantheism that Leslie defends. Is God the creative force behind the existence of our world, our universe, perhaps infinitely multiple universes? The Design Argument for the existence of God and the existence of the world is often thought to entail God, the grand designer, as its conclusion. For there to be design and purpose, there must be a designer with intentions and a plan.

11.1 *Design*

The fact of design in the world, or orders that can be followed and the results of their operations predicted, has had considerable support from scientific evidence amassed about the physical conditions of the existence of life in the world over the last hundred years. This research has fuelled both those who

57 See *Universes,* 165–168.

58 ID, 80–87. We will consider the role of observers in Leslie's theory in the next section on design.

59 ID, 36–39.

would use it to support belief in God, or to buttress the claim that a designer cannot be ruled out as an explanation, and those who would use it to endorse a scientific explanation of the world's existence and its *modus operandi*. Today the former position finds followers who identify with the principle of Intelligent Design. Science alone cannot possibly account for the intricate and unexpected coordination of living species and their environments. The latter group includes philosophers such a Richard Dawkins whose *The Blind Watchmaker* is the benchmark for the resurgence of Darwinian theory—survival through slow, gradual evolutionary changes, and Brandon Carter with whom we associate the Anthropic Principle, according to which the universe is one in which observers can exist. The debates about the universe and the cause of its existence predate Hume and Kant, both of whom suspected the certainties derived from a God-caused universe. Such debates have not disappeared and since the 1950's, the Big-Bang theory cosmologists, physicists, and philosophers have weighed in with prolific fervour.[60] It is not my intent to weigh these debates in this paper but rather to establish John Leslie's position in the discussion.

11.2 *Fine Tuning*

Much research has resulted in the "fine-tuning principle" associated with the anthropic principle.[61] Fine-tuning as a feature of the universe has been supported and furthered by scientists such as R.H. Dicke and S.W. Hawking. Leslie summarizes much of the work amassed hitherto in Chapter 2 of his book, *Universes*. An example will help explain the thrust of the fine-tuning principle. Science has established the fact that the slightest of fluctuations in temperature, gravitational force, and chemical balances of the components of water and oxygen, would have left the earth as a lifeless planet:

> S.W. Hawking estimated that a decrease by as little as one part in a million million when the temperature was 10^{10} degrees, 'would have resulted in the Universe starting to recollapse when the temperature was still 10,000 degrees'. The fine tuning seemingly needed (i.e. for the universe to come to expand at life-permitting speed) must be more accurate, the more one pushes back the time at which it is carried out.[62]

60 An excellent collection containing articles by both supporters of a deity as part of the universe's explanation and advocates of a godless universe can be found in *Modern Cosmology and Philosophy*, ed. John Leslie.

61 This association is noted by Dawkins, *The Blind Watchmaker*, 317.

62 Leslie, *Universes*, 29. Hawking's estimate appears in his article, "The Anistrophy of the Universe at Large Times," in *Confrontation of Cosmological Theories with Observational Data*, ed. M.S. Longair (Dordrecht: Reidel, 1974), 283–286; 285.

What is of interest for our purpose of appreciating the idealist use and contributions to the discoveries of science is the conclusion that Leslie seeks to draw: the scientific evidence of an ordered world that is life-permitting and exists in spite of highly improbable circumstances requires explanation beyond that of chance. Chance or probability theory can establish the proposition that sooner or later conditions that were life-sustaining would have resulted in the universe.[63] For there to be life as a fact about the universe, however, we require an observer of life, or a living being who can bring that particular scientific proposition into existence. This proposition that there is life in the universe is not like an eternal abstract truth which holds beyond the requirement of living beings to sustain it. The observer or documenter of life in the universe is at a place and in time. This fact of being situated will determine the nature of the observations that the observer has. Quantum mechanics might suggest that the likelihood of an observer coming into existence in a very large universe or universes should not surprise us. A commitment to a radically indeterminate universe might yield the same conclusion.[64] Something that is highly improbable but not impossible need be no more surprising than something that happens by chance, provided one believes the universe to be radically indeterminate. If we take chance seriously, then we simply have to accommodate occurrences without fuss. If we take quantum theory seriously, we also have accepted the fact that the improbable could occur. The circumstances for life developing were highly improbable, but given that the fine-tuning of elements that permitted life did occur, our existence can be understood as an unexpected event, as well as the highly probable one. From either perspective, quantum theory supports our coming into existence. The other possible explanation is that the world of living, observing beings came to exist by pure chance. The philosopher/cosmologist has now two possible explanations for our existence in the universe at this place and time. They are not necessarily in contradiction but each does have implications for our understanding of our "situatedness" that are remarkably different.

11.3 *Randomness, Probability, and Observers*

One major implication of choosing a theory which explains our existence concerns the expected demise of the human race. Leslie explores this question in his book, *The End of the World*. It is not my intent to fully review the book's

63 Leslie, *Universes*, 24, cf. *The End of the World*, 16–19.

64 Leslie, *The End of the World*, 18–20.

research but to relate Leslie's observations to his position on the cause of our existence.

The random/unpredictable theory might suggest that some random unpredictable event will hasten our demise, but the weight of observer intelligence gathered thus far greatly weakens that claim. Even a hurtling meteorite would be detected and, though warning might be ineffective, we would have been able to explain the event in which case it would not count as chance. Thus, the demise of the human race seems to be anticipated by probability theory, given the knowledge we have about the universe, and our present conduct on the planet.

Does such a probable event matter? Can probability theory or fine-tuning theory direct us to the answer to that question? Can either theory even shed light on our abilities as observers? Such abilities include participating in the universe's orders, coordinating our observations with knowledge that we develop, and theorizing about the laws of the universe. These are features of being the observational creatures that we are. Why, one can ask, did life not remain unconscious or non-reflective? How would we describe that feature of being an observing species within quantum theory?

Whether we discover the order or impose the order of nature's laws, our particular situatedness is not explained by those orders nor is the fact of our capacity to reflect on our time and place in the universe. Such time and place is not absolute, as even physics continues to debate time and space in an infinite cosmos, and so our speculations about answers to the above questions must seek further concepts than those familiar to us from the scientific domain.

11.4 *Is Explanation of Existence Necessary?*

Inasmuch as science seems to work within the laws of the universe and these laws seem to govern both our means of understanding them as well as the matters to which they apply, we could conclude that the *mere fact of our existence* neither hampers nor advances scientific pursuits. In short, one could not attempt any answers to, nor draw any conclusion about, our happening to exist at this point in space and time and just get on with research. Leslie does acknowledge this possibility:

> Many scientists would say they were not in the business of explaining why there was any universe, why its initial state was such-and-such, and why it obeyed any physical laws at all, let alone ones specifying these or those force strengths.[65]

65 *Universes*, 27.

However, Leslie thinks this is an inadequate response to the kinds of knowledge inquiries that being an observer has aroused: wonder, concerns (about the future of the human race), values that restrain our scientific pursuits, calculations about our past and future, and speculations about other life forms in the universe. These types of inquiries are not amenable to the methods of scientific research as we know them. Why inquiries like this arouse our curiosity is a puzzle in itself.

An easy answer is to say that there must be one law to which all others in the universe have some unifying connection—one which determines not just our calculative capacities but our reflective capacities—and that would be the law (or Mind) of God. Leslie does not conclude that this answer is without alternatives. Instead, he asserts that available alternatives (Darwinism, chance, probability) can be misleading without the further richness of his creative ethically-requiredness principle. For example, to regard our reflective capacities as a product of chance is to devalue their meaningfulness to observers, meaningfulness which is not explained by any documentations of fine tuning, design, or probability theory.[66]

The available answers providing an explanation for our existence give us options. We can be indifferent to life, its future, its past, and its possible extinction. But to adopt a stance of indifference would be analogous to physics rolling back theories of the universe in reverse.[67] Chaos would swamp and obliterate everything that we know at a rate far faster than the rate we associate with the expansion theory of the universe, a rate that is slow and gradual. This rate seems to have application far beyond our chance applications of it. As parts of the universe, our move towards chaos through indifference would undoubtedly hasten our extinction, which proposition should have no effect on us now. But, of course, evidence is increasing that humans do have a vested interest in staying around for a while.

12 Design without a Designer

Leslie is inclined towards design arguments about existence—our place in the scheme of things and all that science may try to ignore—though his interest is tempered by his thorough familiarity with the workings of

66 Much of the rest of Chapter 2 in *Universes* compiles evidence that the conditions for life were astonishingly exact and finely nuanced.

67 *Universes*, 27–28.

quantum mechanics and other scientific theories. (De Koninck relied on the actual workings of science to make his case for the unifying role of values.) Leslie sees no reason to impose a Creator as a requirement of the possibility of scientific endeavours. He realizes that the simple observation that our existence as living beings with the capacity to observe facts about life-supporting universes stands as a confirmation of the claim that the universe could develop life through chance developments or through probability theory. We have documented these developments in spite of the high improbability of life-supporting conditions. Moreover, the possibility of many universes makes the probability of one being life-supporting less surprising.[68] This observation does not negate nor make logically incomprehensible the other possible explanation, however, i.e., that God designed, selected, or chose to establish many universes, including one which was life-supporting. "How could we ever decide firmly whether God did the selecting, or whether it was done by the fact that being alive is a prerequisite of making observations?"[69]

Leslie narrows the options that explain the world as we know it being present to our consciousnesses: a universal mind selecting the life-supporting conditions (maybe called God) or the postulate that multiple universes exist which would increase the chance of one like ours existing. Leslie points out that neither of these hypotheses can be proven to be true.

Neither can we establish that one may be incoherent and false, nor that both have little explanatory power. The multiple-universes postulate has the theory of probability to lean on while the universal-planning-Mind postulate has evidence from fine-tuning, the fact of observational agendas (which cannot be explained by science) and *the fact of goodness* as a working tool in observational explanations.

13 The Good World

Leslie weighs arguments for and against these two positions with studied fairness.[70] He wants the fact of *goodness* to strongly weigh against likelihood or chance arguments as being convincing. We shall arrive at his position through an inquiry that complements but does not repeat his arguments.

68 *Universes*, 153–156.

69 *Universes*, 149.

70 *Universes*, 161–166.

13.1 *Choosing a Theory of Explanation*

Consider the following question: faced with two possible explanations of an event (a world with living observers), each of which explanation may raise objections and neither of which seems completely adequate nor, indeed, inclined to enable further research on its own, how does a scientist choose? Usually theories are valued for their universal applicability and capacity to predict results. Where the multiple-universes theory falls short is in accounting for the fact of goodness in our particular world as we know it. Such goodness is not universal (we do not live in world without evil), so this fact of goodness is in accord with the approximations and probabilities of quantum theory; goodness is not absolute. Furthermore, this fact of goodness is meaningful only to a world of observing living beings. The world of living, observing, reflecting beings, we discovered, was a highly exceptional situation. Had such a phenomenon (our exceptional world) occurred in a lab, its coming into existence would have demonstrated that quantum theory signals approximation, for now the exception would have suddenly happened. The exception (the unexpected world of living beings) would likely have inspired further research to see if present theories needed revising.

This highly unlikely but probably possible world of living beings now exhibits, documented by its own observational "situatedness," behaviours which evoke a need for language and concepts that cannot be empirically observed. This is accompanied by the behaviours and decisions that are not easily addressed by the scientific quantum theories or the random chance theories. If the language and concepts for goodness seem to have evolved to address behaviours that don't accord with Spencerian survival theories, but now try to accommodate aspects of living beings which seem to be not accounted for by survival but move towards Darwinian theory (Darwin acknowledged that morality was an aspect of human development which evolutionary theory didn't address.),[71] then why has that feature, or language concept (goodness), become part of our observational capacities? Furthermore, does our discussion of goodness affect our observations? Does it guide our selective criteria in scientific pursuits? Does it determine research, what shape knowledge takes, and in what manner it is recorded? Why would there be a world with an idea, or ideal, or value, or feature of consciousness, or however one wants to describe this fact of goodness, associated with human beings?

71 See Louis Pojman, *Who Are We? Theories of Human Nature* (New York: Oxford University Press, 2006), 212–213.

The laboratory analogy of quantum fluctuations leaves us without an explanation for goodness; the observation that it was sheer happenstance that goodness appeared robs it of its use, for it might just as randomly disappear and signal the rapid decline of our world. In fact, we can observe that we struggle to maintain what rudiments of order we have, an order which promises at times only approximations of goodness but which we defend against the miserable, pain-filled world that John Leslie postulated at the beginning of this inquiry, a world that ought not to exist.

The most comprehensive explanation for the scientist for our existence in the universe will involve some explanation for goodness as part of the order or design that we experience and struggle to promote. That experience is available only to the conscious observer (and so, in some design platforms, it extends beyond the human world of living things). Such order or design for living the good life requires consciousness on some scale—perhaps a single mind, perhaps multiple universes of infinite minds, but a fact of consciousness as part of the human experience of goodness, evil, and inquiry cannot be denied.

Goodness then pushes us beyond the scientific theorist to the speculative philosopher whose job is to promote understanding of all aspects of our existence; science continues to deny its need to involve itself in explanations involving goodness. Now we can return to Leslie's principle of creative ethical requiredness. He agrees his theory is a postulate:

> An ethical ground for the existence of some possible world would itself lead to that world's actual existence, or else it would not. Neither of these alternatives would be forced by mere logic, yet neither would be a case of sheer happenstance ... Plato's theory of the world's creation could therefore perhaps be right.[72]

Leslie also agrees that his concept of ethical requiredness is one for which the evidence of science will not be helpful. He reasons as follows: a perfectly evil world ought not to exist. That leaves us, the readers, with two possible propositions: 1) that no world is a preferable situation to an evil world and 2) that a good world (a world that comprehends and sustains goodness) is preferable both to an evil world and no world. Now we can choose to escalate our lives towards no world, choosing to believe that nothing is preferable to something, but we would have no ground for doing so except to argue that the world as we know it is perfectly evil, in which case what would be rational (understood

72 ID, 36.

by persons not blinded by theological beliefs that admit no debate) would be to decimate the world. The faculty of rationality usually includes a capacity for deliberative acts, for weighing relevant reasons in making choices. Arguing that the world is perfectly evil admits of no comparator (Does evil mean anything without the idea of good?), and so deciding to decimate the world is not really making a decision unless the maker of the decision has some idea of a better condition of existence. No existence is not a better existence. Thus, world decimation is not a decisive choice but an act driven by some irrational force or fear. Accordingly, if choice is to be meaningful, that is, a deliberative act, we must consider options which reflect "order-recognizing" and "valuing" beings.

A third option can now be considered, i.e., that some kind of world is preferable to no world and that some world can exist that is preferable to an evil world. This is not to make the case that, by definition, some kind of world must be good. It could be good and bad since these are not conditions of existence but properties. Can Leslie make the move that if no evil world should exist, therefore some good world should exist? Not with any logical assurance, for a proposition that states no evil worlds should exist does not entail anything. Thus he can claim all good worlds should exist as easily as claim all evil worlds should not exist. The question is, if a good world should exist though it may not be a perfectly good world, how does the selected-with-goodness world of God or the unlikely-but-probable-with-goodness world of observational beings come to exist? If *nihilo ex nihilo*, then the idea of a good world or goodness in the universe must exist and thus must have come into existence for our minds as conscious beings to contemplate and pursue. If observational living beings can grasp abstract truths (e.g. that a world of horrible pain and suffering should not exist), then the truth of goodness must also be available to us in some existential way. The ethical requiredness of there being good for us to experience, so that we can also grasp the truth of the proposition that pain-filled, evil worlds should not exist, is its ground of being as we know it. Creative ethical requiredness brings into existence being-for-us. This ethical requiredness of good is an eternal truth with the power to create the circumstances of its truth, circumstances which are relevant to the selected observational world of human beings, and quite possibly other living beings who exhibit value behaviours as well.[73]

73 In ID, Leslie does subscribe to Platonic realities (abstract and eternal truths), one of which is goodness (see Chapter 2, 16–35). The fact of evil can also be contemplated as a Platonic truth, one dialectically related to good. Leslie would seem to urge that our proclivity as humans is toward good, but neither empiricism nor logic can establish this claim with any assurance.

It is not difficult now for Leslie to argue that if goodness is required for human beings to observe and decide, then, since goodness is not the kind of universal component that science can investigate, its existence must be associated with a conscious mind or minds. Whether the universe has one mind or infinite minds is not the main point of Leslie's writings. The point is that without infinite value-generating mind or minds for us, there would be nothing at all.[74]

Of course the ruthless proponents of a value-free universe, the goodness of which is simply calculated by rational minds that are self-interested, can and will continue to exist in Leslie's world, but their defeat of his proposition will have to lie not in mounting an attack on some person with god-like status, but in demonstrating that his principle of creative ethical requiredness has contradictions that make it incommensurable with other observations about human nature.

The idealism of Leslie is not directed towards the actual operations of science in the lab (inquiries into legitimacy of causal theory or whether truth claims can be made from inductions), but towards the use one can make of science in establishing fine-tuning theories, probability theory associated with the anthropic principle, and design arguments that rely on scientific data, not on presupposed commitments to God. If one chooses God as a convenient explanation for existence, one has opted for an available choice. Leslie's inquiries into value and existence leave open interpretive avenues for conscious beings, but the options available are grounded in our conscious minds. whether through beliefs in a large-scale designer, or in the value of small acts of kindness that may not ward off all evil, but do confirm the presence of goodness in the world.

References

Armour, Leslie and Trott, Elizabeth, *The Faces of Reason*, Waterloo: Wilfred Laurier University Press, 1982.

Bronowski, J. , *Science and Human Values*, New York: Harper and Row, 1956.

Collingwood, R.G., *The Idea of Nature*, Oxford: Clarendon Press, 1945.

Crossley, David, book review of William Sweet's *Idealism, Metaphysics and Community* (2001), *Dialogue*, Vol. XLII, No. 3 (Summer 2003), 607–609.

Dawkins, Richard, *The Blind Watchmaker*, London: Penguin Books, 1986.

Dawkins, Richard, *The God Delusion*, Boston: Houghton Mifflin, 2006.

Dawkins, Richard, *The Selfish Gene*, Oxford: Oxford University Press, 1976.

74 ID, 53–55.

De Koninck, Charles, *The Hollow Universe*, London: Oxford University Press, 1960.
Dupré, John, *The Disorder of Things*, Cambridge, MA.: Harvard University Press, 1993.
Hoyle, Fred, *Encounter with the Future*, New York: Simon and Schuster, 1965.
Jeans, Sir James, "A mystic universe," *The New York Times*, January 28, 1928, 14.
Jeans, Sir James, *Physics and Philosophy*, Mineola, NY: Dover Publications, 1981; first published Cambridge: The University Press, 1942.
Jeans, Sir James, *The Mysterious Universe*, London: Cambridge University Press, 1930.
Kuhn, Thomas, *The Structure of Scientific Revolutions*, Chicago: University of Chicago Press, 1962.
Leslie, John, *Immortality Defended*, Oxford: Blackwell, 2007.
Leslie, John, ed., *Modern Cosmology and Philosophy*, Amherst, NY: Prometheus Books, 1998.
Leslie, John, *The End of the World*, New York: Routledge, 1996.
Leslie, John, *Universes*, New York: Routledge, 1989.
Leslie, John, *Value and Existence*, Oxford: Blackwell Press, 1979.
Lewontin, Richard, *Biology as Ideology*, Concord, Ontario: Anansi, 1991.
McInery, Ralph, ed. and trans., *The Writings of Charles De Koninck*, Vol. 1, Notre Dame: University of Notre Dame Press, 2008.
McMullan, Ernan, "Is Philosophy Relevant to Cosmology?", *Modern Cosmology and Philosophy*, ed. John Leslie, Amherst, NY: Prometheus Books, 36–56.
Murray, John Clark, "The Dualistic Conception of Nature," *The Monist*, Vol. 6: No. 3 (April), 1896, 382–395.
Pojman, Louis, *Who Are We? Theories of Human Nature*, New York: Oxford University Press, 2006.
Powers, Jonathan, "The Absolute World," in *Philosophy and the New Physics*, Metheun and Co. Ltd., 1982, 89–93.
Schiller, Bill, "A crisis rooted in two Chinas," *Toronto Star*, September 21, 2008, A4.
Smith, C.S., "Structural Hierarchy in Science, Art and History," *On Aesthetics in Science*, ed. Judith Weschsler, Cambridge, MA.: MIT Press, 1978, 9–53.
Sweet, William, ed., *Idealism, Metaphysics, and Community*, Burlington, Vermont: Ashgate Publishing, 2001.
Trott, Elizabeth, "John Clark Murray: Religion, Science, and the Unity of Nature," *Religion and Science in Early Canada*, ed. Douglas Rabb, Kingston, Ontario: Ronald P. Frye and Co., 1988, 43–74.
Whitehead, A.N., *Process and Reality: An Essay in Cosmology*, London: Macmillan, 1929.
Wilson, E.O., *Sociobiology*, abridged edition, Cambridge: Harvard University Press, 1980.

CHAPTER 7

Idealism and Naturalism: a Really Old Story Re-Told with Variations

Fred Wilson

The topic to be addressed is idealist responses to materialist philosophies of science. Materialism, taken literally, as meaning that nothing exists save matter, is like metaphysical behaviourism in denying the reality of things mental, and is therefore one of those philosophies that, as C.D. Broad once said, can be dismissed as silly. I assume that an idealist is like any other serious philosopher, and that his or her response to materialism is to dismiss it as silly. That does not leave much to discuss. In order to have something to talk about, I am therefore going to assume that by "materialist" is meant something like 'naturalist,' and, indeed, that the latter can be understood broadly enough to include such philosophies as empiricism, logical atomism, early (I emphasize *early*) logical positivism, and the new realism and critical realism of the early 20th century. What these have in common and what differentiates them from idealists I shall explain in due course.

Bosanquet devoted an essay to drawing the distinction between philosophy and science[1]—that is, science in the sense of the cognitive discipline or set of disciplines which seeks to understand natural objects like stones or planets. He concluded that philosophy deals with the totality of things[2] while science deals with only some part or parts of this totality.[3] Furthermore, philosophy is the expression of the person as a whole—a feature that it shares with poetry and religion—while science is the expression of more restricted cognitive capacities.[4] Finally, ethics finds its place as a rigorous discipline in philosophy,

1 B. Bosanquet, "Science and Philosophy," in his *Science and Philosophy* (London: Allen and Unwin, 1927), 15–33.

2 "The essence of philosophy lies in the connected vision of the totality ..." He proceeds to observe that philosophy, with this vision, maintains "in every point the subordination of every element and every factor to every other element and factor as conditioned by the totality" ("Science and Philosophy," 25–26).

3 Natural science "restricts itself to a more limited object" than that of philosophy ("Science and Philosophy," 16).

4 Philosophy, with its vision of the totality of things, is "a product of the whole man"—or, one assumes, the whole woman, while natural science, restricting itself to a more specialized object, "springs from a more limited capacity" ("Science and Philosophy," 16).

 | DOI:10.1163/9789004415270_009

whereas it has no place in science, that is, natural science.[5] Philosophy is a matter of thinking, that is, armchair reflection on the vision of the totality,[6] which puts science in a lower, lesser place. Needless to say, this is the attitude behind much idealism and the kind of narcissism that is as much enjoyed by some—the idealists—as it is found distasteful and scorned by others—the naturalists—or, what amounts to the same, the empiricists.[7]

Be all that as it may, ignoring the narcissism that often motivates idealism, and ignoring too the egotism that claims to conquer a world beyond our ordinary world of sense experience and inner awareness, it remains true that the idealists did offer a variety of effective criticisms of science, understood in empiricist terms. They raised in particular the issue of how one was to account for *structure*. The idealists recognized in a way that, until recently at least, naturalists or empiricists did not, that one's ontology needed to include some account of *relations* that could provide the structure or structures which pervade the world. Furthermore, this account of relations had to provide a sense to the claim of Bosanquet (and others) that the task of philosophy requires us to try to come to know not just plain old everything, but everything as a *totality*. One can ask, of course, what is it that makes a totality of everything. It turns out that empiricists are highly critical of this notion of a totality and, indeed, reject this notion, together with the doctrine of relations on which it is based. This leaves them with the problem of structure unsolved and that, in turn, invites a further response by the idealists.

---

5 "Its [that is, philosophy's] object ... is the universe, with all its activities and values, among which the theories of exact sciences with their object form only a certain proportion ... Besides the theory of what concerns the sciences, a philosophy which is to have any claim to deal with the universe is bound prima facie to undertake a theoretical consideration at least of beauty and of goodness" ("Science and Philosophy," 20).

6 "To protest against mere thinking is meaningless. Nothing but thinking can give knowledge; and all thinking without any exception is the endeavour of certain experiences to affirm themselves as a whole" ("Science and Philosophy," 25). What exactly this means is not made clear but this much is evident: there is a teleology such that various experiences are moved as a final cause by the form of knowledge of themselves as a whole, knowledge which is already within that experience, that is, implicitly there to be dragged out by the form as a final cause. This clearly is not what psychology as a natural science would say about these things. For it, we can say at least that any understanding of learning is not to be found in final causes, i.e., causes of a sort which have no place in any science pretending to be natural in the way in which the physics of stones is a natural science.

7 Compare the remark of F.H. Bradley, that "... Common Sense [of which science is but the long arm], taken (as too often it may be seen) at its worst, is in its essence a one-sidedness, which we must not be afraid to mark as stupid or even, perhaps, to denounce as immoral" ("Relations" in *Collected Essays*, 2 vols. Oxford: Oxford University Press, 1935, 2:640).

There are, then, a variety of positions and a variety of criticisms characteristic of what might be called the "naturalist-idealist" dialogue. This dialogue features a variety of positions and a variety of criticisms but many of them are variations on themes to be found already early in this history of philosophy, and in Socrates' engagement with Anaxagoras in particular. We can therefore usefully begin with that interchange.

1 Socrates and Anaxagoras

Empiricism argues that science deals with *a* world, which is *the* world as it is given to us in sensible experience and inner awareness, and that understanding this world consists in locating regularities in it. In thinking about this way of thinking about the world, we can do no better than beginning at the beginning with Anaxagoras.[8] Anaxagoras had a sort of naturalism in terms of which he hoped to explain what happens in the world.[9] The world begins as a mixture of everything with everything; every part contains within itself an infinity of parts. These parts are of many kinds—bright and dark, for example. Whatever else one can say, it is clear that the parts have properties of sensible things: explanation proceeds in terms of these parts, and it is naturalistic in the sense that things in this world are all of sensible sorts.[10] There are no kinds in Anaxagoras that transcend and lie outside the natural world but in terms of which we are expected to explain causally what happens in the world of ordinary experience. Of course, some of the parts—no doubt most of them—are too small for the eye to notice or the fingers to touch, and others, no doubt, too far away, but these entities are *like* the things we sense and are therefore parts of the world we sensibly experience.

---

8 Plato, *Phaedo*, in *Plato, Five Dialogues*, trans. G.M.A. Grube (Indianapolis, IN: Hackett, 1981), 97c 1ff.

9 See G.S. Kirk and J.E. Raven, *The Presocratic Philosophers: A Critical History with a Selection of Texts* (Cambridge: Cambridge University Press, 1957), Chapter XV, 362–394.

10 Anaxagoras writes (according to the fragments): "But before these things were separated off, while all things were together, there was not even any colour plain; for the mixture of all things prevented it, of the moist and the dry, the hot and the cold, the bright and the dark, and of much earth in the mixture ..." (*The Presocratic Philosophers*, fragment 496; 368). Or again, "... this [initial] rotation caused the separating off. And the dense is separated from the rare, the hot from the cold, the bright from the dark and the dry from the moist" (*The Presocratic Philosophers*, fragment 503; 373). So we can get flesh and bones, but the presence of the "dark" allows for shadows and the "bright" allows for rainbows. The world is as much phenomenal as it is corporeal, though this more recent terminology is no doubt misleading.

In answer to Parmenides, the world consists of a real diversity of parts, each of which is in itself unchanging. What changes are their relations; the parts move relative to one another. The world as we experience it is a world of sensible parts changing in discernible patterns: plants grow, animals are born and grow and die, and our thoughts are fleeting. Anaxagoras proposes that the variety of sensible parts receives an initial motion. Once these parts are set in motion, they begin to accumulate to form bodies as we know them, including our own bodies. In fact, everything contains infinitely many parts. Among this infinity are parts gathered together as bits of bone, and an infinitely of such bits. As well, infinitely many groups are bits of flesh. Moreover, every bit of flesh contains not just bits of flesh but also bits of bones and bits of everything else. After an initial starting rotation, the parts started to come together. The initial movement was caused by mind or *nous*. Mind was itself a sort of substance. To be sure, it is a special sort of substance: it is fiery. However, it is of a sort of which there is no question that we can become aware of it. It too is part of the natural world of things we can know in our ordinary experience of things.

Now, ordinary things have parts of the initial mixture concentrated or gathered together in various special ways: ordinary things arise out of the original mixture through the motions of its parts. Originally at rest, the mixture is put into motion through a push from mind or *nous* which gives rise to an initial rotary motion. Mind or *nous* is concentrated before its initial push,[11] but then divides itself among the individual things that began to emerge.[12] After the initial push, the world emerges through changes which seem to have been more or less mechanical, their patterns determined by the natures of the bits and of the initial jolt. Mind now plays only a minor role, organizing and directing some among the individual complex substantial things that come to be.[13] Still, mind itself is, as we said, a natural thing.[14] On Anaxagoras' account, the person

11 "And the things that are mingled and separated and divided off, are all known by Mind. And all things that were to be, all things that were but are not now, all things that are now or shall be, Mind arranged them all ..." (*The Presocratic Philosophers*, fragment 503; 373).

12 Thus Anaxagoras writes (so we are told): "... Mind, which ever is, is assuredly even now where everything else is too, in the surrounding mass and in the things that have been either aggregated or separated" (*The Presocratic Philosophers*, fragment 506; 374).

13 This is how Kirk and Raven (375) read this fragment: "And when Mind initiated motion, from all that was moved separation began and as much as Mind moved was all divided off; and as things moved and were divided off, the rotation greatly increased the process of dividing" (*The Presocratic Philosophers*, fragment 504, 373).

14 It is perhaps misleading to suggest, as Kirk and Raven do, that "mind, like matter, is corporeal ..." (*The Presocratic Philosophers*, 375). The term 'corporeal' suggests that mind is something more bodily than simply consciousness. Their earlier remark that "Anaxagoras ... is striving ... to imagine and describe a truly incorporeal entity" (374) seems to be

seems to have been a body composed of skin, bones, and sinews—a complex, substantial whole structured and moved by mind. This mind of the individual person organizes those things which are persons so that they survive as wholes for as long as they can. Thus construed, it appears, persons are not distinctively different from dogs, and perhaps not even from oysters.

Socrates objected to Anaxagoras' account and Plato gives Socrates' argument. The scene is set in the neatest of Plato's dialogues, the *Phaedo*. Socrates is sitting in prison. According to Plato's earlier dialogue, the *Crito*, his friends offered to arrange things so he could escape to Thebes. Socrates, however, refused. Following the demands of justice, he determined that it would be wrong to escape his native city and run off to Thebes. Instead, he is determined to remain in his cell and drink the hemlock. So we have one event—Socrates sitting in his cell. There are two events that could follow that event, namely, running off to Thebes and remaining in Athens and drinking the hemlock. There is nothing in the first event, Socrates sitting there, that indicates or determines that it will be followed by one rather than the other. But in fact it is the drinking of the hemlock that follows. Why? Socrates objects that Anaxagoras cannot answer this question, and his naturalistic metaphysics is therefore inadequate.[15]

If Anaxagoras were correct, then, like a dog seeking only bodily well-being and survival, he would have run off to Thebes. But he is not doing that, so Anaxagoras must be wrong. The difficulty is the role of mind or *nous*. Socrates has already established that mind or *nous*, the soul, moves the body. Thusfar, he agrees with Anaxagoras. But what is needed to explain Socrates' sitting in jail being followed by his drinking the hemlock is the fact that he is moved by a *vision of the just*. This leads him into his alternative explanation. Earlier he has established two things. *One* is that anything in the world of sensible experience that moves or changes, and one's body in particular, is moved by a soul. This active entity is simple—an indivisible pure activity— and is therefore unlike the things of the world of sense which are divisible and separable into parts. For example, Socrates' sitting there in his cell is separable from the events that will succeed it. The *second* thing that he has established is the existence of the forms as perfect exemplars of the things we observe in sense. Having secured agreement that all equals in our ordinary experience are imperfectly equal, he then argues that in order to judge that something is not-F one needs to have the concept of F—in order to know that something is not-red one must know what

more just. The point is that these things which are incorporeal ("fiery") can *causally affect* and *causally interact with* the things of flesh and bones that are more fairly described as material and are in that respect *like* things *corporeal*.

15 *Phaedo*, 97c and following.

it is to be red. So, to judge something to be imperfectly equal (= not-perfectly equal), we must have the concept of perfect equality. The form of perfect equality must therefore exist, not in the world of ordinary experience, but in another world. What holds for equality holds for other things: perfectly straight lines, perfect triangles, and, of course, perfect moral virtue or justice.

Socrates uses these metaphysical principles to give an explanation of the sequence of events that Anaxagoras is unable to give. Socrates sitting in his cell is followed by his drinking the hemlock because the soul—his soul—is striving to imitate in its outward sensible appearances the form of perfect human justice. This is the ideal good that moves his soul in one direction and not another; the vision of which moves him to do what is right or just and not what the doggie would do.[16] Skin and sinews and bones may explain, as Anaxagoras suggests, what doggies do, but, by themselves at least, they cannot explain what Socrates is doing. However, Socrates being moved by his soul with its vision of the good—the form of the good—*does* explain what Socrates is doing, i.e., why the events of Socrates' life hang together in a sequence rationally intelligible in terms of the form of the good as the real reason for these events happening as they do.

But there is a problem:[17] why does the soul of Socrates strive after the form of human justice rather than the form of doggie justice or oyster justice? Separation is to be explained in terms of a connection being affected by a soul imitating a form. So the soul striving after human justice is explained by the soul striving after the super form of justice, from which it is separate, in which case its striving after that needs another form to strive after—that of the super-duper form of justice—and so on without end. Since there is no end, nothing ever gets explained. This is the Third Man Argument of Aristotle.

To solve the problem in a way that saves the basic form of explanation, Aristotle ends the separation of the forms from the souls. Each soul, or as Aristotle comes to say, each "substance," has an intrinsic and inseparable relationship to its own form or essence. Thus, patterns in the world of ordinary experience which are separate from one another, or at least apparently separable so far as

16 Here we have what has been called, not unjustly, the "natural philosophy" of the *Phaedo*. For more on this, see F. Wilson, *Socrates, Lucretius, Camus* (Lewiston, N.Y.: The Edwin Mellen Press, 2001), Chapter Three. The basic argument that what we have here is the pattern developed for the explanation of ordinary events. For further reading, see G. Vlastos, "Reasons and Causes in the *Phaedo*," *Philosophical Review*, 78 (1960), 291–325, and R.G. Turnbull, "Aristotle's Debt to the 'Natural Philosophy' of the *Phaedo*," *Philosophical Quarterly*, 8 (1963), 131–143.

17 See *Socrates, Lucretius, Camus*, 121ff.

our senses are concerned, are not really separate. Rather, they are connected by the activity of the substance in a direction determined by its form or essence. The sequences of the separable events of sense, of ordinary experience, are explained by virtue of their being a tie between them such that one event necessitates the succeeding event through the activity of the soul in which those events occur where the observable pattern reflects the necessary structure or form or essence of the active substance. The sequences of ordinary events in the world of sense are explained in terms of entities outside this world. The sequence is really one in which events are not after all separable but connected by the striving of the active substance *qua* having a certain form that determines the direction of the striving. Moreover, the explanation is in terms of *best*: the form of human is the form of human justice, and the way things are is explained by the soul aiming at the best. *What is* is explained in terms of *what ought to be.*[18]

The pattern for explanation is clear: *to explain is to unify*. The changing events in the world we know by sense experience or inner awareness are in themselves distinct and separable; there is nothing about one event that requires or necessitates the event that succeeds it. To explain what happens in the world of ordinary experience is to unify the events of that world. On Socrates' scheme, to explain is *to unify by means of entities*. One of these entities is an *active soul*. The soul *unifies by so acting that one thing or event becomes another that succeeds it*. But the activity has a direction; it is *for an end*. This end is determined by a *form*. *Neither the form nor the soul is an entity given to us in our sensible experience of the world*; they are *transcendent*. *The observed events are explained in terms of the soul striving after a form so as to imitate in its outward sensible appearances that form*. In other words, it so strives that the sensible things and events *in* the everyday world are in appearance like the form that *is outside* or transcends this world. The soul is active but the form as it were pulls the soul towards it. To use a metaphor Plato employs elsewhere, the soul is a sort of lover and the form that which is loved, drawing the lover towards it. *The explanation of things in the ordinary world is teleological with the form after which the soul strives being a final cause*.

Now contrast this to the world of Anaxagoras. In that world, there is no unity, only separable parts. There are sequences of events, and even regular patterns; after all, stuffs are stuffs and each kind behaves in its own ways and combines and separates in its own ways. So there is regularity and patterns.

18 Cf. F. Wilson, *The Logic and Methodology of Science in Early Modern Thought: Seven Studies* (Toronto: University of Toronto Press, 1999), Study One, and F. Wilson, "Science and Religion: No Irenics Here," *Metaphysica*, 7 (2006), 159–170.

However, these patterns are contingent. The events that exhibit these patterns do not do so with any sort of necessity; they are separable and without any *real* or *necessary* connection to one another. Even mind or *nous*, while of a unique sort of substance and not mixed with the other stuffs, is not itself a unity. It too consists of parts. What is not a unity cannot unify and therefore cannot explain *in conformity to the Socratic pattern that to explain is to unify*. Mind, moreover, is of *this world*—the world that exists in space and time and the world we know by sense. Mind is not something that is outside this world and certainly not something that grasps the eternal forms and therefore partakes of that eternal being. There is therefore no explaining of what is in terms of the best; there is no teleology that determines the way things are in terms of striving for what ought to be. Socrates and Anaxagoras agree about all the events in the sensible world. In the world of Anaxagoras, however, there are no necessary connections as there are in the world of Socrates Thus, from the viewpoint of Socrates, everything in the world of Anaxagoras is unexplained; with no necessary connections anywhere, everything is quite inexplicable.

What we have in the world of ordinary events, accordingly, is that these events are explained by being instances of patterns: all humans are animal, Socrates is human, hence, Socrates is animal. On this, everyone agrees, Anaxagoras, Socrates, Plato, Aristotle, and everyone since then, including Anaxagoras' naturalist successors! In fact, if we ignore the initial shove, *this is all there is to explanation* on Anaxagoras' account. Naturalists, certainly those who can be called empiricists, agree with Anaxagoras on this point: this is indeed all there is to explanation. In a way, then, it meets the proposal that to explain is to unify. Clearly, on this proposal, there *is* a unity. However, it is a unity provided *not by an entity but by a pattern*. In particular, one should note how this fits the naturalist standard of restricting one's ontology to the world of sense experience. The pattern which explains is a pattern of entities *in* the ordinary world, unlike the unification provided by the Socratic pattern, which is in terms of entities *outside* the ordinary world. For the Platonist and the Aristotelian, however, the pattern is not enough. There is not yet any real unity—no unity provided by an entity that provides a necessary connection. For the Platonist and the Aristotelian successors of Socrates, the major premise of the argument form, recording the pattern, "All human is animal," must do more than merely record a pattern observed in ordinary experience if there is to be an explanation. It must, of course, represent that but further, it must represent a *real connection*—something *necessary* and not merely contingent. It must represent a *necessary structure* that so binds the characters that H in its very being is inseparable from A in its very being. H and A must be so intimately connected that neither the contrary, stating their incompatibility, "No man is animal," nor the contradictory, "Some man is not animal," is conceivable. The connection of H and A must be a timeless structure that rules these out as possibilities.

As Aristotle worked these things out,[19] the required structure is that of a species as part of a genus—the species man within the genus animal—with the connection between affected by the specific difference, in this case the specific difference of being rational. This necessary structure is *displayed* in what Aristotle referred to as a *scientific syllogism*: "All rational is animal, all human is rational, all human is animal." This syllogism displays the logical and ontological structure of the species human. This form is given equivalently in the *real definition* of the species human: "Human is rational animal." This real definition is a *necessary truth, a timeless truth about the logical and ontological structures of the forms of substances.*

More generally, and putting it in terms of a more recent logic, we have characteristics F and G, and the observable pattern, "All F are G" or "(x)(Fx then Gx)." We have this connecting the two events Fa and Ga: (x)(Fx then Gx), Fa, so Ga. According to the experience of sense, being F and being G are separable events. The pattern represented by "All F are G" is merely a regularity, a pattern amongst events that is merely contingent. There is no deeper or stronger connection, and therefore no genuine connection. In order that there be a genuine explanation, there must be a *necessary connection.* We are aware in ordinary experience of things which are F and things which are G but the necessary connection that provides the reason—the real reason—why an F we observe is G is not given in sense. F being necessarily connected to G, their real unity is not apparent to sense and, indeed, transcends ordinary experience. The world of ordinary experience is a world of separable events. For Anaxagoras, this is all there is; he does not have any such necessary connections which, since they are not given in sense, must be connections made by transcendent entities. So in Anaxagoras' world, nothing is ever explained. But that is an unreasonable view of the world. We ought, therefore, as reasonable persons, to accept Socrates' view of the world rather than that of Anaxagoras. This, in effect, is Socrates' argument against Anaxagoras, or part of that argument anyway.

The world as we experience it consists of separable parts: parts that are logically and ontologically separable, parts that are conjoined but merely conjoined. If Anaxagoras is correct, then there is nothing more, only these parts and the patterns that describe how parts of just these sorts are associated with parts of just these other sorts. If Socrates is correct, then this account of the

19 See F. Wilson, *The Logic and Methodology of Science in Early Modern Thought*, Studies One and Two.

world misses something, i.e., the *tie* that provides the *reason* why these parts are associated with those parts. Why is there this pattern rather than that? Socrates answers that there is a reason, an ontological ground, for this pattern being more than a mere pattern—indeed, one that *must* hold of the world as we experience it, a ground or tie which makes of this pattern more than a "mere" pattern.

Recall that Socrates put his point in terms of his own situation. Here he is, Socrates, sitting in prison. According to the earlier dialogue, the *Crito*, his friends offered to arrange things so he could escape to Thebes. Socrates, however, refused. Following the demands of justice, he determined that it would be wrong to escape his native city and run off to Thebes. He is, instead, determined to remain in his cell and drink the hemlock. So we have one event: Socrates sitting in his cell. There are two events that could follow that event, namely, running off to Thebes and remaining in Athens and drinking the hemlock. There is nothing in the first event, Socrates sitting there, which indicates or determines that it will be followed by one rather than the other. In fact, it is the drinking of the hemlock that follows. Socrates complained that Anaxagoras' account of things could not explain the sequence that actually occurs. The explanation must be in terms of Socrates "aiming at the best." But Anaxagoras' scheme has no model of the "best" for Socrates to aim at; there are no ideal forms and in particular no ideal form of human justice. Lacking an objective standard of virtue, Anaxagoras cannot explain Socrates' action. So at least Socrates argues.

However (to give Anaxagoras a helping hand), Anaxagoras need not deny that Socrates is aiming at the "best." Nevertheless, this best is simply what Socrates values as the best. Anaxagoras can argue that Socrates is in fact moved towards the goal that he identifies as the best, but that there is nothing to this beyond that valuing. There is no objective standard in the ideal forms of things. At least, so Anaxagoras could argue: certainly that is the way that Hume was much later to argue when he developed a similar case against the Platonists and other moral objectivists of his own day. There are no objective values but *there are* relative values.

Anaxagoras thus has a reply to Socrates: he *can* explain the events that Socrates wishes to explain without any appeal to transcendent entities, either simple souls or ideal forms, which somehow effect a tie of necessity between events in the ordinary world. All he needs is regularity: Socrates values what he feels to be just, and we know that he regularly acts to make the world, himself in particular, be as justice, so envisioned, requires. Relative values and regularities about things in the world, including people, suffice. There is no need for objective values or forms nor is there any need for a tie or causal nexus or objective necessary connection.

Socrates, Plato and Aristotle, and then, as we shall see, in our own age, the idealists, have all sought for such a tie. Their claims notwithstanding, none has passed the test established by the empiricist's *Principle of Acquaintance*: introduce something into your ontology only if we are acquainted with it in sensible experience or inner awareness.

Nevertheless, I must say that there is a certain point to what Socrates was arguing against Anaxagoras. I do not mean the point above about value. Anaxagoras' ontology, unlike Socrates', lacks objective value. Anaxagoras' ontology does have souls and they do move bodies, just as do Socrates' souls. Unlike Socrates' souls, however, the minds or portions of mind that move individual things on Anaxagoras' scheme do not transcend the world of ordinary experience. They consist of separable parts and we can see Anaxagoras struggling with language to describe them. They are fiery, or like fire, he says.[20] Their parts are in rapid motion—we might recall Hume's comment on selves that "… they are nothing but a bundle or collection of different perceptions, which succeed each other with an inconceivable rapidity, and are in a perpetual flux and movement."[21] Recall Moore's point that mental acts are diaphanous or transparent.[22] In whatever way the fiery bits, which are the separable parts of mind or *nous*, go together, they do separate themselves from other bits of *nous*, and direct some at least of the corporeal groupings or bodies that emerge from the primordial mixture. Like Socrates' souls, these groupings of fiery bits give bodies direction; they are purposive. But there are no transcendent forms that constitute a set of objective values, no transcendent forms that provide an objectively structured teleology. For Anaxagoras, if one may put it this way, there are valuings but no values. There are no objective values but there are things that are valued and these are the goals towards which mind or *nous* directs the bodies that it moves. In other words, there are in Anaxagoras no objective values but there are relative values—the ends towards which the bodies are directed. The teleology is driven from behind as it were by subjective values that derive from antecedent material conditions; the teleology is one of efficient, not of final, causes. It is not a matter of an active form pulling a substance

20 As Kirk and Raven note, "Anaxagoras in fact is striving, as had several of his predecessors, to imagine and describe a truly incorporeal entity" (*The Presocratic Philosophers*, 374).

21 Hume, *A Treatise of Human Nature*, ed. L.A. Selby Bigge (1888); 2nd ed. P.H. Nidditch (Oxford: Clarendon Press, 1978), 252.

22 G.E. Moore, "The Refutation of Idealism," *Mind*, New Series, Vol. 12: No. 48 (October 1903), 433–453; reprinted in Moore's *Philosophical Papers* (London: Kegan Paul, Trench, Trubner and Co., 1922), 23–44.

from in front into a state of being which, in its outward appearances, replicates that form.

The metaphysics of forms deriving from Socrates has often claimed for itself the sole right to the use of the term "reason." This is unfair to Anaxagoras and, indeed, to all versions of naturalism that are defined by the Principle of Acquaintance.

Reason is knowledge of causes; who would disagree? In the world of the ontology of forms, however, there is more to reason than this.[23] *Reason in this world is the grasp of forms.* These forms are the *essences* of things, that is, of particular things; the latter fall under these essences as universals or general kinds. These forms are known by reason. They are not given in sense and reason, therefore, is the capacity to grasp these forms that are outside the world of sense. As we have indicated, these forms are also *ends*—ends towards which things strive to be or strive as early as possible to imitate in their sensible appearances. Thinking in this way, one is naturally enough led to a purposive world-view or a world-view in which the task of reason is to reveal the place of everything, especially the place of human being and of each human being in the "grand scheme of things" or the "totality" as Bosanquet called it. The form or essence, furthermore, is a *law*, unfolding as it were into the necessary structure of the thing of which it is the form. It is the law that describes and explains the changing appearances of that thing. As an end, however, the form not only describes how the thing *is* but also how it *ought to be.* As a natural law, it is not simply descriptive but normative.

Any particular form unfolds further beyond its own internal structure. It also unfolds externally, again of necessity, into the structures that relate it to other forms of other things, unfolding ultimately into the whole which is the totality. Reason thus aims to grasp the whole in its necessary internal structure and to reveal within that whole the station which is appropriate to every thing. Reason aims to grasp the ontological structure of things, not only to understand it. Additionally, it aims to grasp the necessary and objective moral structure of things and, in particular, to locate the stations of things—where they are in the grand scheme. Reason aims with regard to each person to locate the station in which that person is to be located. Since this station is where the person ought to be, reason aims to locate the duties incumbent upon that person, his or her objective duties relative to the wholes and to the ultimate whole or totality in which that person is.

23 Cf. the discussion in Brand Blanshard, *Reason and Analysis* (London: Allen and Unwin, 1962), Chapter II.

Reason in this sense is understood to be *certain*. It is certain because it is *independent of sense*. Sense experience may give us indications about what lies beyond sense, but the entities towards which sense thus points are perfect exemplars of the imperfect things of the sensible world; the forms are not in any way dependent upon any sensible object for their being nor for their existence. In reaching beyond the changing world of sensible appearances, reason is knowledge of the unchanging reality that accounts for those changes (if the world of forms were changing, the problem of the third man would arise, leading to a vicious regress that makes impossible any explanation of change). Reason, in other words, is knowledge of the *eternal*. Reason is absolutely certain because, as it reaches outside the changing world of sensible appearances to grasp the forms, it is grasping something which it sees in its unchanging completeness and therefore sees with complete clarity. Reaching beyond sense, reason, as the grasp of forms, is inerrant. In reaching beyond sense, reason yields *a priori* knowledge of things. Moreover, in grasping forms as the essential truths of things, reason is knowledge of the universal: the ties in things are necessary ties and, as unchanging, they are the same for all persons.

We may, of course, not grasp or grasp only partially, the necessities that lie behind natural processes, so the regularities we observe in the sensible appearances of things may have exceptions. If we do grasp the necessary ties that lie behind the regularities of sense experience, however, then there are no exceptions.[24] In grasping the forms, reason not only grasps their internal structure but also the structures which of necessity relate them to other forms. The forms constitute a *system* and reason is the grasp of this system—ultimately the grasp of this system as a totality. This system, this totality, is one in which each form has its place, its station. Thus, in grasping a form, we not only grasp it in its own self-evident truth but also grasp *why* it is true.

To repeat, the world of ordinary experience is a world of patterns; to know what things are is to know such patterns, the patterns that define their empirical reality. For the naturalist, to know the things of this world, we must grasp these patterns. But the patterns are general patterns, the holding of a population, while in general all we ever are given in experience are samples. A causal judgment is therefore always inductive, going beyond the evidence we have from the sample to draw a conclusion about a population.[25] This is why, for the

24 This is not quite true of Aristotle; see F. Wilson, *The Logic and Methodology of Science in Early Modern Thought*, Study One.

25 It is certainly not clear that Anaxagoras thought that the investigation of nature proceeded inductively. Nor does the surviving evidence indicate whether his successor as a naturalist, Epicurus, thought in terms of inductive inference (though he should have,

naturalist, causal judgments are all fallible. If we think of knowledge in terms of certainty, therefore, we never know for sure the truth about the ordinary world; in this sense, the truth about this ordinary world is beyond our grasp. The case is otherwise for Socrates. Causal patterns are those that hold of necessity, that is, of necessity deriving from the forms. Since we grasp the forms, we grasp the truth of our world. If our cognitive aim is to know the truth of things that we encounter in the world of sense, then Socrates can claim to know the truth. But he knows the truth of this world only by going beyond it to another world that transcends it, the world of the forms. The naturalist, in contrast, cannot claim knowledge in the Socratic sense of inerrant certainty. However, the naturalist also argues that the forms are an illusion and so is the search for inerrant certainty which presupposes that the forms are there to be known; absolute truth is not attainable. However, we do not for that reason give up the search after truth but rather simply do the best we can.

Naturalism eliminates forms and necessary connections. This is a consequence of the Principle of Acquaintance. Thus, all that exists is the world we know by sense experience and inner awareness; it is the world of Anaxagoras rather than that of Socrates. As for the claim that, in grasping the forms, reason transcends the world of ordinary experience, naturalism responds that there are no forms or transcendent world, in which case there is no reason capable of grasping such a world. This is not to say there is no reason. Naturalism is still interested in causes but all that we can know empirically of such causes is regularity which is all that is left of causation after the forms and objectively necessary connections disappear. *Eliminate the forms*, Socrates suggests, *and you eliminate reason*. Anaxagoras' world and *the world of any naturalist are irrational through and through*; such is Socrates' argument. With no forms to define one's station in the grand scheme of things, *the world of the naturalist is everywhere meaningless and everything in it is devoid of any real value.*[26] The naturalist's response to Socrates' argument is simple rejection.

The point is, of course, that the naturalist still takes reason to be the grasp of causes, *except that* it is a different kind of reason. For the naturalist and the empiricist, this grasp of causes is nothing more than the grasp of regularities, and the way to come to grasp a regularity is the inductive method which generalizes from a sample to a population. Since there are no objectively necessary connections, the inferences of reason as the grasp of causes is never certain. For

given his general position about knowledge of matters of fact). In Roman times, there were at least two Epicurean naturalists who thought that way: Lucretius and Philodemus.

26 Cf. the discussion in Brand Blanshard, *Reason and Analysis* (London: G. Allen and Unwin, 1962), Chapter I.

Socrates, this means that the reason of the naturalist is second best—that is, second best relative to the inerrant knowledge of the forms.[27] But there are no forms and thus no inerrant knowledge.[28] However fallible it may be, the reason of the naturalist in the world as delimited by the Principle of Acquaintance is the only reason we have and therefore the best reason available; indeed, there is nothing relative to which it could be second best.

It follows from this that the world of the naturalist, to the extent that it is pervaded by causal regularities, *is, after all, rational.* The judgment that it is irrational derives from an illusion, a deep philosophical illusion but an illusion nonetheless. It derives from a comparison with the illusory transcendent world of the forms. Contrary to the Socratic argument, reason—the reason of the naturalist—does pervade the universe into which we are, willy-nilly, thrown. One cannot even charge that value is beyond the rational. To be sure, it is devoid of the value supposed to be conferred by the forms which are the reasons for things as they are and as they ought to be. It is devoid of any value discoverable by the reason, the spurious reason, that purports to be some sort of rational intuition of the transcendent inner structure of things. There are no such structures and no values that the forms could confer or rather constitute. It is a matter of ends. That which ought to be is what is worthy of being an end. In the world of Socrates, or better, Aristotle, what is worthy of being an end is determined by the objective forms of things, and, in particular, the objective form of human being. In the world of the naturalist, there is no such objective structure determining what is worthy of being an end. But one can say this much: something cannot be worthy of being an end unless it is capable of being an end. Nothing can be desirable, or worthy of desire, unless it is desired. Reason demands that this be the case; to propose an end that cannot be sought after is to propose the impossible. So reason demands that what is worthy of desire be capable of being desired. Happiness can be desired, so it is something that can be worthy of desire. But if X *and only* X can be desired, then X alone can be worthy of desire. Now, *as a matter of fact about human beings, happiness alone can be desired.* Whatever else is sought, is sought as a means. Moreover, there are various kinds or species of pleasure: it may be my pleasure to collect stamps, as it is the miser's pleasure to amass money,[29] and the pleasure of many, if not most, to act kindly towards others and to show

27 If one thinks in terms of a sort of Platonic divided line, inerrant knowledge and the forms are on the top half of the divided line, while sensible objects and our fallible judgments about them are on the lower half of the division.

28 And thus no top half of the divided line.

29 This is one of Mill's examples in chapter 4 of his essay on "Utilitarianism."

gratitude for acts of kindness done to or for us.[30] Rationally, then, happiness alone can be worthy of desire.[31] In other words, contrary to what is suggested by Socrates, Plato, and the other transcendentalists, *reason can, after all, in a naturalistic world, determine what ought to be valued as an end, and what therefore our desires—the desires which as natural facts we experience in our inner awareness—should move us towards.*

This is the view of Hume, of John Stuart Mill, of the (early) logical positivists, and of empiricists in general. The idealists I propose now to talk about were reacting to this version of naturalism.

The difficulty that Socrates found in Anaxagoras was also found in this version of empiricism or naturalism by the British idealists such as T.H. Green, F.H. Bradley, and Bernard Bosanquet. One might include the American Brand Blanshard in this company.

Let us start with Green. Naturalism, he argues, does away with genuine morality:

> The discovery … that our assertions of moral obligation are merely the expression of an ineffectual wish to be better off than we are, or are due to the survival of habits originally enforced by physical fear, but of which the origin is forgotten, is of a kind to give us pause. It logically carries with it the conclusion, however the conclusion may be disguised, that, in inciting ourselves or others to do anything because it ought to be done, we are at best making use of a serviceable illusion.[32]

Getting rid of transcendental standards and human beings is, as Socrates argued, mere skin and bones and, as he also argued, little better than dogs: duty

30 Anaxagoras did not produce an ethics—at least, none has survived, but Epicurus provides us with an ethics that is certainly contrary to the transcendental ethics of Socrates, so we may look upon what he said as Anaxagorean in spirit. As Epicurus put this point, "… every pleasure is a good by reason of its having a nature akin to our own, but not every pleasure is desirable. In like manner every state of pain is an evil, but not all pains are uniformly to be rejected" ("Letter to Menoeceous," in *The Philosophy of Epicurus*, translated with Introduction and Commentary by George K. Strodach (Evanston, Ill.: Northwestern University Press, 1963), 182).

31 See Mill's "proof" of utility in chapter 4 of his essay on "Utilitarianism." This was also the argument offered by Epicurus: "… we speak of pleasure as the starting point and the goal of the happy life because we realize that it is our primary native good, because every act of choice and aversion originates with it, and because we come back to it when we judge every good by using the pleasure feeling as our criterion" ("Letter to Menoeceous," 182).

32 T.H. Green, "Can There be a Natural Science of Man?," *Mind*, Vol.7: No. 25–27 (1882), 1–29 (No. 25); 161–185 (No. 26); 321–348 (No. 27); 25:6.

and obligation thus construed are an illusion. Like Socrates, Green is arguing that *to explain is to unify* and, more specifically, that when events are explained, "the truly good and 'binding' binds and holds them together."[33] It is the form of the good—the good for the thing changing—that binds sequential events into a unity. The form of the good, as what the thing will become, draws it out of its past into this form. The cause as form—that is, as the formal cause—is also the final cause. The form does not so much push the thing into its being of that form but pulls the thing into its being of that formed future. *The unity which alone explains is a teleological unity*, and *the explanation of why a thing*, e.g., Socrates, *becomes to be what it is, is a teleological explanation*. Thus, we clearly have it that *the form as final cause is the Reason for the thing being what it is and what it becomes, and one has an intelligible understanding of why the thing is as it is and as it becomes only when one grasps intuitively the form that is the Reason why it is as it is and as it becomes.*

To repeat, for Socrates, to explain, or to render rationally intelligible, is to unify. However, the unification must be effected by a form of the good or the just acting as a final cause. To explain is to unify, but the unification must be essentially teleological, in terms of active forms.

It was Green's argument that the world of the naturalists was a world of bits and pieces—a pile of pieces but pieces unjoined and not the world as we know it—not the structured world as we experience it. The world, Green correctly argued, is a structured world of *things in relation*. "It is our cognizance of the successiveness or transitoriness of feelings that makes us object intuitively to any [metaphysics] which is understood to imply an identification of the world with the feelings of men."[34] For Anaxagoras, for Hume, for Mill, the entities we perceive are as they appear to be. What we experience are sensible appearances. These appearances or impressions we distinguish one from another. As forming a world, they must, though distinct, be connected to one another. The appearances or impressions are given in sense, Green agreed, but, he also argued, the structure must be cognized as well and the structuring relations are not given in sense. No feeling, as such or as felt, is a relation. The relation is not known by sense, neither is it felt, because it is only for something that distinguishes itself from feelings that is can subsist. The world as delimited by the Principle of Acquaintance to the sensible appearances of things, needs, but cannot admit, those relations. One can suppose that the relations are given in sense only through a confusion. We can suppose a feeling to be a relation only

33 *Phaedo*, 99c.

34 Green, "Can There be a Natural Science of Man?," 25:29.

through a confusion between it and its conditions, or between it and the fact of its occurrence, which is no doubt related to other facts, but, as so related, is not felt. The relation cannot be reduced to a feeling or several feelings as related; for example, the relation of succession connecting two events cannot be a third felt event for then the related events would not be related after all, nor can it somehow be reduced to the two events it relates:

> Even a relation between feelings is not itself a feeling or felt. A feeling can only be felt as successive to another feeling, but the terms of a relation, ... even though the relation be one of succession, do not succeed one another. In order to constitute the relation they must be present together, so that to constitute a relation between feelings there must be something other than the feelings for which they are equally present.[35]

Naturalism, therefore, cannot admit structure. But the world is structured; it is a world of things in relation. So much the worse for naturalism or empiricism. Socrates is correct— restricting ourselves to the world of sense will not do. We need relations and to account for relations we need to go beyond the world of sense: "... nature in its reality, or in order to be what it is, implies a principle which is not natural." It is a principle which is "supernatural" and is reasonably called "spiritual."[36] To understand the world as an intelligible unity, we need to suppose that we cognize relations. We need to suppose that we have an intuition which goes beyond the world of sense to grasp relations as the reasons for things. It thus turns out that we cannot escape idealism. When we seek to understand the facts of the world—such things as matter and motion—then "... to give them [these facts intelligibility], except such as [are] derived from experience and, if from experience, then from relations that have their being only for an intelligent consciousness,"[37] we do so via a spiritual principle.

However, everything is in fact related to everything else; the world is a unity. Thus, the metaphysics to which we are driven is one "... which interprets facts as relations ...;" it is one in which we "... understand relations as constituted by a single spiritual principle ..."[38] Nor does anything lie outside consciousness. For something to be outside consciousness would involve its being external to the kind. But *externality* is a relation, the relation of *being outside*, and this, being a relation, "... like any other relation, exists only in the medium of

35 "Can There be a Natural Science of Man?," 25:28–29.
36 "Can There be a Natural Science of Man?," 26:175.
37 "Can There be a Natural Science of Man?," 25:29.
38 "Can There be a Natural Science of Man?," 25:29.

consciousness."[39] The critique of Anaxagoras' sort of naturalism leads to an idealism which is an absolute idealism: our perceptions are of a reality that we cannot conceive of as passing from outside to within consciousness; our perceptions are awarenesses of facts in relation and, being in relation, "require a consciousness to present them as facts and to unite them in relation."[40]

2 Can There Be a Natural Science of Man?

Green elaborates this point:

> We must hold then that there is a consciousness for which the relations of fact, that form the object of our gradually attained knowledge, already and eternally exist; ... the growing knowledge of the individual is a progress towards this consciousness.[41]

In other words, the knowledge we acquire, including the knowledge we acquire by perception, is, in effect, within us the sense of being *always within* us; it is innate. Consciousness, as that which knows comes to know, is in reality a final cause bringing itself to consciousness. As a knowing thing, consciousness comes to know and this state of knowing is the form attained. The form attained—the knowledge we acquire—we acquire from within us; this knowledge as the form the consciousness attains is implicitly in consciousness as the thinking thing. It is consciousness as a substance moved by knowledge as its active form. This knowledge that we acquire is innately there within us, and this knowledge gradually draws the mind towards itself. It is the knowledge that moves the mind, which in itself is passive. It is the knowledge—the clear and distinct ideas— that moves the mind to assent to itself, i.e., to the propositions known, until at the end of the process, the knowledge is not merely *in potentia* but in act. When the mind comes to know, it is not a matter of being pushed from behind, as in mechanistic causation (or in learning by association *à la* Locke and Hume), but rather a matter of the knowledge as a final cause pulling from in front. The explanation of consciousness is not a matter of mechanical or external causes, as in natural processes such as billiard balls, where Newton discovered the relevant laws, or, to use the example suggested by Socrates, the motion of dogs, where Darwin discovered the relevant laws

39 "Can There be a Natural Science of Man?," 26:181.
40 "Can There be a Natural Science of Man?," 27:325.
41 "Can There be a Natural Science of Man?," 27:325.

(those of natural selection though even here, natural selection is not sufficient), for "... the constituent elements of an organism can only be truly and adequately conceived as rendered what they are by the end realized through the organism."[42]

As for our awareness of the natural world and, indeed, of ourselves and our past history, this is an awareness of things as related. It is an awareness *of* phenomena but it cannot be itself a phenomenon of consciousness, contrary to what is claimed by the naturalist for whom any awareness is a phenomenon among phenomena, generally analyzable into sensory parts to be understood in the same way in which stones and other natural phenomena are understood. For such a one, there is a natural science of psychology, a natural science of man.

Anaxagoras' view of mind as a fiery substance was developed in detail (as were other aspects of his metaphysics) by the Epicureans. As Lucretius makes clear,[43] the ancient Epicureans held a view of the self (which later became of a piece with Hume's view of the self) according to which, on empiricist grounds, i.e., by appeal to ordinary experience of sense and inner awareness (a Principle of Acquaintance), they argued that the self we know through inner awareness is a changeable whole consisting of a series of awarenesses as parts, each of what are causally tied to a state of the body—in Hume's metaphor, a "bundle of perceptions." "The mind," argued Lucretius, "which we often call the intellect, the seat of guidance and control of life, is part of a man, no less than hand or foot or eyes are parts of a whole living creature."[44] The mind, and therefore our awarenesses of things, including our inner awarenesses, is a material thing. This is proved by the interconnection of mind and body; what happens in the one always influences what happens in the other.[45] Perception as our awarenesses of things consists in "replicas or insubstantial shapes of things"[46] being present to the mind.

Plotinus argued in the *Enneads*[47] that this view of the self and of knowledge is mistaken. According to him, there can be no perception without a unitary soul. The Epicurean doctrine that held that perception consisted in having

42 "Can There be a Natural Science of Man?," 27:332.

43 Lucretius, *On the Nature of the Universe*, prose translation by R.E. Latham (Harmondsworth, Middlesex: Penguin, 1951).

44 *On The Nature of the Universe*, 99.

45 *On The Nature of the Universe*, 100–101.

46 *On The Nature of the Universe*, 131.

47 Plotinus, *Enneads*, trans. S. McKenna, 4th edition revised by B.S. Page, Foreword by E.R. Dodds, Introduction by Paul Henry (London: Faber and Faber, 1959). References will be to Ennead, tractate, chapter, and page.

images failed to account for the unity of consciousness. There can be no perception without a unitary percipient whose identity enables it to grasp an object as an entirety. But the various senses yield different perceptions—any one object by itself gives different perceptions. There must be some central unity to which all senses report. The lines of perception converge on a point which unifies them:

> This there must be, as there is a centre to a circle; the sense-impressions converging from every point of occurrence will be as lines striking from a circumference to what will be a true centre of perception as being a veritable unity.[48]

What is perceived is a multiplicity. If the awareness of this multiplicity is an idea which manifests the same multiplicity as the object, then consciousness will be divided into the multiplicity of the idea. There will be a consciousness of each part of the object of perception, but no consciousness of the whole. In order to achieve a unity of consciousness, in which the perceiver is conscious at one moment of each *and all* of the parts, the perceiving consciousness must be a unity. We cannot have an isomorphism of idea and impression, as the Epicureans (or the early Wittgenstein)[49] would have it, but rather we must have a coordination of the complex impression into a unity. We do not have lines running from the various parts of the perceived to the various parts of the perceiver but rather lines from the perceived to a point, as all the radii of a circle converge on the dimensionless and single unified central point. The unity of consciousness in perception requires a simple entity, the soul, as perceiver.

The argument for the simplicity of the soul from its own self-awareness of the multiplicity within itself is simply a variation on this point about the unity of consciousness. The argument for the unity of consciousness generates the argument for the identity of the self throughout the complexity of its own being. Plotinus held, quite correctly, that we are beings capable of reflection upon ourselves and that when we are in this way self-reflective, we are aware of the whole that is the self. He then argued that a part that is within the series

48 Plotinus, *Enneads*, IV, vii, 6, 346.

49 See also the suggestion to this effect in F. Wilson, "The Aboutness of Thought," in T.M. Lennon, ed., *Cartesian Views: Papers Presented to Richard A. Watson* (Leiden: Brill, 2003), 151–164; reprinted in F. Wilson, *Acquaintance, Ontology and Knowledge* (Frankfurt: Ontos Verlag, 2007), 593–608, and in a revised version in F. Wilson, *The External World and Our Knowledge of It: Hume's Critical Realism* (Toronto: University of Toronto Press, 2008).

cannot be aware of the whole series. If a perception is not the whole soul, then it is not soul or soulless. In that case,

> Would not such a soulless addition be subject to just such loss and gain of substance, in fact to the non-identity, which marks the rest of our material mass? And if this were so, how [could we] explain our memories or our recognition of familiar things when we have no stably identical soul?[50]

Plotinus concluded that the self-reflective self which is aware of itself as a whole must be a simple entity that is not part of the series: "no one of the parts can be identical with the entire being."[51] Rather the soul that wholly knows the whole of itself is outside the parts of the series that constitute itself:

> ... surely no honest mind can fail to gather that a thing in which part is identical with whole has a nature which transcends quantity, and must of necessity be without quantity: only so could it remain identical when quantity is filched from it, only by being indifferent to amount or extension, by being in essence something apart.[52]

The soul, in other words, is outside the world of time and change that we know in ordinary experience. The self-reflective self that is aware of the self as a whole is an unchanging simple entity, eternally present to each part of the whole that constitutes the self. Since the soul must be a simple unity, it may be inferred that it is immortal for the simplicity of the soul implies that it is indestructible:

> ... the Soul ... is not a mass, not a quantity. May it not change and so come to destruction? No: the change that destroys annuls the form but leaves the underlying substance: and that could not happen to anything except a compound. If it can be destroyed in no such ways, it is necessarily indestructible.[53]

If it is indestructible, furthermore, then, contrary to the Epicurean, it is immortal.

---

50 Plotinus, *Enneads*, IV, vii, 5, 345.

51 *Enneads*, IV, vii, 5, 345.

52 *Enneads*, IV, vii, 5, 346.

53 *Enneads*, IV, vii, 12, 356.

Kant was later to advance what was essentially the same argument. "The soul, or the thinking 'I'," he argues, "is a [simple] being":

> For suppose it be the composite that thinks; then every part of it would contain a part of the thought, and only all of them taken together would contain the whole thought. But this cannot be consistently maintained. For representations (for instance, the single words of a verse), distributed among different beings, never make up a whole thought (a verse), and it is therefore impossible that a thought should inhere in what is essentially composite. It is therefore possible only in a *single* substance, which, not being an aggregate of many, is absolutely simple.[54]

Green takes up essentially the same point: the self, if it is to be a consciousness of sensible entities as distinct parts of, but connected by, relations into a perceptual whole, cannot itself be complex. As an awareness of a whole of connected parts, it must be simple:

> … the presentation of sensible things, on occasion of sensation, implies the action of a principle [the self] which is not, like sensation in time, or an event in a series of events, but must equally be present to, and distinguish itself from, the several stages of a sensation to which attention is given, as well as the several sensations attended to, which are referred to a single object.[55]

Green rejects any mind-body dualism. There is regularity between mind and body, but for the naturalist, that is all that there is which means for the idealist that the mind-body connection is merely accidental, not genuinely causal. The natural world is connected in perception *non-accidentally* to the consciousness of it for the consciousness of an object, i.e., the feeling that is the appearance of the object, and the natural events which are the conditions of the being of that feeling *are related*, and this relation is constituted by the consciousness that unifies the feeling and its conditions into a connected whole. The same holds for human action. Moved by its passions, the mind wills and the body is moved. This connection has to be non-accidental, indeed genuinely causal, for otherwise there can be no moral responsibility. ("Why not?," responds the naturalist. "Do you suppose that I do not draw a distinction between regularities which are causal and those which are accidental, and that I can thereby

54 Immanuel Kant, *Critique of Pure Reason*, trans. Norman Kemp Smith, 1929; unabridged edition (New York: St. Martin's Press, 1965), A352.

55 Green, "Can There be a Natural Science of Man?," 27:337–338.

account for moral responsibility?" "Why," he says to the idealist, "do you think that yours is the only way that a philosopher can make sense of the causal generality/accidental generality distinction?")[56] But, Green continues, the willing and the bodily movement which is its upshot *are related* into the whole which is the human action. This relation, like all relations, is constituted by consciousness. There is no dualism of separate mind events and bodily events. So Green, with his account of relations, takes it that he has solved important problems which remain unsolved by the naturalist.

3 Bradley's Idealist Account of Structure

That is not the end of the story, however. Upon reflection, one can see that there is a real problem with Green's account of relations. On his view, there are, on the one hand, the sensible entities, the sensations or feelings or impressions, and, on the other hand, the relations that are supposed to connect these into structured wholes. The former are there at the bottom, while the latter at a higher level, as it were, float above these. Our understanding of things is obtained through our reason that knows the relations which are the reasons for things, but it is sense, not reason, which knows the appearances related. How could relations which are spiritual and known by reason define the reality of the facts we encounter in sensible awareness but which are not known to reason? As Bradley once put it, "… the notion that existence [i.e., reality] could be the same as understanding strikes as cold and ghost-like as the dreariest materialism: … the sensuous curtain is a deception and a cheat, if it hides some colourless movement of atoms, some spectral woof of impalpable abstractions, or unearthly ballet of bloodless categories."[57]

Bradley attempts to solve this problem. His solution was to take seriously once again the point that *to understand is to unify*. He solved the problem by bringing into a single whole the sensible impression or appearance and its relations, unifying the relations and the appearances into a seamless sensuous whole. The world, the whole, the totality is constituted by the sensible appearance. This appearance is, as it were, full of everything. Its relations to all other appearances are contained within it; indeed, all the other appearances to which it is related are contained within it. It is not that the appearance contains the totality, but that the appearance is already the whole. The

56 This idealist challenge and the naturalist response is taken up in Section VIII below.

57 F.H. Bradley, *The Principles of Logic*, 2nd ed., revised, 2 vols. (London: Oxford University Press, 1922), 2:591.

appearance is simple, to be sure, but also the whole of reality; indeed, it *is* the totality. This sensuous appearance is directly experienced, or, rather, it *is* direct experience—it is not distinguishable from the experiencing of it, though its relation to the experiencing of it and to the I which is experiencing is contained within it. Since we have a unity here, the appearance is also contained within the consciousness of it. The appearance is already the whole, the totality, and so also the consciousness of it is the whole, the totality. The sensuous appearance contains within itself the whole and therefore turns out to be the Absolute. In experiencing that appearance we have within that experience, we have the intimations of that whole—an implicit sense of the relation—which leads one, when guided by its necessity, from the appearance through the world and the consciousness of it to that totality which is the Absolute. As Bradley put it, "That the glory of this world in the end is appearance leaves the world more glorious, if we feel it is a show of some fuller splendour."[58]

Bradley rests his case on a unity-in-diversity which is given as "immediate experience" or "feeling." "That on which my view rests is the immediate unity which comes in feeling ..."[59] The unity of feeling is not made up of distinct objects; the feeling is the only object, and the apparently related objects within it are not separate or separable objects but merely diverse aspects of the feeling. " ... the unity of feeling contains no individual terms with relations between them, while without these [that is, the terms, not as real individuals, but only as aspects of the feeling], no experience can be really relational."[60]

To have a fact (@ such that a is R to b) that is genuinely relational, then the relata a and b are *not separable*. For, if we consider the properties predicated of a, then we have the non-relational predication in which the non-relational property "R to b" is predicated of a: a is (R to b). If b ceases to exist, then a ceases to have this property. Thus, if a stands in a genuine relation to b and b ceases to exist, then a changes in its very being, in its identity. Upon the traditional naturalist account of things, such as that of Locke, explicitly,[61] and, implicitly, of Anaxagoras,[62] however, the individuals a and b are distinguishable and are therefore separable. That means that the relational fact (@) must be analysed into two separable facts which, to be separable, must be *non-relational*: (&)

58 *The Principles of Logic*, 2:591.

59 F.H. Bradley, *Essays on Truth and Reality* (Oxford: Clarendon Press, 1914), 230–231.

60 Bradley, "Relations," in *Collected Essays*, 2:643.

61 Consider Adam and Eve. Adam existed before Eve and did not change his properties (if we ignore the rib story in Genesis 2:20b-23) upon the creation of Eve. Adam is therefore separable from Eve.

62 One presumes that the pieces of the original mixture are independent of one another.

such that (a is r_1) & (b is r_2) and the two predicates are non-relational properties. Only in this way can the independence of a and b be retained. To achieve independence in such a way is to abandon all real structure. Instead of the structured fact (@), we have a mere conjunction (&), or, to use the metaphor Hume employs in the case of the self, a mere "bundle" or heap.

What is needed is an *objective* relation, somehow connecting the objects related into a unity. As Bradley puts this point,

> … to understand a complex *AB*, I must begin with *A* or *B*. And beginning, say, with *A*, if I then *find B*, I have either lost *A* or I have got beside *A* something else, and in neither case have I understood. For my intellect cannot simply unite a diversity, nor has it in itself any form or way of togetherness, and you gain nothing if beside *A* and *B* you offer me their conjunction in fact.[63]

So much the worse, then, for the empiricist account of relations; since this account of relations makes it unintelligible how any two things can be genuinely related, that position is just wrong. (But as we shall see, the empiricist has a response: William James and Bertrand Russell developed an account of relations that attempted to meet this criticism which effectively rules out the account of Locke (and Hume).)

Bradley proposes what one does not find in Green, namely, a detailed ontology of relations. For our present purposes, three features are important. The *first* important feature is this: the apparent relations of things are not denied or explained away as *merely* apparent, as the empiricists were prone to argue. Rather, the apparent relations of things are construed as genuine relations since their relata are not independent in the sense that the being of one relatum is not separable from the being of the other relatum:

> If it [a relation] is to be real, it must be so at the expense of the terms, or, at least, must be something which appears in them or to which they belong. A relation between *A* and *B* implies really a substantial foundation within them. This foundation, if we say that *A* is like to *B*, is the identity *X* which holds these differences together. And so with space and time—everywhere there must be a whole embracing what is related, or there would be no differences and no relation. It seems as if a reality possessed

63 Bradley, *Appearance and Reality*, 2nd ed. (with an Appendix), (Oxford: Oxford University Press, 1897), 509.

> differences *A* and *B*, incompatible with one another and also with itself. And so in order, without contradiction, to retain its various properties, this whole consents to wear the form of relations between them.[64]

Bradley concludes, against the empiricist account, that there must be a whole embracing what is related.

The *second* important feature is this. Not only does Bradley insist that there are genuinely relational wholes but he also proposes a definite ontology for these relational wholes. On his account, the relata a and b are different things within a whole (a, b)—or, rather, not so much things within such a whole but simply aspects of it, and no longer independent of one another. This whole then "consents to wear the form of" a relation. Thus, if a and b stand in the relation R, the correct representation of this fact consists in attributing a property corresponding to R, say r, to (a, b). According to Bradley's account, then, the correct way to represent the fact reported by (@ such that a is R to b) is given by (&& such that (a, b) is r) or r (a, b). The whole (a, b) is itself a particular thing,[65] of which the two terms a and b are but aspects, and where the arrangement r characterizes the whole. "Relations are unmeaning except within and on the basis of a substantial whole, and related terms, if made absolute, are forthwith destroyed."[66] But this whole consists of the relata as parts—parts that are but appearances of this substantial whole. Thus, the relation holds of the relata, not separately as in the empiricist account, but jointly taken together as one,[67] "... where the whole, relaxing its unity, takes the form of an arrangement; there is a co-existence with concord."[68]

Bradley applies this account to all relations. He mentions something being *like* something else, which is to say, he applies it to resemblance. He applies it to space and time. He applies it to quality orders and to causation. He applies it to contrariety among qualities, so that general facts of the sort, "Whatever is F is not G," which record such contrariety, become relational facts among qualities. Indeed, he applies the account to all predication. We have seen all this above. At each stage, newer and more embracing unities appear to support the relations that structure the world. In the end, all the apparently separable particulars, and all the apparently separable qualities or properties of things that we are aware of in ordinary experience, disappear into one all-embracing

64 Bradley, *Appearance and Reality*, 18.
65 Cf. Bradley, "Relations," *Collected Essays*, 2:635–636.
66 Bradley, *Appearance and Reality*, 125.
67 *Appearance and Reality*, 18.
68 *Appearance and Reality*, 19.

unity. As we have noted previously, the ultimately real thing is one substantiality.[69] This is the Absolute.[70]

Third, Bradley conceived the substantial particular that constituted the relational fact to be active. It *develops* into greater wholes. Any given "this" that might be is something that "… exists within, and by virtue of an embracing unity, and apart from that totality both itself and its terms would be nothing":

> … the self-transcendent character of the 'this' is, on all sides, open and plain. Appearing as immediate, it, on the other side, has contents which are not consistent with themselves, and which refer themselves beyond. Hence the inner nature of the 'this' leads it to pass outside itself towards a higher totality.[71]

Bradley, of course, understands the activity of the substantial particular in terms appropriate to his idealist philosophy.[72] It is a *cognitive* activity; it is a grasping of the reasons for things, and is therefore reason. But this reason is also a form of conscious experiencing. It is a form of experiencing that develops teleologically towards an experience which, at the ultimate end in the Absolute, is the experiencing of the reasons for all things and the truth of those things. It is the whole truth and the whole truth of the totality of things, the ultimate whole:

> Feeling … supplies us with a positive idea of non-relational unity. The idea is imperfect, but is sufficient to serve as a positive basis. … [The ultimate] Whole [is] qualified, and qualified non-relationally, by every fraction of experience. But this unity of all experiences, if itself not an experience, would be meaningless. The Whole is one experience then, and such a unity, higher than all relations—a unity which contains and transforms them—has positive meaning.[73]

---

69 *Appearance and Reality,* 124–126.

70 *Appearance and Reality,* 151.

71 *Appearance and Reality,* 201–202.

72 This view is not exactly Christian. On the latter, God, with His or Her free will, could create Adam but not Eve; that is, the parts of God's creation are independent of one another and, indeed, God is independent of His or Her creatures. On Bradley's account of relations, however, each part implicates every other part; taken as a totality, there are no independent parts (nor, for that matter, could God be independent of His or Her creatures, a decidedly non-Christian view).

73 Bradley, *Appearance and Reality,* 470.

There is a movement from the apparently unstructured "this" of immediate experience, through partial structures to structures that are more and more complete, less and less inadequate, towards the fully complete structure of the Absolute: "... to the ultimate whole nothing can be opposed, or even related."[74] This ultimate Whole "... is the final destiny and last truth of things."[75] Bradley's ultimate Whole is the totality of which Bosanquet spoke and made the object which philosophy aimed to come to know.

The Socratic view of things seems thus to triumph over the naturalistic view of Anaxagoras, *but only at a cost*: the price of structure seems to be absolute idealism, and an Absolute is too high a price for many. For these latter, their aim is to have an account of the world that is essentially that of science: they prefer that science not be put into its (insignificant) place by the Absolute or the Totality of things.

The problem was structure. It was there in Socrates and absent from Anaxagoras. A naturalist can follow Anaxagoras to the extent of rejecting, on plausible grounds, Socrates' account of the pursuit of the good. There is, that is, a plausible naturalist account of teleology in the sense of a pursuit of human ends, including noble ends such as justice. But we still have not got structure. Socrates' account or its idealist descendent seems still to be the only game in town. However, this account of relations brings in the Absolute as that which gives structure to the things in the world of ordinary experience, and that clearly violates the Principle of Acquaintance which defines the empiricist ontology of the naturalist.

There is more than just ontology at stake here. There is also an important issue in the social sciences.

4 Althusser: Marxism as Idealism

To see what this issue is, we can do worse than consider the Marxist theory of society. According to Marxists, the classical economics of Ricardo was a reasonable theory concerning the economic workings of society, i.e., the economic workings of capitalist society, especially the capitalist society of Great Britain during the first two thirds of the nineteenth century. The laws that Ricardo discovered were not unconditional. To the contrary, these regularities were conditional,[76] holding only for a society structured by capitalist social

74 *Appearance and Reality*, 201.

75 *Appearance and Reality*, 469.

76 See John Stuart Mill, *System of Logic*, in *The Collected Works of John Stuart Mill*, ed. J. Robson, Volumes 7 and 8 (Toronto: University of Toronto Press, 1973–1974). Mill dealt

relations of individuals to the means of production.[77] The regularities Ricardo discovered were in fact imperfect, omitting social relations to the means of production as relevant variables. Where Ricardo and his followers went wrong was in holding that these imperfect laws were truly causal laws in the sense of unconditioned regularities.[78]

Now, if the laws were truly unconditional, then there was nothing one could do to change them. Attempts to reform society in ways that violated these laws of economics were bound to fail; a violation would be a counterexample to the law. Since the regularity was a law without exceptions, the attempt to act contrary to it would simply be to try to act in a way that contradicted natural necessity. Not only was such an attempt bound to fail but it was bound to make conditions worse: one cannot with impunity attempt to violate one of nature's causal laws. For example, legalizing trade unions or enacting state regulations governing the conditions of labour for children, say, or, women, while perhaps well-intentioned, could not in fact alleviate the conditions they aimed to remedy. If these regularities about the economics of society held only conditionally, then there was nothing inevitable about them. As regularities, they did describe how society worked but how society worked could be changed by effecting changes in the conditions under which alone the regularities were valid.[79] Calling Ricardo's regularities "causal laws" when they were in fact imperfect, holding only under certain conditions,[80] was not a piece of science but rather ideology, an ideology aimed at deflecting as impossible attempts to achieve the reformation of capitalism. In other words, it was an ideology disguised as science, the point of which was to keep the conditions of capitalist society unchanged—to keep in power those owners of capital who benefited from the current social relations at the expense of the majority whose only capital was their own labour and who did the work, usually hard and ill-paid, from which the capitalists benefited.

---

with this distinction between conditional and unconditional regularities in his *System of Logic*, Volume 7, Book III, Chapter v, Section 6 (hereafter III, v, 6, etc.).

77 Mill discussed political economy in the *System of Logic*, Volume 8, VI, ix, 3, 900–904, where he noted that the regularities of this science are conditioned regularities.

78 Mill noted that "it has been a very common error of political economists to draw conclusions from one state of society, and apply them to other states in which many of the elements are not the same ..." (*System of Logic*, Volume 8, VI, ix, 3, 903).

79 This was Mill's criticism of many political economists of his day.

80 For a good discussion of imperfect or gappy laws, see G. Bergmann, *Philosophy of Science* (Madison, WI: University of Wisconsin Press, 1956), Chapter II; see also F. Wilson, *The Logic and Methodology of Science and Pseudoscience* (Toronto: Canadian Scholars Press, 2000).

The Marxist aimed to change the system, to benefit the masses who did the labour by changing the conditions, the social relations, under which alone Ricardo's regularities were valid. All is fine so far. There is nothing wrong with the logic of the Marxist's position.

But Marx, in attempting to come to grips with the social relations he had to emphasize, drew on the same Hegelian heritage as Green and Bradley. Recall how Bradley did these things. Corresponding to the relational fact (@ such that a is R to b), we have a single r-ish individual of which a and b are aspects that we represented as (&& such that (a, b) is r) or r (a, b). The whole (a, b) is itself a particular thing,[81] of which the two terms a and b are but aspects, mere appearances of the whole which is the only substantial reality, and where the arrangement r characterizes this substantial reality, the whole, or, as Bosanquet would say, the "totality." Now, let a and b be two individuals related to one another by R, the relation jointly exemplified by persons selling their labour. They form a whole (a, b) which is of the relational kind, "related as sellers of labour" or "related as workers." The schema (&&) such that r (a, b) thus represents, in a simplified way of course, *the working class. The working class* is thereby a single thing metaphysically, the real thing and sole substantial reality of which a and b, the individual workers, are not genuinely individuals but rather less than fully real aspects. If we fit this to the discussion of Ricardo's economics, where the imperfect regularities are conditioned by the social relations characteristic of capitalist society, then we can schematically take it that (@ such that a is R to b) represents these conditioning facts.

So far so good; according to the idealist account of relations, however, this fact, while apparently involving two individuals connected by a relation that is a causally relevant variable in a process in which they together participate (as in Newtonian mechanics, Mars and the Sun are two individual things with a certain distance between them, where this relation is a variable relevant to the process they jointly undergo; where this process is described by what has been called a process law,[82] and where, historically, this unconditioned law was discovered by Newton), actually involves only one individual (a, b) undergoing a process for which the non-relational entity r is the relevant variable. What we have is not so much a relation relating two individuals as an r-ish single whole. Since we are taking the individuals *qua* related as sellers of their labour, what "r (a, b)" represents is *the working class as a unified whole or totality. It is a totality superior to, different in kind from, and indeed more substantial and*

81 Cf. Bradley, "Relations," in *Collected Essays*, 2:635–636.

82 Cf. Bergmann, *Philosophy of Science*, Chapter II.

more real than—but causally determining the being, such as it is, of what we take in the ordinary way to be the members of—the working class. Moreover, on the idealist account, a and b are phenomena given to us in sense, in our ordinary sensible experience of the world, whereas both the relation relating them and the whole into which the relation makes them but aspects, are not given in sense but are grasped by a higher cognitive faculty. Though a Marxist would not speak this way about the working class and the property of being the working class, *they are non-empirical transcendent entities known by a reason higher than sense which discerns the real reasons for things*. Moreover, of course, there are several classes, or, rather, two classes, the workers and the capitalists. These are two wholes which are not yet the ultimate totality, since they are related to each other. This more grand totality is what has been called History—History as a material process which unfolds dialectically towards a certain end, the classless society of the Marxists being a final cause which teleologically draws the social process towards itself, the goal of History.

One can recognize this idealist pattern of thought in Marxism, and in particular, the Marxism of Louis Althusser. I will use his *Reading Capital*[83] as a source, but in that work one can find sufficient references to Marx himself to make the case that he too exemplifies the idealist pattern.

Althusser tells us that "A science is a systematic theory which embraces the totality of its object and seizes the 'internal connexion' which links the ... essences of *all* economic phenomena."[84] It was Ricardo's merit to have delineated clearly and fully the "physiology" of bourgeois capitalist society—its essence.[85] According to Althusser, what Ricardo took to be unconditioned laws of economics were in fact imperfect, conditioned laws. As Althusser puts it, "classical economics had not an historical, but an eternalist conception of its economic categories ..."[86] So far no empiricist is likely to disagree—certainly not John Stuart Mill, as we have noted. According to Althusser, however, what is crucial to understanding economic science is not the empirical investigation of the regularities of our economic behaviour but rather "the Marxist conception of the social totality."[87] This is not the "spiritual totality" of Hegel nor that, of course, of Hegel's successors, the British idealists. In other words, the relations relevant to constituting the Marxist totality are not such relations as

83 L. Althusser and É. Balibar, trans. B. Brewster, *Reading Capital* (London: Verso Editions, 1979).

84 *Reading Capital*, 84.

85 *Reading Capital*, 84.

86 *Reading Capital*, 93.

87 *Reading Capital*, 97.

those of patriotism (which constitutes the nation as a totality) or religious sentiment (which goes to constitute the church as a spiritual institution). Rather, the relations to understanding the social process are those defining economic relations. It is these latter that constitute "the unity of a structured whole"[88] and it is such structured wholes to which we must turn for scientific understanding. The totality which is the object of the science is one in which the parts are all related one to another, as they are in the idealist totality. The Marxist totality, we are told, is

> ... a totality all of whose parts are '*total parts,*' each expressing the others, and each expressing the social totality that contains them, because each in itself contains in the immediate form of its expression the essence of the totality itself."[89]

Taking this seriously, we must recognize that the entities Marxism attempts to use to conceive the world it aims to investigate are "totalities," entities which are "structured wholes,"[90] and that such totalities are of a sort incompatible with empiricism. Like the totalities discussed by Bosanquet, they are structured according to the Hegelian concept of relations made articulate by Bradley. Furthermore, the Marxist position is that these structures are not known through empirical observation. This is to be expected: the structures transcend the world of ordinary sensible appearances, so they must be grasped or known by a sort of thinking that grasps them as the transcendent reasons for things of this world being as they are. Althusser quotes Marx himself: "The whole, as it appears in the mind as a thought-whole is a product of the thinking mind."[91] Althusser goes on to note:

> ... the thought-concrete belongs to thought and not to the real. The process of knowledge, the work of elaboration ... by which thought transforms its initial intuitions and representations into knowledge or thought-concrete, takes place entirely in thought.[92]

No empiricist, nor any naturalist who adopts the empiricist stance, would agree; economics, and indeed any social science, is an empirical science, and its

88 *Reading Capital*, 94.
89 *Reading Capital*, 94.
90 *Reading Capital*, 94.
91 Marx, *Grundrisse*; quoted in *Reading Capital*, 87.
92 *Reading Capital*, 87.

method consists, roughly put, in testing hypothesized regularities against the empirical facts as we encounter them in our sense experience of the world.[93]

However, it is still the argument of Marxists that the relations that yield the social whole or totality are not spiritual but material in the sense of economic. But the critic must insist that these relations are taken by the Marxists—some anyway—to have the same form as idealists like Bradley attach to relations, where one part is connected to all the others to form a totality and is thereby connected back to itself. *The relations that are taken to form the Marxist totality are not spiritual in the way that the Hegelian takes his or her structuring relations to be, so the Marxist wholes are not idealistic like those of Hegel; the relations are economic and, if one wishes, materialistic.* So goes the argument of the Marxists: Marxist science, they conclude, is materialistic. One presumes this could be one of the philosophies of science, the idealist response to which we have been asked to investigate. However, *Marxist totalities, while based on "material relations," are conceived in terms of the idealist account of relations.* It follows that *Marxism, however much it claims to have a naturalist or materialist philosophy of science, is in fact at its core structurally idealist.*

We said above that the issue given to us by Bradley's idealist account of relations was ontological but also more than ontological. We now see that the issue created by this doctrine of relations is central to how we conceive a social science, either as empirical, as Hume and Mill conceived it, or, *à la* Hegel, Bradley, Bosanquet, and Marx, as something that must penetrate to a world that is at once beyond the world of ordinary experience but also the world of reasons why the ordinary world is as it is.

5 John Stuart Mill on the Naturalist Science of Man: His Defence and Its Problems

John Stuart Mill confronted neither Bradley nor the Marxists but he did confront much the same view of social wholes in the thought of Coleridge.[94] Of course, these social wholes were taken to be structured by relations essentially "spiritual" rather than "material" in the sense of the Marxists, but basically they were relations conceived on an earlier and less articulate form of the idealist account given by Bradley. Despite the nascent form of the view of social wholes

93 Cf. Mill, *System of Logic*, in *Collected Works*, Volume 8, VI, ix, 3, 903–904.

94 See, for example, Coleridge, "On the Principles of Political Knowledge," Essay III (from *The Friend*, First Section), in *The Portable Coleridge*, ed. I.A. Richards (New York: Viking Press, 1961), 332–338.

he confronted, Mill was clear: these transcendent entities were contrary to, and excluded by, the Principle of Acquaintance. Accordingly, they were illusions which, if left unchallenged, could do nothing but so muddle things as to frustrate progress in any social science that was aiming to be empirical. He made his case using the Principle of Acquaintance:

> The laws of the phenomena of society are, and can be, nothing but the laws of the actions and passions of human beings united together in the social state. Men, however, in a state of society, are still men; their actions and passions are obedient to the laws of individual human nature. Men are not, when brought together, converted into another kind of substance, with different properties; as hydrogen and oxygen are different from water, or as hydrogen, oxygen, carbon, and oxygen are different from nerves, muscles, and tendons. Human beings in society have no properties but those which are derived from, and may be resolved into, the laws of the nature of individual man. In social phenomena, the composition of causes is the universal law.[95]

Put men and women together as sellers of labour and they remain just men and women. There does not emerge a new entity, the working class, different in kind from, and superior to, the individuals which have been reduced, as the Marxists argue, to mere aspects of this class, swept along willy-nilly as this latter substance develops in interaction with another super-individual, the capitalist class, according to special laws for social wholes or totalities. These new super-entities or substances simply do not exist. Neither does any super-super entity such as History exist. We are not acquainted with any such entities in our ordinary sensible experience of the world. Based on the empiricist Principle of Acquaintance, then, the idealist account of social wholes or totalities, whether Hegelian or Marxist, will not do.

However, Mill, in presenting his argument here, in fact makes a grave error. His appeal to the Principle of Acquaintance may well be sound. One may *not* conclude from this that "the Composition of Causes is the universal law," however. By the "composition of causes," Mill means that one can take the laws for individual persons and from these, *directly and with no further premises*, deduce the laws for the social aggregate, i.e., deduce deductively, as from 'p' and 'q', that 'p & q'. From this, Mill infers that the method appropriate to the

95 *System of Logic*, Volume 8, VI, vii, 1, 879.

social sciences is *deductive* from the laws of psychology for individual persons making up the social aggregate.[96]

This method for the social sciences that Mill proposes can work only if the deductions he describes really are valid, however. Mill argues that they are and, in fact, occur elsewhere in science, particularly in physics.[97] He points to mechanics, where it is in fact possible to deduce the laws for a complex system from the laws for simple systems. The procedure is this.[98] If we have a three-body system, we can conceptually divide it into three two-body systems. Knowing the forces that would operate in the two-body systems were they isolated, we can deduce what the forces are that are operating in the three-body system. Mill holds that this deduction proceeds *a priori*[99] in these cases, as opposed to those such as chemistry and psychology where the effect is "heterogeneous" with its causes, "the joint effect of the causes being the sum of their separate effects."[100] While we know the law of the separate causes by induction, the inference to their joint effects involves no further induction but only "ratiocination."[101]

In fact, he is wrong on this point.[102] In order for the deduction to proceed, one must take into account *the relations by which the simpler systems are constituted into the more complex system*. However, and this is the crucial point, *there is no a priori reason for assuming that a given relational structure will yield one sort of law for the complex system rather than another*.[103] This means that the deduction of the law for the complex system depends not only upon the laws for the simpler system but also upon *another factual assumption* that relates the laws of the complex system to *both* the laws for the simpler systems *and* the relational structure that constitutes the complex system out of the simpler

96 Mill does qualify this in certain ways. Owing to the complexity of social phenomena, the "physical or concrete deductive method" will not be successful in discovering laws or in validating hypotheses concerning such laws; one must use instead the "inverse deductive, or historical" method (*System of Logic*, Volume 8, VI, ix and x, 1–8; 895–930).

97 Mill, *System of Logic*, Volume 8, VI, ix, 1, 895.

98 Cf. Bergmann, *Philosophy of Science*, Chapter III.

99 Mill, *System of Logic*, Volume 7, III, vi, 1, 370.

100 *System of Logic*, Volume 7, III, vi, 1, 371.

101 *System of Logic*, Volume 7, III, xi, 1–2, 454; 458–460.

102 Cf. Bergmann, *Philosophy of Science*, Chapter III; also B. Russell, *Principles of Mathematics*, 2nd ed. (London: Allen and Unwin, 1937), 484.

103 See G. Bergmann, *Philosophy of Science*, Ch. III; E. Madden, *Philosophical Problems of Psychology* (New York: Odyssey Press, 1962); F. Wilson, *Empiricism and Darwin's Science* (Dordrecht, Holland: D. Reidel, 1991), Ch. III, and F. Wilson, *Psychological Analysis and the Philosophy of John Stuart Mill* (Toronto: University of Toronto Press, 1990), Chapter II, section 2.

systems. This factual assumption relating the laws for the complex system to both the laws for the simpler system and the relational structure, is itself a *law*. It is not a specific causal law but rather *a law about such laws*. Since it is a law, the step from the causal laws for the simpler systems is not one of pure ratiocination or pure deduction but also involves an inductive feature. This law—this inductive step the existence of which Mill denies—has been referred to as a "composition law."[104] When Mill asserts that the inference is a deduction that proceeds wholly *a priori* without any inductive step beyond those that provided the laws for the simple systems, he is neglecting to take into account this additional factual premise. In effect, *this amounts to neglecting the causal role of the relations which constitute the whole out of the parts*. Mill, then, is wrong in his claim that, in mechanics, the deduction of a law for the complex system can be deduced *a priori* from the laws for simpler systems; what he calls the "deductive" method does not in fact have any place in mechanics.

When Mill makes his claim that in the social sciences "the Composition of Causes is the universal law," he is making a similar error. When he claims that the deduction of the laws for the complex social wholes can be deduced *a priori* from the laws for the parts, that is, from the laws for persons taken individually, he is claiming in effect that there is no need for a composition law, or, what amounts to the same, *no need to take into account the social relations* which, by virtue of holding among individuals, constitute the social whole out of those individuals. Mill suggests that in the social sciences, the individual cases act "conjunctively" in just the way that they act in mechanics:

> The Social Science … is a deductive science; not, indeed, after the model of geometry, but after that of the more complex physical sciences. It infers the law of each effect from the laws of causation on which that effect depends; not, however, from the law merely of one cause, as in the geometrical method; but by considering all the causes which conjunctively influence the effect, and compounding their laws with one another.[105]

Mill says "conjunctively," by which he means that this deduction is the same as when one deduces 'p & q' from 'p' and 'q'. But a conjunction is merely that, and not a relational whole. He indicates the same neglect of relations when he speaks of "computing the *aggregate* result of many co-existent causes."[106]

104 Cf. Bergmann, *Philosophy of Science*, Chapter III and Russell, *Principles of Mathematics*, 2nd ed., 477.

105 Mill, *System of Logic*, Volume 8, VI, ix, 1, 895.

106 *System of Logic*, Volume 8, VI, ix, 1, 896; emphasis added.

Mill also indicates that the inference from the laws of the co-existent causes to the "aggregate" effect is something that we can "calculate *a priori*;"[107] the inference will, of course, be *a priori* if it proceeds on the basis of a *conjunction* of premises, but not if it requires additional factual premises concerning the relational structure and a composition law. He also suggests that the total social effect is merely the "sum" of the individual effects. He makes this point when he explicitly compares social phenomena to those of mechanics, when, as we have seen, he argues that "in social phenomena the Composition of Causes is the universal law," where the Composition of Causes is "the principle which is exemplified in all cases in which the joint effect of several causes is identical with the *sum* of their several effects."[108] While it is true that, if the connection is merely that of a sum, then the deduction does proceed *a priori*, like any other inference based on the mathematical notion of addition,[109] it is also the case, as we know, that a mere sum is not a relational whole, which means that the deduction cannot be wholly *a priori*. One can argue, as Mill does, that the inference is wholly *a priori* only if one neglects the need for a premise concerning relations (and also the need for a composition law). In short, *when Mill proposes the "deductive" method for the social sciences, he is neglecting to take into account social relations as relevant factors*. It is much as if Newton failed to take into account the relative positions of the planets when he inferred the forces acting in the solar system from the assumption that gravity would act among the planets and the Sun, taking them pair-wise, but then, Mill's account of mechanics implies that Newton did just that![110]

We may therefore conclude that Mill did not provide an adequate account of the scientific method appropriate for the social sciences. More to the point with regard to the problem of structure, Mill, when he (implicitly) denies the causal relevance of relations, is in effect denying the existence of social relations; he is proposing that *social structure is nothing real*. But of course, as the idealists insisted, social structure *is* real. Even Mill must accept this. Consequently, he must also accept the idealist position on social relations. While

---

107 *System of Logic*, Volume 8, VI, ix, 1, 896.

108 *System of Logic*, Volume 7, III, vi, 1, 371; emphasis added.

109 It is worth noting a further confusion on Mill's part, between the *logical* notion of "conjunction" and the *arithmetical* notion of "addition." The inferences of logic and mathematics both proceed *a priori*, but they are for all that very different sorts of inference. This was finally made clear to empiricists by Mill's intellectual heir, Bertrand Russell.

110 Samuel Hollander, *The Economics of John Stuart* Mill, 2 vols. (Toronto: University of Toronto Press, 1985), fails to note this central defect of Mill's proposed methodology of social science. To be sure, he does mention the "Composition of Forces" (1:91), but he does not notice how Mill lacks an adequate understanding of this principle.

the empiricist argues against this account of social relations, any realistic view of social relations, according to which they *do* exist, seems to drive him into accepting this very account. Socrates seems after all to be correct in rejecting Anaxagoras' naturalistic account of things.

Things ought not to proceed so quickly, however. As is so often true in philosophy, there is a reply. Consider once again the relational state of affairs (@ such that a is R to b). To obtain an objectively real connection, as the idealists (and Bradley, most articulately) argued, we must make it the predicate of something substantial, some one individual thing. For there to be such an individual, it had to be one thing that incorporated within itself both the individuals a and b. The problem is that if a and b are to be incorporated into one individual, then they could themselves be independent individuals. Thus, they had, as it were, to lose their reality as independent individuals and become absorbed into the substantial reality of the R-ish individual. They were indeed a and b but they were less than fully real; they were simply aspects of the more real individual that connected objectively into an R-ish structure. This more substantial thing of which a and be are simply aspects we represented by (a, b), and we represented it as being of the R-ish sort by the subject-predicate proposition r (a, b).

This is how the idealist got the empiricist into the problems about structure. There is another way out, however. It argues that the idealist in fact gets himself or herself—as well as philosophy—into the problems we have noted by insisting that all propositions are of the form *one subject, one predicate*. It is this that led Socrates and the idealists who came after him into accepting entities that are contrary to the Principle of Acquaintance and into proposing a non-empirical method in the social sciences. This, of course, was traditional from Aristotle to Hobbes and Leibniz; recall the remark of the latter to the effect that predicates do not have feet in two substances. However, if one holds to this position, then *either* one retains the independence and uniqueness of individual things only by giving up the reality of relations—denying, in other words, the obvious fact that things come structured—*or* one ensures the reality of structure, of relations, by guaranteeing that they are genuinely predicated of substances which, unfortunately, are not the ordinary things of everyday experience but rather wholes not given in everyday experience and in which the things of everyday experience are but aspects lacking independent reality.

In spite of the problems with relations that Mill had when dealing with the composition of causes, he did in fact see the way out: *not every proposition has only one subject*. Some propositions do have this form. Others, however, have a form not considered by the tradition in logic: *some propositions have two or more subject terms*. The former sort represents states of affairs in which the ordinary thing denoted by the subject term exemplifies the non-relational quality connoted by the predicate. The other sort of proposition, those with two or more subject terms, represent states of affairs in which the objects *jointly exemplify* the relation connoted by the (multi-term) relational predicate; they jointly exemplify the relation in the sense that they remain distinct from one another, equally real, and equally as real as the state of affairs that they and the relation form through the relation relating.

What is needed for this way out of the problems confronting the idealists is a new ontology: an ontology with 1) *objective* relations, connecting the objects related into unities, but where 2) in these unities, the objects related retain *objectively* their being as genuine individuals. The former is contrary to the traditional account of relations given by empiricists like Locke and Hume; the latter is contrary to the account of relations propounded by idealists like Green, Bradley, Bosanquet (and, we might add, Althusser).

Mill did not himself confront his idealist critics; they came later in the century. However, one can see two things. First, it is clear on empiricist grounds that there are no substantial forms and therefore especially no Absolute. Second, there are, contrary to Locke but in keeping with what the idealists were to assert, genuine relational unities—unities which are in fact given to us in our ordinary sensible experience of things and in which the objects related retain their reality as individuals. Peter and Paul are related by the relation of kicking into the state of affairs of Peter kicking Paul; the relation of kicking is really there but Peter and Paul remain fully the individuals they were before the unfortunate incident of the kicking. Just ask them. Such states of affairs, upon clear reflection, are compatible with the Principle of Acquaintance and are therefore as surely admissible into the empiricist's ontology as substantial forms are excluded. Moreover, it is a contingent matter that Peter kicked Paul; the world could have been otherwise. Contrary to what Socrates held and contrary to what Green and the other idealists held, *a relation does not imply a necessary connection*. Mill's ontology is clearly one that admits relations in the anti-idealist, anti-Bradleyan sense just indicated. He has therefore a reply to the idealist that enables him to escape the challenge that one can admit objective structure into one's ontology and one's philosophy of science only if one is prepared to accept idealism. In particular, he can do this even as he, no doubt inadvertently, denied (wrongly, as a matter of fact) that there are in the social sciences social relations as relevant variables.

He does allow that some relations are susceptible to a Lockean analysis. Furthermore, he agrees that a relation cannot be a third thing existing alongside the relata. "Dawn and sunrise announce themselves to our consciousness by two successive sensations. Our consciousness of the succession of these sensations is not a third sensation or feeling added to these …" On this point, Mill agrees with the idealists. However, there is a unity to these that cannot be further analyzed:

> To have two feelings at all, implies having them either successively, or simultaneously. Sensations, or other feelings, being given, succession and simultaneousness are the two conditions, to the alternative of which they are subjected by the nature of our faculties, and no one has been able, or needs expect, to analyse the matter any further.[111]

Resemblance, too, is a basic, unanalyzable feature of things and of properties.

Mill is here arguing, in effect, that there are certain facts in the world that cannot be represented by propositions consisting of a single subject and a non-relational predicate. "In the same manner, as a quality is an attribute grounded in the fact that a certain sensation or sensations are produced in us by an object, so an attribute grounded on some fact into which the object enters jointly with another object, is a relation between it and that other object."[112] Mill's point is ontological: there are relational facts in which the relation is an attribute jointly of two objects. Recall Green's objection to the empiricist,[113] or at least the Humean, account of relations: the relation of succession connecting two events cannot be a third felt event for then the related events would not be related after all, nor can it somehow be reduced to the two events it relates. Mill's point is that the relation is not a third felt event and is not something that can be reduced to the two events which are its relata. It is not something exemplified by one entity but rather its own kind of entity, something that is exemplified by the two relata jointly. That it might be this sort of entity does not even occur to Green, but insisting on it, as Mill does, in fact provides an adequate naturalistic response to Green's idealism.

The point here is ontological, but it has implications—radical implications—for logic, as De Morgan was to recognize. Unfortunately, Mill does not examine, or even really notice, the implication for logic that, besides sentences of the form, "This is F," logic must include sentences of the form, "This is R to that." The traditional logic does not admit sentences of the latter sort. It was De Morgan who drew attention to the fact that the traditional logic could not account for the validity of the inference, "A horse is an animal; therefore a head of a horse is a head of an animal." The relational predicate "x is of y"

---

111 Mill, *System of Logic*, Volume 7, I, iii, 10, 69.

112 *System of Logic*, Volume 7, I, iii, 10, 68.

113 See the passage from "Can There Be a Natural Science of Man?" quoted above, note 34.

adds something with which logic ought to deal—after all, it ought to be able to show why this argument is valid (which it clearly is)—though it has no place in the traditional logic nor, indeed, in the formal logic that Mill allowed. Mill had an insight, an ontological insight, which had implications for formal logic, but he failed to develop them. Instead, it fell to De Morgan to work out a newer logic that admitted of dyadic as well as monadic predicates. De Morgan's work, along with that of Boole, opened up logic into the broad science that it has become, a science within which the formal logic of Aristotle and Mill is but a fragment. It is also safe to say that Mill was not fully aware of the implications of his views for ontology. It was William James and Bertrand Russell who developed the ontological insight into a systematic response to the idealist. But the essential point, both for logic and for ontology, was given by Mill.[114]

There is a second aspect to Mill's views on relations, specifically relations in the social sciences. Mill in fact gives an account of the relations that define social institutions that is rather different from the ontological analysis we have just noted. He views social institutions as constituted by inter-related roles, with people acting in ways characteristic of the institution. He refers to this when he speaks of a "consensus" in social groups and society in general, "… similar to that existing among the various organs of man and the more perfect animals, and constituting one of the many analogies which have rendered universal such expressions as the 'body politic' and 'body natural'."[115] The point is the same: an institution is not a heap of individuals any more than an animal is a heap of organs. Rather, an institution is constituted by *relations among the parts*, in this case, coordinated action and behaviour. These ways of acting and behaving are determined by the norms that define the roles, with people moved to conform to these norms by their moral sentiments. They have learned to value those characteristic ways of behaving.

One of the simplest, yet most fundamental, of social institutions, and one to which Mill gave considerable thought, is that of promising.[116] There are two roles here, that of promisor and that of promisee. Two persons, ego and alter,

114 It is worth noting that Herbert Spencer also recognized that relational structures exist, that they cannot be analyzed as Locke and Hume had suggested, and that the relata retain their individuality in the relational state of affairs. He argued that our capacity to recognize relational structures, such as spatial relations, is innate in humans, but that these innate capacities are developed through an evolutionary process. See his *Principles of Psychology*, 2 vols. (New York: D. Appleton, 1902), 2:195.

115 Mill, *System of Logic*, Volume 8, VI, ix, 2, 899.

116 Cf. Mill's account of promising to be found in Chapter V of his essay on "Utilitarianism" in *The Collected Works of John Stuart Mill*, ed. J. Robson, Volume 10 (Toronto: University of Toronto Press, 1969), 240–259.

enter into the roles of promisor and promisee when ego says to alter, "I promise you that I shall do x." Humans early acquire the moral sentiment that people ought to keep their promises. That is, there is a general pattern of behaviour (gp) such that every promisor does for the promisee what was promised. This general pattern of behaviour has attached to it moral sentiments such that people feel that it ought to be that every promisor does for the promisee what was promised (*). These moral sentiments shared by all people are, of course, learned or acquired through association, rather than innate as thinkers in the idealist tradition, such as Mill's great opponent, William Whewell, had claimed,[117] but they are very deeply embedded in our human nature.[118] The point is that the sentiment that people ought to conform to the norm (*) moves people to conform their own behaviour to it. Thus, when the promisor and promisee enter into their roles by ego promising alter to do x, ego comes to feel the moral obligation to do x, what was promised, and also feels that alter has the moral right to expect that x be done. Similarly, alter comes to feel that ego is morally obliged to do x and to feel that he or she has a moral right to expect that x be done by ego. Since we feel that it is obligatory that all conform to this norm, we take care to raise children so that they conform to the rule (*); we put them in situations where they come to internalize this standard,[119] i.e., in situations where pleasure comes to be associated with behaviour conforming to this rule, both their own behaviour as well as others'. Moreover, since we feel that it is obligatory that all conform to (*), when a promise is not kept we are liable to take steps to punish the violation, or at least we feel that it is morally appropriate to punish the violator.[120]

There are of course other rules besides (*) which are involved. There is, for example, the pragmatic rule of language that makes 'I promise …' a performative utterance that moves ego and alter into the roles of promisor and promisee. Moreover, it is clear that most institutions have many more roles than two and that the norms governing those roles are much more complex. Mill did not subject these institutions to detailed analysis and, although sociologists since have gone into greater detail on occasion, it is still true that most institutions have not been carefully mapped. What, for example, were the differences

117 Cf. F. Wilson, "Mill on Justice," in *J. S. Mill on Justice*, ed. L. Kahn (London: Palgrave Macmillan Press, 2011).

118 Cf. Mill, "Utilitarianism," Chapter V.

119 I am not suggesting that "internalization" is an *explanation* of this event; it merely describes it. The explanation is to be found in the learning theory which provides a law or generalization that relates learning experiences to the fact of internalization.

120 Mill, "Utilitarianism," Chapter V.

in the norms defining the roles of Chief Examiner as opposed to clerk in the East India Company? Often enough, no doubt, there would be very little to be gained by providing an abstract description of the rules, formal and informal, that define social roles. The point here is that if one is to think clearly about social relations, then one must have at least a basic outline for how one proposes to think about them. Mill has provided himself with this in his account of promising.

This basic account, or "model" if you wish, can be extended to other social relationships, e.g., kinship relations, where, corresponding to the norm (*), we have the norm (**): it ought to be that every person in the bear clan marries a person who is not in the bear clan. What is important about norms like (*) and (**) is that, provided the moral sentiments are in fact efficacious in motivating people, then these sentiments will bring it about that the pattern deemed obligatory will in fact hold in the group. Thus, if the sentiments attaching to (*) are efficacious, then the generalization (gp) will *as a matter of fact* truly describe the people in the group. Similarly, if the sentiments attaching to (**) are efficacious, then it will be a true generalization about the group that (gp2): every person in the bear clan marries a person who is not in the bear clan. Rules such as (*) and (**) provide models that people have of their social relations. If these normative models are efficacious, then they will in fact be true models.

Note that (*) and (**) are models in the minds of those *in* the social group. They are normative for these people. Insofar as they make descriptive claims (gp) and (gp2), these generalities are, upon the empiricist account, explanatory of the observed behaviour of members of the society.[121] These prescriptive norms, which provide an explanatory model for those in the group, can also constitute an explanatory model for the scientist who is studying the group; the patterns they prescribe are explanatory not only for those in the group but for those studying the group. However, those studying the group must also take these patterns as the object of study. The social scientist thus not only uses the model but also studies it.[122] Often he or she studies it in order to understand how change might be effected in the institution being studied. One's cognitive interest in the institutional structure might be the disinterested concern of the research social scientist or the pragmatic interests of the reformer or administrator, but in any case does not require one to have internalized the norms governing the institution. Thus, it has been suggested that, with regard to the

121 Cf. Claude Lévi-Strauss, *Structural Anthropology*, trans. C. Jacobson and B.G. Schoepf (Garden City, N.Y.: Doubleday [Anchor], 1967), 273–274.

122 *Structural Anthropology*, 121; cf. Levi-Strauss, *Elementary Structures of Kinship*, 2nd ed., trans. J.H. Bell, J.R. von Sturmer and R. Needham (Boston: Beacon, 1969), Preface.

social institutions of India, John Stuart Mill lacked "any special regard for existing institutions or traditions, except that they formed the given, the datum line in any particular case."[123] For those who are participating in the institution, however, the structure is not, or at least not merely, an object of study but a set of norms to which their moral sentiments move them to conform.

In recognizing this, we also recognize a further explanatory element. The moral sentiment expressed by the normative statement, "It is obligatory that p," brings it about that p. That is, there is a causal relationship to the effect that (c): the sentiment that it is obligatory that p brings it about that p. In this sense, (*) and (**) *explain*, respectively, the patterns (gp) and (gp2) by means of regularities like (c). In other words, we can explain the observed patterns by appeal to the causal efficacy of the moral sentiments that makes such behaviour felt to be obligatory. The model in the minds of the members of the group thus not only explains insofar as it is descriptive but also insofar as it is prescriptive. In fact, the latter explanation, since it is based on a more comprehensive causal principle (c) which takes into account more relevant factors, provides a fuller explanation of the observed behaviour.

Yet more comprehensive explanations are possible. (c) by itself is imperfect, holding only under certain conditions; specifically, it holds only because of certain learning situations. What we in fact have is something like the following: whenever a person is in a learning situation of such and such a sort, then that person comes to be such that he or she feels the sentiment that it is obligatory that p (L). Mill argues that sentiments such as this are *learned*. In order to explain why persons feel the sentiment, we turn to their past history, the learning experiences they have had. These experiences, together with the law (L), provide a (covering law) explanation of why persons feel the sentiment. This, when conjoined with (c), explains (via the covering law model) why they act as they do. The law (L) yields a yet more comprehensive explanation than (c). Of course, (L) as stated is just a sketch of a law or theory, specifically a sketch of a psychological theory of learning. For Mill, of course, the relevant theory of learning is associationism—or, behaviouristically re-stated, classical conditioning—with more than a touch of re-enforcement theory.[124] As Mill puts it, "The laws of mind ... compose the universal or abstract portion of the

123 R.J. Moore, "John Stuart Mill at East India House," *Historical Studies*, 20 (1983), 497–519, 518.

124 For a discussion of the connections between John Stuart Mill's theories of psychology and more recent theories, now stated in the language of behaviourism, see F. Wilson, *Psychological Analysis and the Philosophy of John Stuart Mill* (Toronto: University of Toronto Press, 1990), Chapter VIII.

philosophy of human nature; and all the truths of common experience, constituting a practical knowledge of mankind, must to the extent to which they are truths, be results or consequences of these."[125] The point which needs to be emphasized here is that Mill's account of social theory and of the methods for justifying its acceptance for purposes of explanation and practice do not presuppose the commitment to any specific theory of learning. Associationism or classical conditioning is one possible theory. There are others, such as that of Piaget, which one could hold. Mill, of course, was prepared to defend associationism, and the movement of his own thought, before, during, and after his mental crisis, remained within this context. The account of social theory, its justification, and its relation to psychological theory do not depend upon the details of Mill's own theory of learning, however. All that is required for these is that there be some learning theory or other—some account of how the norms for social roles become internalized and how our moral sentiments become attached to these patterns of behaviour and not others.

If the structural patterns (*) and (**), or what, for purposes of explanation, is much the same, the patterns (gp) and (gp2), are explanatory, then at the same time it is essential to recognize that the events that these patterns describe are the evidential basis for any claim or thought that those patterns in fact describe correctly the social reality. The observed social relations provide the data for the scientist attempting to ascertain the social structure.[126] At the same time, those relations are a consequence of that structure. Lévi-Strauss seems to put the relationships in pretty well the correct order: observed social relations are the raw material in which social structural relations "inhere"[127] and out of which non-statistical models are "constructed." Lévi-Strauss distinguishes "mechanical" from "statistical" models; he characterizes this distinction as one of level, but it is clear that with a "mechanical" model on the level of structure, there is associated a "statistical" model which describes, with some degree of precision, the actual distributions of behaviours that conform and do not conform with the non-statistical or mechanical model.[128] Such models explain structural features exemplified in concrete social relations and are "translatable" into statistical models.[129] The structural models serve to explain

125 Mill, *System of Logic*, Volume 8, VI, v, 1, 861.

126 "... social relations consist of the raw material out of which the models are built, while social structure can, by no means, be reduced to the ensemble of social relations to be described in a given society" (Levi-Strauss, *Structural Anthropology*, 271).

127 Cf. Lévi-Strauss, *Elementary Structures*, 483.

128 Lévi-Strauss, *Structural Anthropology*, 275–276.

129 Lévi-Strauss, *Structural Anthropology*, 292.

the social relations that can be observed, with the approximation to reality of those models measured by the probabilities in the statistical model. To be sure, it is an approximation, but if the approximation is close enough, then a degree of social control is possible. As Mill makes clear in his discussion of natural scientific explanations in the social sciences, the empirical generalizations which are the starting point of any social science are only "approximate generalizations,"[130] but these laws which cannot be used to give firm predictions are nonetheless useful: "… whenever it is sufficient to know how the great majority of the human race, or of some nation or class of persons, will think, feel, and act, these propositions are equivalent to universal ones."[131]

Such is Mill's empiricist account of explanation in the social sciences. It is evident that he has a clear doctrine of social relations. Of course, we did see earlier that in laying out the basic framework of his philosophy of the social sciences, he implicitly denied that there are any social relations and therefore that there are any such relevant variables. Since social relations do exist and are relevant variables, this left room for the idealists to move in with their account of the objective reality of relations. The result was a holism which Mill rejected on the grounds that the social wholes it asserted to exist did not pass the test of the Principle of Acquaintance. We now see that Mill did have a perfectly reasonable account of social relations and that we are therefore not forced back into an idealist account of the explanation of social phenomena. We do need to recognize, however, that *it is possible that* in social groups of a smaller size, say less than n, the relations causally relevant to the social process are different from the relations that are causally relevant to the social processes in groups of larger size (larger than n as we are suggesting). Thus, some have suggested that mob behaviour is different and could not be predicted from behaviour in smaller groups. People in mobs sometimes do things they would not instinctively do given the moral principles by which they are moved in everyday normal contexts. Can the laws of learning that apply in the latter sort of case also apply in the former? Perhaps. Whether this be so or not, it is at least *possible*. However, one would wonder, I think, how the patterns of behaviour that occur only in mobs were learned or acquired—just how do the laws of learning differ in the two cases?—but that sort of question need not detain us. Whether behaviour in large groups is different in kind from that in smaller groups is an empirical, not a philosophical, question as is the issue (supposing there to be such different patterns of behaviour) of how the

130 Mill, *System of Logic*, Volume 8, VI, v, 1, 862.
131 *System of Logic*, Volume 8, VI, iii, 2, 847.

patterns characteristic of the larger groups are acquired. Perhaps behaviour patterns for large groups might be a sort of emergent behaviour not derivable from, nor, therefore, reducible to, behaviour patterns characteristic of smaller groups; perhaps. In Mill's way of putting these issues, perhaps the laws of learning have a *chemical* aspect, yielding for large groups a person *different in kind* from the sort of person found in small groups. None of this commits one to any sort of holism of the idealist sort.

We can conclude that Mill, although he has a reasonable empiricist account of social relations, missed these sorts of issues because he failed to recognize the real relevance of social relations. Once we become clear on the nature of social relations, the existence of these issues can, we now see, fit into a naturalistic philosophy of science; they do not force upon us any sort of idealistic philosophy of science. In particular, on the Millian empiricist account of social relations, the individuals that are the relata do not lose their reality as separate individuals; in no sense are they absorbed, or their reality dissolved, into social wholes, the totalities characteristic of idealist, including Marxist, philosophies of science.

6 Green Again: Can There Be a Natural Science of Man?

There is another issue, however, that is raised by other idealist critics of naturalist accounts of the explanation of human being. This is the argument, discussed above, that the mind must be, as Socrates argued against Anaxagoras, a simple unity, and not a multiplicity, as Anaxagoras proposed—not a "mere" bundle of perceptions, as Hume argued much later. The argument runs as follows: as the world of appearances is a multiplicity of diverse things, to recognize a group of these appearances to be unified as an ordinary perceptual object such as a chair, the consciousness that recognizes the unity must itself be a unity. We have seen how Plotinus developed this argument against any naturalistic science of mind which takes its subject matter to be a sequence of inner events. Platonists continued to defend it against the naturalists down through the centuries.[132] And so we find it in T.H. Green.

Green developed the point in detail in his "Introduction" to Hume's *Treatise*.[133] He argues that, for Hume, the object of perception is complex. It

132 Cf. F. Wilson, "The Impact on Psychology of Bradley's Humean View of the Self," *Bradley Studies*, 5 (1999), 5–44.

133 T.H. Green, "Introduction to *Treatise of Human Nature*," in Volume I of *David Hume: The Philosophical Works*, ed., T.H. Green and T.H. Grose (London: 1886; re-print, Aalen: ScientiaVerlag, 1964).

consists of a bundle of impressions: visible, tactile, olfactory and other sorts of impressions.[134] Green goes on to ask "… whether a multitude of such impressions can be present at once, i.e., as one impression.[135] He concludes that "… we have to suppose the presentation of the succession of feelings to a consciousness [which is] not in succession."[136] Similarly, when it comes to the self as a bundle, it really is a succession of perceptions or feelings. But we are in fact aware of this as a whole. This "… implies that a feeling, which is one among a succession of feelings, is at the same time a consciousness of these feelings as successive and under that qualification by mutual relation which implies their equal presence to it."[137]

Green elsewhere[138] resisted the requirement that Hume's position seems to entail with respect to the self: a sense of unity which is itself a feeling, and, as a feeling, part of the bundle which is the self. In other words, the self that Hume's (and Anaxagoras') philosophy requires is a self in which a part could represent the whole, and thereby provide the unity any account of the self must provide. That special feeling which is of the whole, and provides the unity of the latter, Green argued, is not a feeling but rather the "thinking subject," a unitary substance which is presupposed by Hume's unity felt in self-consciousness but which, at the same time, he is compelled to deny. Green argued that a whole consisting of separable parts could have no unity; no part of such a whole could provide that unity. Where such a whole did have unity, the latter derived from a timeless self-consciousness that comprehended the whole. "To be conscious of it [one's personal history]," Green tells us, "we must unite its several stages as related to each other in the way of succession; and to do that we must ourselves be, and distinguish ourselves as being, out of that succession":

> It is only through our holding ourselves aloof, so to speak, from the manifold affections of sense, as constant throughout their variety, that they can be presented to us as a connected series, and thus move [us] to seek the conditions of connection between them.[139]

Green's argument for this is not so much to deny Hume's point as it is to insist that the *relational structure* of the self presupposes a self that is outside

134 "Introduction," 198.
135 "Introduction," 204.
136 "Introduction," 219.
137 "Introduction," 296.
138 T.H. Green, *Prolegomena to Ethics* (Oxford: Clarendon Press, 1906).
139 *Prolegomena*, 92.

time. According to Green, both selves and material objects have as their basic constituents entities given in sense experience, or, as he calls them, "feelings." On this point he is in agreement with Hume. However, "… feelings *are* facts; but they are facts only so far as determined by relations, which exist only for a thinking consciousness and otherwise could not exist.[140] Structure can be provided only by a thinking, conscious, self-conscious, unity or substance in which the whole as such exists, or, rather, the unifying activities *of* the whole.

The unity that Green insisted must be there was fairly traditional; it was that of an active substance. In his "Introduction" to his edition of Hume's *Treatise*, Green praised Berkeley for, in effect, keeping the simple self that Hume attempted to eliminate. How else indeed than by the activity of such a self can one account for structure and order?

Ordinary things are structured bundles of sense impressions. Each impression in the bundle is itself but a fleeting thing. A fleeting impression is made permanent, made into the thing that it *is*, by the relations in which it stands to other things. "It is permanent with the permanence of the combining and comparing thought which alone constitutes it. …"[141] The relation is not given in sense, but the thing does not exist in its being apart from the relation. This relation is contributed to the thing by something which is outside the thing:

> … just so far as they are qualified by likeness or unlikeness to each other, they must be taken out of that succession by something which is not itself in it, but is indivisibly present to every moment of it. This we may call soul, or mind, or what we will. …[142]

Thus, on Green's account, if there are indeed any genuine relations—which, of course, there are, then we end up in an idealism that requires, contrary to Hume, a doctrine of a substantial self:

> … any such modification of Hume's doctrine of the singleness and succession of impressions as will entitle us to speak of their carrying with them, though single and successive, the consciousness of their resemblance to each other, will also entitle us to speak of their carrying with them a reference to that which is not itself any single impression, but is

140 *Prolegomena*, 53.
141 *Prolegomena*, 151.
142 *Prolegomena*, 176.

> permanent throughout the impressions; and the whole ground of Hume's polemic against the idea of the self or spirit is removed.[143]

Things sensible are unified into wholes by the activity that accounts for them and connects them. These activities that unite things into relational wholes are not given in sense. Nor, of course, is the substantial self to which these activities belong. Green's account of relations is idealist in the sense that, since the structure in the world produces a genuine connection among the structured entities, then that structure enters into the very being of those things. Indeed, his vision of those relational structures is very much in the substantialist tradition: the unity achieved is achieved by the activities of a substantial self.

Green thus restores to philosophy the substantial self. It plays the role that it had played ever since Plotinus: that of accounting for the unity of consciousness, including the unity of self-consciousness. "… feelings *are* facts; but they are facts only so far as determined by relations, which exist only for a thinking consciousness and otherwise could not exist.[144]

To this the empiricist or naturalist does have a response. Green came too late for Mill to confront directly, but he did face up to the argument. The argument derived from Plotinus and Green took it from this tradition, a tradition to which Mill attempted an empiricist response.

Mill had developed an account of things as groups of sensations—structured groups of sensations. Most of the relations were lawful, that is, upon the empiricist account, matter-of-fact regularities. Needless to say, none of them were understood in idealist terms of the sort proposed by Green and, more articulately, by Bradley. But however that went, there remained the argument that to perceive the set of sensations as unified, it must be present to a unified consciousness which is outside, not only the sensation, but also the multiplicity of the stream of perceivings of the sensations. Such a self, transcending the world of entities given in sensation and in inner awareness, violates the Principle of Acquaintance. It therefore has no place in any world proposing a naturalistic account of, among other things, human beings. Mill did in fact take up the argument developed by Plotinus and Green.

Mill argues in this way:

> If … we speak of the Mind as a series of feelings, we are obliged to complete the statement by calling it a series which is aware of itself as past

143 *Prolegomena*, 176.

144 *Prolegomena*, 53.

> and future; and we are reduced to the alternative of believing that the Mind, or Ego, is something different from any series of feelings, or possibilities of them [Green's idealist position], or of accepting the paradox, that something which *ex hypothesi* is but a series of feelings, can be aware of itself as a series.[145]

The first alternative is wholly unacceptable. The empiricist or the naturalist can therefore do no other than accept the second—accept it *as a fact about the world* as we experience it in sensation and inner awareness, puzzling though that fact may be. It is a fact about the mind, a fact among other facts. But how can we think about what does not exist? What do we say, if anything, about the nature of *intentionality* (to use Brentano's term) that accounts for our capacity to contemplate things that are not immediately present to us? The puzzle is, as Mill puts it, how is it that "… something which has ceased, or is not yet in existence, can still be, in a manner, present …?"[146] How, indeed, can we think of things which could never be present to us: things that are too far away, perhaps, or are too small, or do not exist or even those that could not exist? There is a real puzzle in this last item. How can we think what is contradictory? How could the impossible be there before consciousness (as it must sometimes be, as when we consciously deny that a contradiction is true)? While it is a puzzle, as Mill indicates, it is also a fact—*a matter of empirical fact about our world and ourselves in the world*. It is a fact which is there, simply to be accepted, a fact which we cannot, or cannot yet, explain in terms of the only empirical theory of mind that we have, namely, associationism. So we accept it as a fact that "a series of feelings, the infinitely greater part of which is past or future, can be gathered up, as it were, into a single present conception accompanied by a belief of reality."[147]

Let this be emphasized: *it is accepted as a fact, perhaps a puzzling fact, but not as a puzzle that demands that it be solved at any cost, that is, "solved" by inventing an entity like a transcendental ego which is even more puzzling*, and, worse, interferes with the progress of the empirical or naturalistic science of mind.

If we do seek a theory about how one acquires a feeling of personal identity, then this theory has to be in terms of the empirical science of learning, i.e., associationism. It has to be a theory like in kind to the associationist theory

145 John Stuart Mill, *An Examination of Sir William Hamilton's Philosophy*, in *The Collected Works of John Stuart Mill*, Volume 9, 1979, 194.

146 *An Examination of Sir William Hamilton's Philosophy*, 194.

147 *An Examination of Sir William Hamilton's Philosophy*, 194.

of perception—the “Psychological Theory of Matter, of the Non-ego,” as Mill puts it[148]—that had been generally accepted by associationists since Berkeley, including Mill. We do not have such a theory of how the learning proceeds but we should not mark this state of unknowing by trying to introduce some sort of transcendental soul or self or ego to make it seem that we do know after all. “I think,” says Mill, “by far the wisest thing we can do, is to accept the inexplicable fact, without any theory how it can take place; and when we are obliged to speak of it in terms which assume a theory, to use them with a reservation as to their meaning.[149] This is sage advice.

7 A Naturalistic or Empiricist Response to Bradley’s Account of Structure

We now see that the naturalistic philosophy of science, and in particular the naturalistic understanding of the sciences of human being, can be defended against the variety of idealist responses deriving from Socrates’ response to Anaxagoras. The problem Socrates posed was that of *structure*. It was argued, as Socrates argued against Anaxagoras, that the naturalist could not account for structure, whereas the idealist had an account of relations which did provide an ontologically adequate account of the structure of the world. Bradley’s ontology of relations was the most articulate of the idealist proposals.

Now, we have seen that there are in fact defects in this idealist position. Most importantly, individuals it counted as related ceased to be fully individuals and became but aspects of the whole which is the relational fact. Mill produced an ontology of relations which did not suffer from this defect. Instead of giving up the individuals, as Bradley did, Mill gave up the idea that every proposition was subject-predicate, and allowed that some propositions could have two or more subject-terms, and a multi-placed relation predicated of them jointly. The core notion was that, when related, individuals did not lose any of their reality or any of their individuality. The point was fundamentally a phenomenological one: when experienced as related, individuals retained their distinctness and did not blur one into the other or vice versa, as was required by the idealist view of structure, or at least as required by Bradley’s view of structure.

148 *An Examination of Sir William Hamilton’s Philosophy*, 194.
149 *An Examination of Sir William Hamilton’s Philosophy*, 194.

William James was to carry on Mill's phenomenological argument based on the Principle of Acquaintance. Thus, in an early philosophical essay, "On Some Hegelisms," he notes:

> The world of qualities is a world of things almost wholly discontinuous *inter se*. Each only says, "I am that I am," and each says it "on its own hook" and with absolute monotony. The continuities of which they *partake*, in Plato's phrase, the ego, space, and time are for most of them the only grounds of union they possess.[150]

Characteristics of things, taken to be entities that satisfy the Principle of Acquaintance, and the things of which they are the characteristics, are wholly self-contained, as it were; each is what it is in itself and there is nothing about its being that implicates some other entity. This means that such an entity may be related to other entities, but those relations are not ties that mean its being is necessarily connected to any other such entity; connections among these entities are contingent, not necessary. James, in effect, is making the point we made earlier: where 'F' and 'G' are empirical concepts satisfying the Principle of Acquaintance, then any regularity to the effect that "All F are G" or, in language made familiar to us by Russell, (x)(Fx then Gx), is contingent, not necessary. This is not so for the idealist, however. According to such a philosopher, any empirical regularity of this sort that is non-accidental will be justified as more than contingent, indeed, as necessary, by virtue of a necessary connection linking in their very being the forms F and G.

We see this claim about necessary connections—that is, objective necessary connections, already in Socrates where the idea developed against Anaxagoras was that the connections effected by the active forms among the things of the world of sensory appearances are *necessary*, not contingent and accidental. So far as we are acquainted with things in the sensible world, however, there are connections but there is no necessity to them since they are all contingent. "What feeling shall fill *this* time, what substance shall execute *this* motion, what qualities combine in *this* being, are as much unanswered questions as if the metaphysical axioms never existed at all."[151]

Also, contrary to the idealists, being related to other things does not undo the individualities of the relata. Each is complete on its own; it may be connected to other things, but such a connection does not compromise the reality of the thing as truly and fully individual. Contrary to Socrates, one event does

150 William James, "On Some Hegelisms," *Mind*, o.s. 7 (1882), 186–208; 189.

151 "On Some Hegelisms," 189.

not necessitate another; rather, many possibilities are there of which only a few become actual. Just which ones become actual is not determined as a matter of necessity by the event or events to which the first event turns out to be connected. There are connections, but these are only contingent. "The transitions are abrupt, absolutely, truly shot out of a pistol, for while many possibilities are called, the few that are chosen are chosen in all their sudden completeness."[152]

Certainly, so far as things appear to us in our ordinary sensible experience, there is no necessity to the connections among things:

> The existence of such syntheses as there are does then but slightly mitigate the jolt we get when we pass over the facts of the world. Everywhere indeterminate variables, subject only to those few vague enveloping laws, [are] independent of all besides—which seems the truth.[153]

Socrates is wrong in insisting that one thing necessitates another to which it is related. In the world of appearances, there is nothing about the entities we experience to indicate connections effected by, or from, something that transcends the sensible realm and somehow necessitates what goes on in the latter; the work of the forms is *not apparent* and, accordingly, is not allowed in the naturalist's ontology.[154]

As James was developing this phenomenological argument, Bertrand Russell was developing the implications for logic of Mill's new ontology of relations, and then using this new logic to attack the idealists and Bradley in particular.[155] On Bradley's view, as we have seen, a relational states of affairs, a is R to b, ontologically consists of a property r corresponding to the relation being predicated of a complex individual whole (a, b) in a proposition r (a, b) of the traditional subject-predicate form—one subject and one predicate. We must here note what Russell observed but Bradley did not—namely, that the relation of a to

152 "On Some Hegelisms," 190.

153 "On Some Hegelisms," 189–190.

154 This is the crucial matter-of-fact about the world that forms the basis for the metaphysics of logical atomism. See F. Wilson, "Bareness, as in 'Bare Particular': Its Ubiquity," in H. Hochberg and K. Mulligan, eds., *Relations and Predicates* (Frankfurt: Ontos Verlag, 2004), 81–112; reprinted in F. Wilson, *Acquaintance, Ontology and Knowledge*, 329–362. See also F. Wilson, "Placing Bergmann," in Laird Addis, Greg Jesson and Erwin Teigtmeir, eds., *Ontology and Analysis: Essays and Recollections about Gustav Bergmann* (Frankfurt: Ontos Verlag, 2007), 185–276, and the discussion of the holistic philosophy of W. Sellars in F. Wilson, "Acquaintance, Ontology and Knowledge" in F. Wilson, *Acquaintance, Ontology and Knowledge*, 1–108.

155 B. Russell, *Principles of Mathematics*, 2nd ed.

the whole (a, b) is the same as the relation of b to that whole. That is, the role of a in this whole is symmetrical with the role of b. Thus, (*) such that r (a, b) represents indifferently *both* the fact that (f) such that a is R to b *and* its converse, the fact that (fc) such that b is R to a. Where R is a *symmetrical* relation, i.e., one for which we have for any two things, if the first R is the second, then the second R is the first, then we have no problem: if (f) obtains so does (fc), and (*) can represent the two indifferently. But the same does not hold for *asymmetrical* relations.[156] Where R is asymmetrical, we have (+) such that for any two things, if the first R is the second, then the second is *not* R-ed by the first. In this case (f) obtains while (fc) does not. In the case of an asymmetrical relation, there is a difference—an ontological difference—between a relational fact and its converse that is not captured in an account, like the monistic, which requires both facts to be represented indifferently by the same notation. Since the Bradleyan account of relations was introduced precisely to solve the problem of relations and it cannot do that, it must be rejected.[157]

Russell had another argument against the idealist account of relations.[158] According to the latter, everything is related intrinsically to everything else. One can know what a thing is, in itself, only if one knows all these connections. One cannot know anything until one knows everything—until, that is, one knows the totality of things, i.e., the Absolute. But as finite beings, we can never know this totality in its infinity. The search after the truth of things inevitably fails, and we know beforehand that it will fail. In short, scepticism is the direct consequence of the idealist account of relations. Setting out to know something with absolute certainty, it turns out that we know nothing and can know nothing. Worse still, judgments falling short of the final grasp of the totality are all basically false because truth lies in the totality and only there. Setting out to know, we discover that what we think we know is no more than ignorance. We don't know anything to be true, and what we do know is that everything we think we know actually is false.

Contrast this to the naturalist position. Judgments may, upon the naturalist's view of things, all be fallible, but if they are wrong then that can be discovered. For a judgment to the effect that "All F are G" to be discovered to be wrong, one needs but one counterexample, an individual which is F and also

156 *Principles of Mathematics*, 221.

157 See F. Wilson, "Burgersdijck, Coleridge, Bradley, Russell, Bergmann, Hochberg: Six Philosophers on the Ontology of Relations," in F. Wilson, *Acquaintance, Ontology and Knowledge*, 287–328.

158 See B. Russell. "The Monistic Theory of Truth," in *Philosophical Essays*, Revised Edition (London: Allen and Unwin, 1966), 131–146.

not G. A judgment like "All F are G" is always in effect an hypothesis, repeatedly to be tested against the world as we experience it in everyday life. If a judgment is thus discovered to be wrong, then our judgments can be revised and a new hypothesis formed—one that accommodates the counterexample to our former judgment. The new judgment will still be an hypothesis, and will still be fallible, but we do know that this new hypothesis is closer to the truth, whatever ultimately that may be, in at least not being falsified by observations that eliminated our former hypothesis. In this sense the naturalist can grasp, albeit tentatively, the truth about the ordinary world, whereas the idealist, who locates the truth of the ordinary world in another world which transcends this world, finds that discovering the truth is impossible since it is beyond our cognitive capacities.

Yet things get even worse. Our knowledge of the world takes the form of propositions. Any proposition (the idealist holds) is of subject-predicate form, where the subject is asserted to have a certain property. But this involves a distinction between the subject and what is predicated of it. This, in turn, means that any proposition involves a *relation* between subject and predicate. Upon Bradley's account of relations, these two terms must turn out to be aspects of a greater whole, the totality. That knowledge, then, cannot be propositional, and therefore if we do know this entity, then we cannot *say* it. If we do know the totality, the Absolute (which we could not, since it is beyond our finite minds), then we could never say or express that knowledge; *the Absolute, the supposed truth of everything in the world of ordinary appearances, is ineffable, and in that sense not knowable.* We may, as Bradley held, in some sense experience the Absolute, but that will never count as "knowledge" in any ordinary sense of the word.

The downward spiral of the idealist account has not come to an end. If one knows something, then there is a distinction between the knower and the known. But that means that there is a *relation* between the knower and the known. According to the Bradleyan model, relata as related cease to be individual things and become merely aspects of a larger whole; the distinction between knower and known disappears into a deeper unity. It follows that *you*—the individual person you are and experience yourself as being—cannot in truth know the Absolute. The distinction between you and the Absolute disappears for *all there is to you knowing any truth at all is a mystical unity with the unitary totality of things.*

Thus, the idealist, aiming to know inerrantly the truth of our world, ends up knowing nothing, at least in any ordinary sense of "know." The idealist response to the naturalist ends up in a radical skepticism which is a good reason for not taking it seriously.

8 Once More Necessity: Empiricist Response and Idealist Counter-Response

The story does not end here, however.

8.1 *Hume: an Empiricist Account of the Necessity of Causal Laws in Science*

We must recognize that there is a point to what Socrates says, and which the naturalist has not (in our telling of the tale, anyway) yet addressed. I am thinking of the point about the need for a tie, for a *necessary connection.* Without such a tie, without objective necessities, nothing can be explained. This is the point that Socrates, or Plato's Socrates, was trying to make. This is the point that Aristotle made when he insisted that the syllogism (s), "All human is animal, Socrates is a human, therefore, Socrates is animal," could not be explanatory if the major premise was understood to be nothing more than a matter-of-empirical-fact regularity. He argued that it could be explanatory only if the major premise could be demonstrated in a *scientific syllogism* (ss): "All rational is animal, *all human is rational,* all human is animal." Such a syllogism is taken to display the logical and the ontological structure of the species human. This form is given equivalently by the *real definition* of the species human: "human is rational animal." This real definition is a necessary truth—a timeless truth about the logical and ontological structures of the forms of things or "substances" as Aristotle would call them. These structures are not empirical but in fact transcend the empirical. That is what makes these connections *necessary*; we discern these structures, these ontological structures of being, not through sense but through a rational intuition. What makes the syllogism(s) explanatory is the fact that the empirical generalization which is its major premise can be demonstrated in a syllogism (ss) as a necessary truth about the forms or kinds mentioned in the major premise of (s). The necessary connections among the forms provide the necessary connections which make the major premise of (s) necessary, and therefore make (s) explanatory, rather than being merely contingent and accidental, in which case (s) is not explanatory.

This sort of objectively necessary tie is absent from Anaxagoras, or at least from what we know of Anaxagoras. In any case, this was the reason that formed the basis of Socrates' criticism of Anaxagoras and so on throughout the history of naturalist philosophy. Such a tie is likewise absent from the purely natural world of Hume and the (early) logical positivists. The tie or necessary connection which is missing from the worlds of these philosophers is just the real connection that distinguishes those patterns or regularities which are laws and those that are "mere" regularities—mere "accidental" generalities. In fact, as we have argued, these are entities with which we are not acquainted in the world of our ordinary sensible experience. They are excluded from our ontology by the Principle of Acquaintance. At the same time, we seem to need them in order to distinguish those patterns which are lawful or necessary and those which are accidental and merely contingent. We seem, after all, to be driven, like Socrates, to find a world outside the world of ordinary experience—a world that transcends our ordinary world, but a world in which we find the tie that we must grasp if we are truly to have the reasons which explain things in the ordinary world. Otherwise, all things are loose and everything is an accident, including, we should add, our knowledge of the world The connection between knower and known can only be accidental too, a point, as we saw, taken up by Green.

But the critic must be careful. This is more rhetoric than careful analysis. For something to happen by accident is for it to happen without apparent cause, or for its happening to be unintended. Some things in the universe of the naturalist do happen without apparent cause, of course, but most do not. Thus, to say that everything happens by accident is simply false and the notion that the whole somehow turns out to be a matter of accident makes sense only if one supposes that the whole as a whole might have a cause, a supposition which, for the naturalist, is devoid of sense. To suggest that, for the naturalist, every thing is unintended, therefore, is either simply wrong since many events in the world are intended (though of course most are not), or simply wrong since the idea that the universe as a whole might somehow be either intended or unintended lacks any cognitive sense.

Still, as we have suggested, there is a point to the criticism. What can be the response of the naturalist or empiricist to the challenge of the idealist—or, if you wish, of Socrates—that the former cannot account for the necessity of causal connections? We can best approach this issue, I think, by way of the philosopher who first thought it through clearly, David Hume.

In his discussion of causation, Hume comes to a conclusion about its nature in which he gives two definitions of "cause."[159] The first definition is this: "… an object precedent and contiguous to another, and where all objects resembling the former are plac'd in a like relation of priority and contiguity to those objects, that resemble the latter." In other words, one event causes a second where the two are subsumable under a matter-of-fact generality. Particular instances of causation are to be understood in terms of *de facto* regularities. This

159 Hume, *Treatise of Human Nature*, 172.

means that any proposition stating a causal regularity has the logical form "(x)(fx then gx)," where the descriptive predicates are logically and ontologically independent of each other.[160] This renders in the language of *Principia Mathematica* and in the empiricist's language the Humean point that cause and effect are logically independent of each other.

The idealist point follows: so far as propositional content is concerned, the assertion of a causal regularity involves no difference in logical form from the assertion of an accidental generality. Both are of the form, "(x)(fx then gx)," from which it follows directly that *objectively there is no logical difference between causal regularities and accidental generalities*. Objectively, then, there is no difference of a logical sort between *post hoc* and *propter hoc*. Yet, the idealist insists, such a distinction is to be drawn. Hume has been careful enough to note this point earlier in his discussion,[161] so here he states the question, "*What is our idea of necessity, when we say that two objects are necessarily connected together*."[162]

What he finds is that when two sorts are causally connected, upon the appearance of an object of the one sort "the mind is *determin'd* by custom to consider its usual attendant" and that it is this "*determination*, which affords [one] the idea of necessity."[163] This yields the second definition of "cause" as "… an object precedent and contiguous to another, and so united with it in the imagination, that the idea of the one determines the mind to form the idea of the other, and the impression of the one to form a more lively idea of the other."[164] What this means is that, if in a propositional content of the form "(x)(fx then gx)" the form of a regularity according to the first definition of "cause" is such that, as *a matter of psychological fact*, it is used to support assertions of subjunctive or contrary-to-fact conditionals ("the idea of the one determines the mind to form the idea of the other"), *then* the assertion of that proposition is the assertion of a casual generality. A generality is lawlike or causally *necessary* just in case it is in fact used to predict and to support subjunctive conditionals; otherwise it is a statement of "mere regularity."

There has been some controversy about this. How can one give two different definitions of the same notion without contradiction? J.A. Robinson has suggested that the first definition is the *real* definition and that the second is *not* really a definition but is a psychological comment on the relation so

160 *Treatise*, 139.
161 *Treatise*, 77.
162 *Treatise*, 155.
163 *Treatise*, 156.
164 *Treatise*, 172.

defined.[165] On the one hand, it is evident that there is a point to this, since it clear that the second definition introduces an element of psychology derived from and explained by the associationist theory. On the other hand, it does fly in the face of Hume's explicitly saying that he is giving two definitions. More seriously, however, if one construes Hume as giving only the first definition of "cause," then one is attributing to him what A.C. Ewing for one has called the "regularity" theory of causality, which holds that there is nothing more to causal assertions than assertions of *de facto* regular connection. It attributes to Hume an analysis of causation "as nothing more than an instance of a general uniformity of concomitance between two classes of particular occurrences."[166] But this theory may be criticized, as it is by Ewing, for being unable to distinguish causal associations from those that occur by chance—in other words, for being unable to distinguish laws from accidental generalities.

Another who makes this charge is Jonathan Harrison, who quite wrongly asserts that "He [Hume] also fails to distinguish between believing that whenever there is smoke there is fire, and that smoke is caused by fire."[167] The difficulty is that Hume never held this "regularity" theory.[168] The *Treatise* fully recognizes the distinction Ewing's "regularity" theory denies between *post hoc* and *propter hoc*. As it points out, "an object may be contiguous and prior to another, without being consider'd as its cause. There is a NECESSARY CONNEXION to be taken into consideration; and that relation is of much greater importance, than any of the other two above-mention'd."[169] Most of the rest of Book I, Part III is devoted in one way or another to an investigation of the nature of this necessary connection. Robinson thus commits Hume to a theory Hume explicitly denies. In fact, Robinson simply misses the whole point of Hume's second definition, which is to introduce the element of necessary connection that a "mere regularity" theory fails to capture. What needs to be done is to make clear how the two definitions are related to each other.

165 See J.A. Robinson, "Hume's Two Definitions of 'Cause'," Thomas J. Richard, "Hume's Two Definitions of 'Cause'," and J.A. Robinson, "Hume's Two Definitions of 'Cause' Reconsidered," all in V.C. Chappell, ed., *Hume* (Garden City, NY: Anchor Books, 1966), 129–147, 148–161, and 162–168 respectively.

166 A.C. Ewing, *Fundamental Questions of Philosophy* (London: Routledge and Kegan Paul, 1951), Chapter Eight.

167 J. Harrison, *Hume's Moral Epistemology* (London: Oxford University Press, 1976), 54.

168 See F. Wilson, "Hume's Theory of Mental Activity" in D. F. Norton *et al.*, eds., *McGill Hume Studies* (Los Angeles: Austin Hill Press, 1976), 101–120. For greater detail, see F. Wilson, *Hume's Defence of Causal Inference* (Toronto: University of Toronto Press, 1997), *passim*.

169 Hume, *Treatise*, 77.

Fundamental to understanding Hume's account of causation is the distinction that can be drawn with respect to our beliefs between the propositional attitude and the propositional content. If one person believes that Toronto is west of Hamilton and a second disbelieves it, the propositional content of both beliefs is the same, namely, the proposition that Toronto is west of Hamilton. But the propositional attitudes in the two beliefs are different. In one case, the attitude is that of believing or asserting; in the other, it is that of disbelieving or denying. The propositional content is true or false. Its truth-value depends only on the facts it is about. In particular, it does not depend upon what propositional attitude one has with respect to the content. This independence is important since it is precisely such independence which enables truth and falsity to function as *standards*, justifying the attitude one has. The attitude of believing or asserting with respect to a proposition is *objectively justified* if and only if that proposition is true while the attitude of disbelieving or denying is objectively justified if and only if that proposition is false. Clearly, if truth-value depended on attitude, the former could not provide an objective standard for evaluating the latter.

As Barry Stroud has said, "To believe that something *must* happen is to reject any possibility of its not happening."[170] Thus, any account of causal necessity must show how laws apply not only to actual events but also to, as it were, unrealized possibilities. The law that unsupported bodies fall not only describes the regular behaviour of actual things, but also the behaviour of possible things, that is, of supported bodies which are possibly unsupported, asserting that if they *were* unsupported they *would* fall. Of course, no empiricist could countenance the literal introduction into his or her ontology of unrealized possibilities, especially Hume, who accepts the point made by the Megaric position against the view of Aristotle—and in effect, that of Socrates—that there are no unanalyzable powers and that there is, therefore, no distinction between a power and its exercise.[171] To suppose that there is such a distinction is, in Hume's view, "entirely frivolous."[172]

What Hume does to accommodate this point is to make it a matter of *inference*. A prediction is an inference of the sort, "All F are G, a is F, so a is G," where the major premise is our asserted regularity and the minor is asserted as a statement of fact. A counterfactual inference is of the same logical form, and

170 B. Stroud, *Hume* (London: Routledge and Kegan Paul, 1977), 88.

171 Hume, *Treatise*, 171: "the distinction ... betwixt a *power* and its *exercise* is ... without foundation." Cf. F. Wilson, *Hume's Defence of Causal Inference*, 30ff; also F. Wilson, *The Logic and Methodology of Science in Early Modern Thought: Seven Studies*, Study One.

172 Hume, *Treatise*, 311.

the major is the same asserted *regularity*, but the minor premise is a statement of fact which is not asserted, but only expresses, as Hume puts it, an idea of such a fact, which idea could be false or, perhaps, even is false. A counterfactual thus does not make an assertion about an unrealized possibility; rather, it simply makes a statement of fact which is *not* asserted and then makes an inference from it by means of precisely the same statement of regularity that is used to make a predictive inference. Predictive and counterfactual inferences differ neither in logical form nor in the form of the regularity asserted but only in the *attitude* adopted towards the minor premise: in the prediction it is asserted while in the counterfactual it is suspended or perhaps even denied. Counterfactuals do not differ from predictions in being about another sort of entity—unrealized possibles—but rather in being another sort of inference. In particular, the matter-of-fact regularity that is the basis of the prediction is precisely the same regularity that is used in the counterfactual inference. Thus, when the Humean argues that laws are, objectively, nothing more than matter-of-fact regularities, it is *not* true, as Stroud suggests, that "they are restricted to the actual in a way that we [who use counterfactuals] are not."[173] There is no reason whatsoever why one who defends the claim that, objectively all there is to lawfulness or causation is regularity, should not be able to accommodate counterfactual discourse. As a criticism of Hume, Stroud's point cuts no deeper than Ewing's, that is, not at all.

Now look at what Hume has so neatly done. The connection between law-like generalities and subjunctive conditionals has often been observed;[174] details aside, it is the point made by Socrates against Anaxagoras, and by idealists from Coleridge on against naturalists and empiricists. What Hume does in his defence of naturalism is use this subjective connection between some asserted regularities and the making of inferences to *define* lawlikeness, i.e., causal or nomic *necessity*. The necessity of a law statement thus lies not in the logical form of the regularity but in our *attitude* towards its truth—the propositional attitude we adopt with respect to the contexts in which we assert it and use it in inference. A generality is lawlike if and only if it is not merely believed or asserted but is asserted in a certain more specific way, namely, with a preparedness to take risks with it, that is, to predict and to assert subjunctive conditionals. We may refer to this propositional attitude as the "law-assertion attitude."

Lawlikeness thus becomes a subjective matter. It is not objective, a feature of the logical form of causal propositions. As we have suggested above, there

173 Stroud, *Hume*, 88.

174 Cf. R. Chisholm, "Law Statements and Counterfactual Inference," *Analysis*, 15 (1957), 97–105.

is a thesis in ethics parallel to this, to the effect that value is a matter not of propositional content but rather of psychological attitude. This thesis is often called "emotivism." Hume's account of causation may therefore perhaps not unreasonably be characterized as an emotivist account of causation. But if the analogy is to be pursued, one should be careful not to call Hume's ethical views "emotivist" *simpliciter*. In particular, one must account for our correcting and adjusting our moral judgments, a point he emphasizes many times.[175] This is a point Hume also emphasizes in aesthetics. In his essay, "Of the Standard of Taste,"[176] Hume discussed how we correct aesthetic judgments, giving a very detailed analysis of the process. The same thing holds with respect to causal judgments. They are not merely subjective but can be *corrected*, and the view developed by Hume in response to the idealist criticism is not subjectivist *simpliciter*. With regard to ethics and aesthetics, Hume argues, the mind, guided by certain ideals, can regulate its own activities in terms of those standards.

Hume similarly holds, in a way compatible with his associationism, that the mind, guided by the ideal of truth, can actively correct its causal judgments. There is a difference, however. In the case of ethics and aesthetics, while a defence (as we saw above) can be made of certain standards over others, any such defence is not in terms of an objective moral or aesthetic standard. Contrary to Socrates and idealists such as T.H. Green, there are no such ontologically grounded standards in the case of either ethics or aesthetics. But in the case of causal judgments, there is such a standard, namely, *objective matter-of-fact truth*. To be sure, this is not an objective standard for the *necessity* which we ascribe to the causal judgment—this is the point of Hume's second definition of "cause"—but the objective facts constitute a standard for the truth of the judgment. This is the point of Hume's first definition of "cause."

Of course, given that the judgment is general, about a population, while the available evidence is only about a sample, it follows that we can never be inerrantly certain about our assertions of causation. Those judgments are always fallible and therefore open to correction. We can place the objective truth of facts as our cognitive ideal, but can never be certain that we have attained that ideal, and must allow for correction as new evidence becomes available, whether by accident or by deliberate research.

Let us return to the point where Hume's theory of causal judgment is similar to emotivism: that causality, like value, is a matter not of propositional content

175 See, for example, *Treatise*, 582.

176 Hume, "Of the Standard of Taste," in. T.H. Green and T.H. Grosse, eds., *Hume: The Philosophical Works*, 4 vols. (London: Oxford University Press, 1882–86; reprinted Aalen, Scientia, 1964), 3: 246–284.

but of psychological attitude. As we noted, given this similarity, Hume's account of causation, or, more specifically, of causal necesssity, might be called an emotivist account. Once this is noted, one will also recognize that similar objections can be raised against both the thesis about moral judgments and the thesis about causal judgments. As with the emotivist theory in ethics, the immediate question about the emotivist theory of causation is this: does not emotivism, when it denies the existence of an objective standard, entail that whether the attitude is adopted is not something that permits of justification? In other words, the idealist objects, does not Hume's account reduce causal reasoning to irrationalism?

The reply to this, in both the cases of causation and ethics, is to challenge the presupposition of the objector's question and ask what one might *reasonably* mean by "justification" in this context. In ethics, if the emotivist's arguments that objectivism is false are accepted, then it is not a serious objection to this account that he or she leaves no room for an objective justification of value judgments. *Of course* the emotivist has left no such room. He or she has just finished arguing that such justification is not possible. If it is not possible, furthermore, it is not reasonable to insist upon it. Whatever justification amounts to in ethics, the one thing that cannot *reasonably* be demanded is objective justification. In the case of Hume's account of causal necessity, the same sort of response is required. *Given* Hume's argument that, objectively considered, all causal assertions are assertions of constant conjunction, *then* it is not reasonable to demand an *objective* justification for the adoption of the law-assertion attitude toward some generalities rather than others.[177]

However, from the fact that *every* causal inference lacks, so far as its truth is concerned, complete objective justification, and lacks, so far as its necessity is concerned, any objective justification whatsoever, it does not follow that all causal inferences are equally unjustified, or, what amounts to the same thing, equally justified. Though the demand for objective justification is unreasonable, not all law-assertive attitudes are equally justified, nor therefore are all equally reasonable. Hume clearly recognizes this point and explicitly draws our attention to cases where the law-assertion attitude holds but where it also is not justified. His discussion of credulity,[178] of the often adverse effects of education,[179] of the role of imagination,[180] of unphilosophical probability (to

177 On this point see Hume, *An Enquiry concerning the Human Understanding*, in his *Enquiries concerning the Human Understanding and concerning Morals*, ed. L.A. Selby-Bigge, 2nd ed. (Oxford: Oxford University Press, 1902), 160.

178 Hume, *Treatise*, 112.

179 *Treatise*, 116.

180 *Treatise*, 123.

which a whole chapter is devoted),[181] all make clear that the *Treatise* draws a distinction between these cases where the attitude is unjustifiably held and cases where it is justifiably held—where its adoption is in accordance with the "Rules by which to judge of causes and effects," which rules appear in their own chapter with that very title.[182]

An assertive attitude is objectively justified if and only if the proposition in question is true. This holds equally for laws and for "mere" regularities; provided that the latter are true, they may justifiably be asserted. So truth, while *sufficient* to justify the attitude of assertion, is only *necessary* to justify the attitude of law-assertion. Thus, if a generality is false, one is objectively unjustified in holding toward it the law-assertion attitude. This will be so even if the other necessary conditions of justification (whatever they may be) are all fulfilled and it will be so even if we have *all possible reason* to believe that that necessary condition of truth is fulfilled.

A generality is a statement about a total population. Normally all one ever observes is a sample. Between sample and population there is a logical gap. This logical gap is, at bottom, *ontological*; it is the consequence of the separability of events in the world of sense experience. In Hume's case, it is a consequence of the separability which is the basis of his negative claim that there are no objective necessary connections, i.e., those connections which, if the Aristotelian or idealist is correct, are entities such that, in tying the separable events of sense together, they can bridge the logical gap between sample and population. That properties are constantly conjoined in a sample is, however, a necessary condition for their being constantly conjoined in the population. A necessary condition for being objectively justified in making a law-assertion is that the generality asserted be true. Given the logical gap between sample and population, it is not possible to know whether this necessary condition is fulfilled simply by observing that the regularity obtains in the sample. The best we can do is to know that a necessary part of this necessary condition obtains, namely, that the regularity holds in the sample. That is the best we can do, short of omniscience. Since it is the best we can have, we must make do with it. *If we observe that a regularity holds in a sample, we thereby have every objective reason it is (at that point) possible to have to justify ourselves in believing that the regularity holds in the population.* Subjectively, the only and best objective evidence that a regularity obtains in the population is that we have observed it to obtain among the facts we already know. Subjectively, we may be justified

181 *Treatise*, 132ff.

182 *Treatise*, 173ff.

in asserting a generality, therefore, when objectively, the assertion is not justified. Still, if we have done the best we can do—if we assert only when we are subjectively justified—then we cannot be blamed for not having done more, even where we are objectively unjustified. Fallibility is not a vice; it is rather the human condition.

Of course, for the idealist this falls short. It is second best compared to incorrigible and *a priori* knowledge of the forms. So, for the idealist, to settle for fallibility is indeed a vice, a cognitive vice (and perhaps also a moral vice); it involves settling for something that one ought not cognitively to settle for (and abandoning the search for the forms as objective moral standards). The Principle of Acquaintance of the empiricist excludes the forms, however. They do not exist, nor therefore, naturally enough, could knowledge of the forms be a reasonable cognitive aim, either for matters of fact or matters of morals. Hence, for the naturalist or empiricist, fallibility is not unreasonable; it is not cognitively vicious (nor therefore could it be a moral vice).

The remarks just made, drawing attention to our fallibility, apply equally to both causal and "mere" or "accidental" regularities. People often make law-asserting general claims that clearly are cases of "unphilosophical probability": for example, "Whenever a black cat crosses my path, it brings me bad luck" or, to use one of Hume's examples, "Irishmen lack wit." The issue is what *reasonable* subjective grounds are for law-asserting a generality like, "Water, when heated, boils," which are absent when one law-asserts a case which is one of "unphilosophical probability." An observed constant conjunction is a necessary condition for our being subjectively justified in adopting the law-assertion attitude toward a generality—indeed, any generality, but it is not a sufficient subjective condition, and certainly is not sufficient to justify treating some generality as more than a "mere" regularity or, in other words, a law. What evidence would justify our reckoning some generality as necessary or, alternatively, what evidence would justify our adopting that subjective attitude in some cases rather than others?

What Hume argues is that such evidence is sufficient only if the acquisition of this evidence has proceeded in accordance with the "rules by which to judge of causes and effects." We use law-asserting generalities to predict. What could justify our using them to predict? In the absence of objective necessary connections, holding in the observed sample can be no guarantee that a generality will hold in the population, nor perforce in the absence of objective necessary connections can past predictive success be a guarantee of future success. The absence of objective necessary connections withstanding, past predictive success is *the only evidence we could have*. However, not even just any predictive success will do. It will not do where, *so far as one can see*, such success is accidental, or a matter of sheer luck, quite unintended and not a matter of skill in the game of knowledge.

Hume is clear that his negative argument against objective necessities is, in effect, an argument for the existence of the ontological and therefore logical gap between the assertion of a generality and the observational evidence for that assertion.[183] At one point,[184] he expresses this in terms of the following principles: "*there is nothing in any object, consider'd in itself, which can afford us a reason for drawing a conclusion beyond it*" and "*that even after the observation of the frequent or constant conjunction of objects, we have no reason to draw any inference concerning any object beyond those of which we have had experience*"—that is, reason drawn from those objects "consider'd in themselves" rather than "never any sort of reason." After all, Hume goes on to give us the "Rules by which to judge of causes and effects" where he sketches the conditions under which one can *reasonably* infer a causal judgment about a population on the basis of evidence drawn from a sample, i.e., the conditions under which one can reasonably use an asserted regularity to predict and support contrary-to-fact conditionals. Hume's idea is that the evidence we use to assert the regularity must be gathered in a certain way and have survived certain tests. Such survival, again, can be no guarantee of continued success but, also again, *this is the best we can do* and it is therefore *the only evidence that can count as reasonable*.

The conditions that the evidence must satisfy are those of being justified by the "rules by which to judge of causes and effects." The "Rules" go like this: given Hume's negative thesis, we have a gap between sample and population such that properties regularly associated or constantly conjoined in the observed sample may not be constantly conjoined in the population. Hume notes carefully that in such situations the mind is often confronted with a set of contrary hypotheses, each of which initially fits the data and fulfills the necessary condition for being subjectively justified.[185] Since these are contraries, the law-assertive attitude toward any particular one cannot be justified initially. Now suppose more data is sought, in accordance with the "rules by which to judge of causes and effects." Suppose further that these data ultimately eliminate all but one of the contraries. In that case, the adoption of the law-assertive attitude toward that hypothesis will be (subjectively) justified, that is, justified to the extent that it is reasonable to seek such justification.

183 *Treatise*, 123.

184 *Treatise*, 139.

185 *Treatise*, 131–135; 154.

However, this is not the end of the story for other principles of selection are possible. Hume cites the principle that we choose as worthy of law-assertion the hypotheses we *want* to be true, quoting Cardinal de Retz on the principle that the wish is the father of the belief, "*That there are many things, in which the world wishes to be deceiv'd*."[186] Where the world wishes to be deceived, it can avoid trying to gather together the *evidence* relevant to reasonably deciding among the possible hypotheses.[187] In the chapter "Of unphilosophical probability," de Retz' principle and a number of other such unreasonable principles are mentioned.

How do we choose among these several principles that which it is reasonable to adopt, leaving the remaining to be unreasonable? Also, what could the grounds be for deciding that the rules one chooses really are reasonable? To begin, it is clear that those who desire a *reasonable decision* among contrary hypotheses must go out and *actively collect* additional observational evidence that will permit a decision to be made. The data that are the given, from which one infers the causal judgment, are often collocated only with great difficulty. As Hume puts it, directly after stating the "rules by which to judge of causes and effects":

> There is no phaenomenon in nature, but what is compounded and modify'd by so many different circumstances, that in order to arrive at the decisive point, we must carefully separate whatever is superfluous, and enquire by new experiments, if every particular circumstance of the first experiment was essential to it. These new experiments are liable to a discussion of the same kind; so that the utmost constancy is requir'd to make us persevere in our enquiry, and the utmost sagacity to choose the right way among so many that present themselves.[188]

One makes a reasonable decision among alternative possible hypotheses when one has *actively* sought out such data as would permit one logically to make such a decision, eliminating or falsifying hypotheses precisely until one is rendered subjectively worthy of law-assertion. *The choice among hypotheses may be made* EITHER by collecting relevant observational evidence according to the "rules by which to judge of causes and effects" OR by some other principle. If the choice is made on the basis of the latter, then the resulting law-assertion is unjustified.

186 *Treatise*, 153.
187 *Treatise*, 152–153.
188 *Treatise*, 175.

The issue is, why pick out the rules Hume does as providing *reasonable* justification rather than some other rules? Unless Hume can provide an answer to this question, his empiricist response to the idealists falls short of what philosophy can reasonably demand. So, what makes Hume's rules the *reasonable* ones to adopt? In part, the answer is that the data alone should decide among hypotheses and until sufficient data are available, the law-assertive attitude ought not to be adopted. But this is not the whole story. In order to see what else is involved we must look more closely at the associationist ideas Hume makes use of in his definition and, in particular, at his second definition of "cause."

The relevant laws are, once again, those of learning theory.[189] What these laws of learning describe are (among other things) the conditions under which law-assertive attitudes are acquired. There are a number of sufficient conditions for acquiring this attitude. Frequent repetition by itself is one.[190] So is frequent repetition under the further condition that the "rules by which to judge of causes and effects" have been observed. So is the wish that is parent of the belief.[191] Let us call the acquisition of a law-assertive attitude *successful* just in case the predictions it yields turn out to be true. (Recall the role of prediction in Hume's second definition of 'cause.') Now, the philosopher is motivated by curiosity[192] or love of truth,[193] which, indeed, "was the first source of all our [Hume's] enquiries,"[194] and is the sole motive of the academic philosopher.[195] *If we are guided by this aim of discovering the truth, so far as we can, then we shall be aiming to acquire successful law-assertive attitudes.* What we discover are regularities to the effect that certain conditions bringing about law-assertive attitudes are more likely than others to yield successful law-assertive attitudes. These regularities are inductions about making inductions. The infirmities attaching to all inductive inference thus attach to these, and doubly so, if you wish. Yet we can and have tested them. Thus, of the three conditions mentioned immediately above, the third is much more likely than not to yield *unsuccessful* law-assertive attitudes. The first, too, except in very

189 Actually, all that is relevant is that there are laws of learning which can provide the relevant psychological scientific explanations. The fact that Hume takes this learning theory to be associationism is not essential to the Humean account of how one decides which principles for law-assertion are reasonable.

190 Hume, *Treatise*, 156.

191 *Treatise*, 153.

192 *Treatise*, 270.

193 *Treatise*, 448.

194 *Treatise*, 448.

195 Hume, *An Enquiry concerning Human Understanding*, 41.

special cases, is unreliable. Only the second is in general relatively efficient in producing successful law-assertive attitudes. *This is a lawful fact we discover about ourselves, the world, and our interactions with it.* We discover that if we are serious about the truth, then we should put aside the pieties of religion (it is a "blameable" method of reasoning to condemn a doctrine because it is dangerous to religion),[196] the platitudes put into us by education ("an artificial and not a natural cause" of belief),[197] and rely instead upon the "rules by which to judge of causes and effects," the norms of "experimental philosophy."[198]

Previously we asked why it is reasonable to adopt Hume's "rules by which to judge of causes and effects" as the conditions of subjective justification for adopting the law-assertive attitude, rather than some other rules. We now have our answer to that question. *Given that our end is to come to believe general truths, then what we discover is that, as a matter of lawful fact, conformity to these rules is the best or most efficient means for achieving that end, so far as we are able.*[199] What we must do, then, is so discipline ourselves that we adopt the law-assertive attitude under only these conditions, not the others. Such self-discipline is, of course, difficult: "The utmost constancy is requir'd to make us persevere in our enquiry and the utmost sagacity to choose the right way among so many that present themselves."[200] Such discipline is possible, however. Furthermore, *if our cognitive end or goal is that of truth*, then it is a discipline we *ought* to practice. We earlier emphasized that causal inference involved an element of conative impulse, of mental activity. What we are now seeing is that these impulses can be disciplined, just like other impulses, such as the urge to jump the queue or the urge to take the last candy in the bag, and just like our impulsive appetites, e.g., hunger or the desire for food. As Bishop Butler indicated, Hume concurring, so far as concerns ethically relevant actions, particular impulses to action are neither to be denied nor reduced to self-love, though they are to be disciplined by the principle of self-love and also by other higher principles, such as those of conscience. Similarly, according to Hume, our mental impulses can and ought to be disciplined to higher ends: "Who indeed does not feel an accession of alacrity in his pursuits of knowledge and ability of every kind, when he considers, that besides the advantage, which immediately results from these acquisitions, they also give him

196 Hume, *Treatise*, 409.

197 *Treatise*, 117.

198 *Treatise*, 175.

199 For greater detail, see F. Wilson, *Hume's Defence of Causal Inference*, Chapter Two.

200 Hume, *Treatise*, 175.

a new lustre in the eyes of mankind, and are universally attended with esteem and approbation?"[201]

How else, in other words, than through the careful accumulation of evidence in conformity to the rules of natural science could we justify law-asserting certain regularities, treating them as causal necessities and, in so doing, relying upon them to predict and to support contrary-to-fact conditionals? What else could be more reasonable? Such is Hume's and the naturalist's response to the Socratic and the idealist challenge to give an account of causal necessity that does not presuppose objective necessary connections.

8.2 *Green and Blanshard: Idealist Responses to Hume on the Necessity of Causal Laws*

We now have seen in detail how the naturalist, while denying the existence of objective necessary connections, can respond to the challenge of Socrates and idealists like Green and Bradley that the naturalist must account for the apparent necessity of causal judgments. Idealists have replied to this response, however. T.H. Green countered directly is his "Introduction" to his edition of Hume's *Treatise*. More recently, Brand Blanshard in particular has argued that this position of the naturalist or empiricist will not do: he has argued that, contrary to the naturalist, there are necessities that clearly cannot be accounted for with any sort of Humean subjectivist or emotivist analysis of causal necessity that the naturalist could provide.

Green raises a variety of objections, most of them aimed at showing how Hume's position on causation is inconsistent with Hume's position on other topics, e.g., his account of our knowledge of the external world. Green's objections of the latter sort we can here ignore. Some stand on their own, however. We have seen that Hume has two definitions of 'cause,' the first where he refers to a "philosophical" relation and the second where he refers to a "natural" relation.

As for 'cause' in the *first* sense, Green argues as follows:

> ... it is a surprise to find that the 'philosophical relation' of cause and effect, as distinct from the natural one, should ... appear. ... Such a distinction has no meaning unless it implies a conceived relation of objects other than the *de facto* sequence of feelings, of which one 'naturally' introduces the other. ... [For,] [h]is [Hume's] language constantly implies

201 *Treatise*, 620.

> that knowledge, as a perception of relations, is other than the sequence of feelings ...[202]

Of course, our ordinary discourse implies that causation involves a necessity which is more than simple regularity or sequence. It does not follow that this sort of everyday discourse is correct. Hume's argument is that, in fact, it is not correct that, objectively, causation is indeed nothing more than *de facto* regularity [first definition], that the apparent necessity is really nothing more than the felt psychological attitude when one makes a causal judgment [second definition], and that our ordinary discourse implying an objective necessity is but a blurring—an erroneous blurring—of these two relations deriving from their similarity as relations. Green just takes it for granted that our and Hume's ordinary discourse is correct, and that therefore causation must be something more than mere regularity. *That*, however, is no great argument; indeed, it is no argument at all, but rather a simple begging of the question against Hume and against the regularity account of causation that any naturalist or empiricist defends.

Green's reference to the second definition makes clear how his own thinking goes: a connection is needed, and connections are effected by the mind, by mental activity; hence, Hume really must absorb his first definition, which suggests a connection but provides none, into his second definition, which allows for the mind to do some connecting at least. Of course, for Hume, as for Anaxagoras, mind is a bundle of separable parts, whereas (Green argues) mind must be understood as Socrates understood it, as a simple active entity. In order for there to be unity before one, the separable parts must cease to be separable. "They cease to be so in virtue of the presence to them of the Ego, in common relation to which they become related to one another."[203]

This is why Hume's *second* definition of 'cause' as a natural relation won't do either. One can get something connected by some essentially mental act only by presupposing this Ego. But Hume cannot allow for this Ego since, for him, the world, including the self, consists wholly of separable parts. Hume therefore, on Green's account, falls into incoherence:

> ... by help of the phrase 'natural relation', conveying as it does two wholly different things—the involuntary sequence of one feeling upon another, and that determination of each by the other which can take place only

202 T.H. Green, "Introduction to *Treatise of Human Nature*," 249.
203 "Introduction," 267.

> for a synthetic self-consciousness—he is able on the one hand to deny that the relations which form the framework of knowledge are more than sequences of feeling, and on the other to clothe them with so much of the real character of relations as qualifies them for 'principles of union among ideas'.[204]

But that conclusion depends solely on Green's question-begging argument that causation must involve objectively something more than regularity. He merely assumes right from the beginning, what Hume (and any naturalist or empiricist) denies, that "no philosophical relation is admissible … that is not derived from a natural one."[205] Furthermore, he merely assumes, or takes for granted, the idealist claim, emanating from Socrates but which the naturalist and the empiricist deny, that a connection to be necessary must be effected by a simple active entity.

Green raises an objection to Hume such that one can often quite correctly judge something to be the result of some cause on the basis of a single experiment without, in fact, having previously observed that effect or anything specifically like it.[206] A causal judgment is a habit of inference, however, and a habit of inference can arise only from a repetition of similar events in the past. So a single event cannot, on Hume's view, give rise to a habit nor, consequently, justify a causal judgment. But single events do often and correctly justify causal judgments. Hume's view therefore cannot be adequate. Green infers that one must go to the idealist alternative which allows, through insight into the logical structure of the forms, judgments of causal necessity on the basis of a single event, where that event instances the relevant form.[207]

This idealist inference is too hasty, however. Hume himself allows that there can be causal judgments justified by a single event. As he tells us:

> … the mind, having form'd another observation concerning the connexion of causes and effects, gives new force to its reasoning from that observation; and by means of it can build an argument on one single experiment, when duly prepared and examin'd. What we have found once to follow from any object, we conclude will forever follow from it …[208]

204 "Introduction," 271.
205 "Introduction," 269.
206 "Introduction," 288ff.
207 "Introduction," 289.
208 Hume, *Treatise*, 131.

Obviously this inference, whatever it may be, cannot be an inference from "All observed F are G" to "Therefore all F are G." One might suppose it to be such if one thinks of the judgment, i.e., inference, that F's are G's. One might think that one can hold that "All F are G" only on the basis of there being a reasonably sized sample and one will think that a sample of one is not reasonably sized. However, the inference from "All observed F are G" to "All F are G" is induction by simple enumeration and Hume specifically argues, as we have seen, that this is an unsafe method for forming causal judgments; it is a case of "unphilosophical" probability. For it to be an instance of what we may speak of as "philosophical" probability, that is, for the causal judgment to count as reasonable, we need to make the judgment in conformity with the "rules by which to judge of causes and effects."

It is easy enough to know that this could be so, for some cases at least, that the forming of causal judgments on the basis of single experiment is a reasonable procedure. Take the case of one flipping a switch and a light in the ceiling of the room coming on where one concludes that the flipping of the switch causes the light to come on. Everything else in the situation, so far at least as one can see, remains the same. Let F=the switch unflipped, G=the light being on, H=the other properties that are present. We now have *two* events, the switch (s) at t_1 before flipping, that is, unflipped, and the switch at a succeeding moment t_2, when it is flipped. Then the mentioned properties are present (P) and absent (A) in the two events. With this information, we can use the traditional "method of difference" (Rule 6 of Hume's "rules by which to judge of causes and effects") and infer that F, the switch having been flipped, is sufficient for G, the light being on. We can infer, in other words, that in any situation in which all the other properties H are the same, we have "All F are G=(x)(Fx then Gx)." This regularity states that F is sufficient for G, that is, that F is the cause of G.

Of course, in making the inference to the judgment about the cause of the light being on, one takes for granted a *Principle of Determinism*, i.e., here is a conditioning property sufficient for the conditioned property, and a *Principle of Limited Variety*, i.e., the conditioning property for the conditioned property is in the set consisting of F and H. Hume is clear on the need for Determinism but less clear on the need for Limited Variety. (Bacon, who first stated the methods of eliminative induction, including the Method of Difference which we are using, was clear on the need for Limited Variety but less clear on the need for Determinism.) These two follow from the *Principle of Universal Causation*, i.e., for every property there is a conditioning property sufficient for it and this conditioning property is from the set of occurring natural properties. Note that Determinism, Limited Variety and Universal Causation are *laws about laws*; they are generalizations at the generic level about regularities at the specific level. This law of universal causation is the fourth of Hume's "rules by which to judge of causes and effects," which states, as he puts it, that "like causes have like effects."

In Hume's language, persons in the room, "having formed another observation concerning the connexion of causes and effects," or, in other words, having acquired by means of previous inferences the knowledge that every event has a cause and that the same cause has the same effect (rule 4), can by means of that knowledge "build an argument on one single experiment when duly prepared and examined. What [they] have found once to follow from any object, we conclude will for ever follow from it ..."[209]

Green does advert to Hume's "rules by which to judge of causes and effects."[210] While he does not examine them in detail, he notes that they turn on the thesis that "the connection between all causes and effects is equally necessary and that its seeming uncertainty in some instances proceeds from the secret operation of contrary causes," and that "what the vulgar call chance is but a concealed cause."[211] But he clearly recognizes that the inferences legitimated by the "rules" all presuppose the Principle of Universal Causation. He argues, however, that Hume's account of necessary connection will not do: "The 'equal necessity between all causes and effects' means not that any 'law of causation pervades the universe,' but that, unless the habit of transition is 'full and perfect,' we do not speak of these feelings as related in the way of cause and effect."[212]

However, if the necessity of the law of universal causation is to be carried to the conclusions regarding specific connections, there has to be objective necessity guaranteeing true universality and not just psychological feelings and habits. However, it is not an argument simply to suggest that one cannot use the rules of the scientific method (Hume's "rules by which to judge of causes and effects") since these rules presuppose the universality of causation and, on the Humean account, " 'probability' cannot become 'proof' " because induction is always fallible.[213] This point is part of a more general criticism, but this too is unsupported. Simply to declare that "mere chance, probability, and causation" on Hume's account amount to the same thing and that this is absurd[214] is hardly an argument. The response of the naturalist is rather that probability or fallibility will in fact suffice, and that inerrant necessity is not something that is

209 *Treatise*, 131.
210 Green, "Introduction," 287ff.
211 Cf. Hume, *Treatise*, Book. I, Part III, Section 12, "Of the Probability of Causes," 130–142.
212 Green, "Introduction," 289.
213 "Introduction," 289–290.
214 "Introduction," 289.

really required. The Law of Universal Causation, and specific regularities, all have been confirmed by observation and experiment. Such confirmation can never yield inerrant certainty; the assertion of any matter-of-fact generalization, including the generic law about laws, the Law of Universal Causation, will always be fallible though that, for the naturalist, suffices. Therefore, fallible belief, in the absence of the absolute certainty of the forms, can suffice for our cognitive standards, even though the claim that fallible belief need not be second best is seemingly beyond Green's capacity to imagine. Green has thus failed to establish that Hume's account of causal necessity is not adequate.

Let us now turn to Brand Blanshard, who developed an idealist response[215] to recent "analytic philosophy" (more recent than Bradley, anyway), some of which is empiricist (Russell, the early positivists) but some of which is clearly not (the later positivists, Ryle).[216] Blanshard's critique is thorough and space here hardly permits comment on every point he makes. There are two important points which warrant discussion, however, and an indication of how a naturalist might respond.

Blanshard considers various accounts of *necessity*.[217] In particular, he examines an assortment of views to the effect that necessity is somehow linguistic, and shows them not able to accomplish what their authors hoped they would. Much of his discussion concerns the necessity of logic and of arithmetic, so it is beyond our topic, which is science. But in this context he does takes up the Humean view of necessity, where causal necessity is not objective but rather subjective. Blanshard quotes Huxley on this view:

> [Necessity] does not lie in the observed facts, and has no warranty that I can discover elsewhere. For my part I utterly repudiate and anathematize

215 Brand Blanshard, *Reason and Analysis*.

216 The break of the positivists, those who followed Carnap at least, from empiricism occurred in 1936, when Carnap gave up the notion that abstract concepts are to be introduced by explicit definition in favour of the notion that they can also be introduced by what he calls "reduction sentences" (see Carnap, "Testability and Meaning," *Philosophy of Science*, 3 (1936), 419–471, and *Philosophy of Science* 4 (1937), 1–40; see also the discussion of Carnap in F. Wilson, "Placing Bergmann," 202ff, notes 44 and 45. This discussion emphasizes Carnap's turn to (logical) holism in this 1936 essay. While (logical) holism is not the entirety of idealism, it is certainly an ingredient. Carnap's turn connects him to Heidegger, on the one hand (see "Placing Bergmann"), and W. Sellars and M. Dummett, on the other hand (see F. Wilson's essay, "Acquaintance, Ontology, and Knowledge," especially the Appendices, in F. Wilson, *Acquaintance, Ontology, and Knowledge*, 1–108, and F. Wilson, "Dummett's History," in *Acquaintance, Ontology, and Knowledge*, 683–725). Sellars and Dummett developed the details of their views too late for Blanshard to address, however.

217 Blanshard, *Reason and Analysis*, Chapter VII.

> the intruder. Fact I know, and Law I know; but what is this Necessity [which is] … an empty shadow of my mind's own throbbing?[218]

There is in the tone here a sense of regret at the loss, something absent from Hume who has no qualms that objective necessity has been jettisoned. Huxley's observation is reminiscent of the late Victorian thinker who shakes his fist at heaven and damns God for ceasing to exist.[219] In any case, Blanshard challenges this position on some of the same grounds that Green used. According to this position, Blanshard observes,

> What looks like necessity is only habit, and there is nothing self-evident about that; what has become fixed by mere association might in theory become unfixed. It *may* be that the habits we have acquired in our little corner of the universe are an index to its nature everywhere. But it would be sheer good luck if that were so.[220]

This, however, is simply a repetition of the argument we found in Green. If we in fact have observational evidence that leads us to believe that we have grasped the basic regularities describing things, from the large to the small, from the inorganic to the organic, from the human and to the social, then we do have that evidence. If we have that evidence, furthermore, then it is *not* a matter of luck that we think this to be the structure of the universe. We have the only evidence that we could have which is, the naturalist argues, therefore the best evidence that we could have. If a golfer has good luck in sinking the ball, that means that, as the play unfolded, the sinking of the ball happened, though neither as a matter of skill nor merit. How could the unfolding of the universe be a matter of luck? What or whose skill is lacking? What or whose merit is absent? Is the creator an amateur, lacking the skill to keep things going as he or she had tried to have them go? Or is the creator undeserving for some reason of the way this universe has worked itself out? The notion that things have occurred as a matter of luck is, I suppose, possible but unlikely in the

---

218 T.H. Huxley, *Methods and Results* (London: Macmillan, 1892), 161; quoted by Blanshard, *Reason and Analysis*, 254.

219 Compare these lines from Matthew Arnold's *Dover Beach*: "The Sea of Faith/Was once, too, at the full, and round earth's shore/Lay like the folds of a bright girdle furled,/But now I only hear/Its melancholy, long withdrawing roar … /for the world, which seems/To lie before us like a land of dreams,/So various, so beautiful, so new,/Hath really neither joy, nor love, nor light,/Nor certitude, nor peace, nor help from pain …" One finds this combination of resentment, sadness, and nostalgia not only in Huxley but also in Bradley.

220 Blanshard, *Reason and Analysis*, 254.

universe of the idealist. There is nothing at all about the universe of the naturalist that would allow the applicability of that concept, however. There can be lack of skill only if there is a creator, which there is not. Furthermore, that the way things have gone is unmerited presupposes that there is an objective standard of merit, which again there is not. Neither of these concepts makes any sense in the universe of the naturalist. Accordingly, to suggest that the way the world has turned out is a matter of luck is hardly a criticism of the naturalist or empiricist account of that world. Things in the naturalist's world, i.e., things in the world we live in and know by sense experience, to be sure, are logically and ontologically separable and so everything we believe *might* be wrong. Our judgment is always fallible. But *posse* does not imply *esse*. Everything we believe about the structure of the world in which we live *might* indeed be wrong, but we have no reason to believe that it *is* wrong. Furthermore, the best evidence we could have is that it is *not* wrong but, in fact, is correct. Blanshard's argument simply begs the question, as did Green's; confronting the naturalist with misplaced ridicule is not to offer a criticism.

Much of Blanshard's argument concerns the necessity of logic and of arithmetic rather than that of science, so, while it is interesting, we cannot tarry over it. He does raise an important point about relations, however. Bradley's account of relations, we recall, turned out to be inadequate because it could not handle asymmetrical relations. Blanshard appeals to such relations to try to show that there must be objective necessary connections in the world. Specifically, he appeals to *serial* relations.[221] In the first place, he proposes. "Necessity enters into ... [an] order [that is, an ordered set of things] in the sense that things that are equal to the same thing *must* be equal to each other."[222] But this 'must' is misplaced. Define 'equal in temperature' in the usual way, using thermometers: two things are equal in temperature (ET) just in case that each sends a standard thermometer to the same height. Then it is true that (Z) (For all x, y, z, if x is ET to y and y is ET to z, then x is ET to z).

This means that ET has the following *structural form*: it is transitive. For all x, y, z, if Rxy and Ryz, then Rxz. The point is that not every relation has this form. Consider the scratch test for hardness: if x scratches y, then x is harder than y. So diamonds are harder than glass. Two things are equal in hardness (EH), then, just in case something that scratches the one also scratches the other. If we consider the form of transitivity, for all x, y, z, if x is EH to y and y is EH to z, then x is EH to z, then this *turns out to be false*: not every factual relation is

221 *Reason and Analysis*, 436ff.

222 *Reason and Analysis*, 463; emphasis added.

transitive. Therefore, the mere fact that a relation is transitive does not imply, as Blanshard suggests it does, that its truth is necessary; on the contrary, its truth will be contingent.

Interestingly enough, in thermodynamics, scientists have formulated the First and Second Laws of Thermodynamics. The Second Law gives an additive ordering of thermodynamical objects. It presupposes that these objects could be set in a rank order. This was given by the First Law. Seeing this, it was realized that both presuppose a nominal ordering, one given by an equality relation. Both the Second and First Laws presuppose a more elementary law. This law is given by (Z), that things equal in temperature to a third thing are equal in temperature to each other. Thermodynamicists knew this law to be true as a matter of fact but they had not made this part of their theories explicit. (Axiomatization has something to be said for it, and so does the development of the logic of relations!) They knew that the law (Z) was true but they had not recognized that its role needed a name: it came to be called the "Zeroth Law" of Thermodynamics.

For us the point is that the Zeroth Law is a *law* in the sense of *matter-of-fact regularity*, that is, it is *not a necessary truth*. Blanshard is therefore wrong to claim that because a relation has a certain logical structure, namely, that of an equality relation, it therefore necessarily has that structure, e.g., that it is necessary that this specific relation is a transitive relation. So Blanshard has not found in this example a case of an objective necessity that cannot be fit into the Humean picture of the nature of scientific and causal laws.

The other argument I propose to consider also aims to make a case for necessary connections, or, rather to make at least the somewhat weaker point that Hume's argument that there are no objective necessary connections does not succeed. Blanshard's next argument does not try to establish that, yes, *there are* necessary connections that cannot be accommodated in the Anaxagorean-Humean-naturalistic view of the world. Rather, it tries to show that Hume has *not* made a case that *there are not* any such connections, in which case one is not compelled to renounce idealism in favour of naturalism.[223]

Hume's argument for the contingency of causal connections goes like this: we have one event (e1) such that a is F which causes a second event (e2) such that a is G. If one can consistently affirm the former and deny the latter then there is no necessary connection between the two. The Humean point is that causation is such that, if there were a necessary connection, then denying the effect event (e2) necessitates denying the cause event (e1). Since for all

223 *Reason and Analysis*, 471.

events it is always consistent to affirm the cause and deny the effect; there are therefore no objective necessary connections. Blanshard considers the case of a person who is bitten by an anopheles mosquito and as a consequence gets malaria.[224] There is nothing, it would seem about the bite of such a mosquito that necessitates falling ill with malaria. In fact, sometimes being so bitten does not lead to malaria; the regularity of falling ill with malaria when one is bitten by an anopheles mosquito is, in fact, inconsistent, holding only under certain conditions. This demonstrates that, though we say, speaking loosely, that the bite "causes" the malaria, it is not the real cause, or not the whole of the real cause. There is clearly no necessary connection between the sort of event that we call the "cause" and the sort of event that we call the "effect." If a person falls ill with malaria, the cause is not the mosquito bite, but rather something more proximate. We should therefore say that the cause of the person falling ill is the pouring into the bloodstream, through the bite, of a mass of parasites called *plasmodia*. Again, however, we have not got the "real" cause since it happens at times that a person has the parasites without falling ill. There is more to the process than the presence of the parasites; the parasites must also attack the red blood corpuscles in the bloodstream. Again, that is not enough either; there is still a gap between this event and the falling ill. The parasites must drain the corpuscles of their hemoglobin too before signs of the illness appear. However, once again, this is not enough: additionally, the draining of the hemoglobin must effect the draining of the capacity to carry oxygen to the tissues of the body. Now we seem to have reached a point in the process where we can say that we have the cause of the disease: starved of oxygen the various tissues cannot perform their normal functions. But, of course, this necessitates the disease. Indeed, it *is* the disease—it *is* the factor that is essentially constitutive of the disease.

We have a cause event and an effect event, and there is also a series of events constituting a process that leads from the former to the latter. If the process, as it were, breaks down at some intermediate point—if there is something that interferes with what happens, then the event which is usually a consequence of the cause does not occur. The events in the process become more and more proximate to the effect. So long as there is a separation, there is no necessary connection: Hume is correct in his argument. We reach an event necessarily connected to the effect only when it is so proximate that there is no gap between it and the effect—so proximate that is *merges* with the effect, which is to say when we reach an event that is so proximate that it *is* the effect. We have

224 *Reason and Analysis*, 452.

a genuine case of causation linking cause to effect only when we have a series of events linking the ends points *continuously*, in the sense that between every pair of events there is another event between them. Only when we have such a series can we say that one event merges with what comes after it. In a series like this, while there are successors, there is never an *immediate* successor in which case there are no gaps that could be separated in the way that Hume asserts that cause and effect are always separable.[225] Hence, in the case of continuous causal processes, Hume's argument that there are no necessary connections, depending as it does on the separability of cause and effect, cannot be successful. We can therefore suggest that in the case of continuous processes, since these are not arbitrary in how they work out, we do have examples of non-arbitrary, i.e., necessary, connections. To be sure, the mind cannot grasp the process in detail; it cannot dissect it in sufficient detail, if you wish, to discern any of the specific causal connections. The non-arbitrary patterns of these processes do suggest that these things are not a matter of accident, however, and therefore do suggest the presence of necessary connections.

We may not have established the existence of necessary connections, but we have at least established that their non-existence has not been established. We have shown that the naturalist has not made out the case that implies that "the universe is ... a mere heap of things and events thrown together in hit-or-miss fashion." Since it is implausible that this is how the universe is, we have "reason to believe" that "nature is shot through with filaments of necessity," and that "there is such a filament in causal connection"—"whether at present we can dissect it out or not."[226]

Blanshard is taking seriously Hume's remark that "One event follows another, but we never can observe any tie between them. They seem *conjoined*, but never *connected*."[227] Blanshard's point is that there are causal processes in which the parts do *not* seem conjoined but rather *do* seem connected; they do seem, in Hume's words, to consist of events that *are tied* together.

Now Blanshard has not been the only one to have made this point. S.P. Lamprecht is another who argues that causal processes in fact lack the gaps that imply they are nothing but conjunctions of events, untied events, unhooked to one another:

> There is not merely a fire and then a hot oven ... There is not merely an avalanche and then destruction, but there is an irresistible sweep in

225 *Reason and Analysis*, 453.

226 *Reason and Analysis*, 472.

227 Hume, *An Enquiry concerning Human Understanding*, 74.

> which the avalanche destroys whatever is in its path. There is not merely a storm on the coast and then the loss of many feet of property, but there is the inevitable process in which the stormy waters eat away the soft soil.[228]

Consider one of these cases: that where the avalanche sweeps away a hut. There are two events, one at t_1 and one a little later at t_2. Let "G" be the situation in which the hut is standing whole unharmed, let "F" represent the descended avalanche, and let "H" represent the other properties characteristic of the two events. Then, as before, let P=present and A=absent. Given the assumption, perfectly reasonable and derived from experience, that for every event like the destruction of the hut there is a sufficient condition for the destruction (determinism) and that sufficient condition involves the application of a strong force (limited variety), then, once again, a simple application of the method of difference (Hume's Rule 6 of the "rules by which to judge of causes and effects") leads directly to the conclusion that the following regularity, or something like it, holds: (^) such that whenever the other conditions obtain, then a hut's being struck by an avalanche is sufficient for the destruction of the hut. That is, from the observed facts we can directly infer that there is a causal law of which those facts are an instance: *the causal explanation of the facts is discovered in the facts themselves.*

Moreover, let us also note Lamprecht's argument from the avalanche example: we see in causally connected objects an "active compulsion" which constitutes, like the active Socratic soul, a "necessary relation between" events. "There is not merely an avalanche and then destruction; but there is an irresistible sweep in which the avalanche destroys whatever is in its path."[229] It is easy to see that the Humean naturalist can grant active compulsion a power without granting that this implies any sort of necessary connection. Thus, while Hume accepts the Megaric position rather than the Aristotelian and denies that there are any ontologically unanalyzable powers that yield necessary connections—"the distinction," he tells us, "... betwixt *power* and the *exercise*

228 S.P. Lamprecht, *The Metaphysics of Naturalism* (New York: Appleton-Century-Crofts, 1967), 131. Note that in Lamprecht's view, "naturalism" does not imply that the scientific picture of the world has as a constitutive part the Humean account of causation. In this respect, his naturalism contrasts to that of the present essay. What I am now contending, however, is that his case against the Humean view does not stand up to close examination. I also discuss Lamprecht's anti-Humean argument in F. Wilson, *Body, Mind and Self in Hume's Critical Realism* (Frankfurt: Ontos Verlag, 2008), Chapter 2.

229 *The Metaphysics of Naturalism*, 131.

of it, is ... without foundation,"[230] nonetheless, in ordinary life, e.g., in the "philosophy of our passions," the language of power does have a sense, though it is not ontological, i.e., "*power* has always a reference to its *exercise*, either actual or probable, and ... we consider a person as endow'd with any ability when we find from part experience, that 'tis probable, or at least possible that he may exert it."[231] What Hume is saying here we may put as follows: the predicate 'if F then G' represents a power just in case there is a regularity to the effect that for all x, if x is H, then, if x is F, then x is G, where 'H' represents some categorical property. To put it in language that has become familiar to philosophers subsequent to Hume, (x)[Hx then (Fx then Gx)].[232] Now if we now take the property, "if struck by an avalanche then destroyed," we recognize that this represents a power that avalanches exercise on huts by virtue of the fact that it occurs in the law (^). In short, when we discover in the facts the law that explains them, we therein also discover that the avalanche has the *power* to destroy huts when it strikes them, exactly as Lamprecht recognizes. In thus granting Lamprecht's point, however, we are not saying, contrary to what Lamprecht believes, anything with which the Humean will disagree.

There is another aspect to Lamprecht's examples that should also be noted, one that ties in with Blanshard's argument. This is his point that in a causal process, as in the avalanche-hut example or Blanshard's malaria example, there is not first one event and then another clearly distinct event—first the avalanche as a prior event and then the hut being destroyed or first the mosquito biting and then the disease, but a continuous process in which the apparent "cause" and "effect" are boundaries and not parts and certainly not separable parts. We have been rather coy about this so far, striving to fit these examples and the commentary about them into a Humean framework of clearly separable "cause" and "effect," made clear when we resorted to symbols by our quite separate F and G. However, the issue ought not to be avoided. Perhaps here in the continuity of causal processes there is something which is essentially idealist—continuity can perhaps be found only in something like the Socratic simple soul and its exercise of active powers —and so it may be that there is something to which the Humean naturalist has no adequate response here. Let us see whether this is so.

230 Hume, *Treatise*, 171.

231 *Treatise*, 313.

232 Cf. F. Wilson, "Dispositions: Defined or Reduced?," *Australasian Journal of Philosophy*, 47 (1969), 184–204; see also F. Wilson, *The Logic and Methodology of Science in Early Modern Thought: Seven Studies*, Study One.

8.3 *Russell's Response to Blanshard: Causation in Science Examined and the Contingency of Causal Laws Sustained*

The issue Blanshard raised which we have been discussing was also raised by Bertrand Russell, though in a somewhat different way. Russell argues that thinking of "A causes B" in the traditional way in terms of a sequential regularity actually results in an impossible position, given the continuity of the causal process to which idealists like Blanshard—Russell himself refers to Bergson—and even some like Lamprecht who call themselves naturalists, have drawn our attention. Given this continuity, then between any two parts there is a third in which case it follows that no part is immediately contiguous to another and no cause could have as its effect an immediate successor. The cause itself as an event will have some duration or other, and therefore one part prior to a later part. Since the earlier parts are not contiguous to the effect, they cannot be the cause of that effect. The cause will therefore have in respect to its duration the property to decrease without limit. But however near one approaches the effect, there will always be a space occupied by some part that could be altered so that the effect is absent though any purported cause remains present. This is fatal to the view that continuity implies a necessary connection and to any view that holds that the causal process is a continuous process, though one in which the cause is contiguous to the effect.[233] There are indeed laws in science and in physics in particular, but these do not have the form implied by the "same cause same effect" rule. Where a science is inexact this simple sort of rule does have some place, but in any exact science the principle is otiose. In such a science there are laws quite properly so-called, but these laws, while describing processes rightly seen to be causally determined, are of a form that hardly fits the "same cause, same effect" pattern.[234]

There are three aspects to a causally determined process which need to be noted. Wherever one has a certain system or configuration of objects that undergoes the process, 1) one needs a complete set of relevant variables. There is nothing else inside the system that affects the values of the variables over time. A state of the system is given by the values of the variables at any one time. (2) The system must also be closed. There is nothing else outside the system that affects what happens inside the system. The idea of given boundary conditions, i.e., knowledge of the ways in which what is outside the system affects what happens inside the system, is a generalization of the idea of closure.

233 B. Russell, "On the Notion of Cause, with Applications to the Free-Will Problem," in H. Feigl and M. Brodbeck, *Readings in the Philosophy of Science* (New York: Appleton-Century-Crofts, 1953), 387–407, at 390–391.

234 Russell, "On the Notion of Cause," 392.

(3) One needs a rule or law that permits one to infer deductively from the state of the system at any one time the state of the system at any other time, before or after the time of the given state. A process law makes both prediction and postdiction possible.[235]

Knowledge of this sort has been called "process knowledge."[236] The clear example of something approximating such knowledge is Newton's explanation of the motions in the solar system by means of the law of gravity. (A) There are 10 objects in the system. The masses, positions, and velocities of each object form a complete set of relevant variables. One does not need to know the colours of the objects or their chemical composition or anything else in order to predict or postdict the values of any of these relevant variables. (B) The system is for practical purposes closed. Nothing outside the system affects what is going on inside; other stars are too far away to have any (noticeable) gravitational influence. (C) Newton's law of gravity within the theoretical structure of classical mechanics provides a set of differential equations that jointly constitute the process law.

Whether we have a complete set of relevant variables, whether we know that the system is closed, and whether the differential equations that we have really are a process law are, in each case, something that we know only through inductive inference.[237] Any assumption that we have process knowledge is therefore fallible and hazardous, just as any inductive inference is fallible and hazardous, falling short of any idealist notion that knowledge, in order to be such, must be inerrantly and incorrigibly certain.

Note how, with process knowledge, given the state of the system at any one time, one knows the state at any succeeding time and the state at any preceding time. In this sense, the process is *determined*—indeed, one wants to say *causally determined*. This is what happens to causation. Stones still break windows, apples fall, and water, when heated, boils. Each of these is a reasonably good piece of knowledge, yet, when compared to process knowledge, something is left out: relevant variables are not mentioned, boundary conditions are not given, and details about how the processes unfold over time are not specified. Such knowledge is in a clear sense "gappy" or "imperfect" relative to process knowledge as the standard.[238] It is the task of scientific research to find

235 Russell, "On the Notion of Cause," 395.

236 Cf. G. Bergmann, *Philosophy of Science* (Madison, WI: University of Wisconsin Press, 1956), Chapter Two.

237 Russell, "On the Notion of Cause," 397.

238 "On the Notion of Cause," 395; cf. Bergmann, *Philosophy of Science*, Chapter Two. See also F. Wilson, *The Logic and Methodology of Science and Pseudoscience*.

the facts that fill these gaps and to move closer to what is clearly the cognitive ideal of process knowledge. These gappy laws, characteristic of everyday life and any science in its infancy, describing regular sequences of events, often have the clear cause and effect pattern characterized by the form "same cause, same effect."

As science advances, however, the gaps are filled and the cause-effect form disappears. Even the regularity in a way disappears. It is displaced by, or, perhaps better, is absorbed into, something that more and more closely approximates to process knowledge. Regularity becomes constancy of relation.

When one has process knowledge, then prediction and postdiction occur by means of mathematical functions. It is these functions that permit the deductive inferences in which, from the state of the system at any one time, one infers the state of the system at any other time. It is not the sameness of "same cause, same effect" that is constant in the laws constituting process knowledge. Rather, it is the sameness of functional relations. More precisely, the constancy of the law, its basic regularity, is given by the sameness of the differential equations, the solutions of which yield the functional relations that are the process law.[239]

Then there is the law of universal causation or the thesis that "for every event there is a cause." Allowing "cause" to mean 'causal law', then this is the thesis that, as Mill put it, "it is a law, that every event depends on some causal law" or "it is law that there is a causal law for everything."[240] This is a generic law, asserting that there are specific laws and that these laws have a certain form. Taking "cause" and "causal law" now to be 'process law', as we have just suggested, then this generic law states, in the same way, that there are specific laws and that these specific laws have a certain identity, namely, that of process laws: "it is a law that there is a process law for everything" or "it is a law that, for every thing, that thing is or is part of a system or configuration of things the development over time of which is described by a constant set of constant differential equations."[241] This is more a "principle of the lawful uniformity of nature" than it is a "principle of universal causation," where causation is understood in the older fashion as "regular sequence." This principle as we have come to understand it asserts that the cognitive ideal of process knowledge is something that we can in principle attain; there are process laws, there to be discovered. This has some inductive support (given the universal quantifier, such support will always be fallible and could never be conclusive) and the

239 Russell, "On the Notion of Cause," 395.

240 Mill, *System of Logic*, Volume 7, III, V, 1, 325.

241 Russell, "On the Notion of Cause," 395–396.

generic law definitely has not been falsified (given the existential quantifier, its logical form is such that it cannot be falsified, and in particular, is not to be considered falsified merely by virtue of the fact that one has tried to find a law and has failed). Should the latter happen and one's search does fail, one does not reject the generic law but simply tries harder.

A causally determined process is one that is continuous. There is nothing remaining that could be a simple "cause" followed by a simple "effect." In the continuity of the process, the cause (if one still wishes to speak this way) merges into, and, as Blanshard indicates, in a way becomes identical with, the effect. It is not possible to discern any break between events in the process that requires that it be bridged. Since there are no breaks to be bridged nor, accordingly, any bridges, there is no place for any sort of ontological tie to do the bridging. Properly understood, the universe of Anaxagoras develops smoothly and continuously (though no one then had the mathematical tools that could give a clear idea of what this meant). There was no need of the activities of Socratic souls or unanalyzable powers to provide ties and linkages to provide an ontological basis for the continuity so clearly apparent in ordinary processes and dimly discerned in others.

More generally, given that the only fully clear notion of cause has become that of a causally determined system, there is little indeed to be said for the necessity of causal laws. Blanshard notwithstanding, no one could pretend, I think, that a process law, i.e., a constant set of constant differential equations, is anything we know *a priori*. If we do know such a law, then we know it only empirically by checking it predictively against observational data. The same is true of what has become of the law of universal causation. This too is clearly nothing that can be known *a priori* by pure reason but rather an empirical, matter-of-fact generalization, arrived at inductively from a number of laws which are themselves empirical, matter-of-fact generalizations.[242]

From Anaxagoras through Hume and Mill, cause and effect were argued to be logically and ontologically separable terms. This claim, itself based on the Principle of Acquaintance, was the basis of the claim that causation was simply regularity—that all causal regularities were contingent, not necessary. Separability was the basis of the denial of the metaphysical claim of Socrates, Aristotle, and their successors down to the present, that there are objective necessary connections. This argument, based on acquaintance, still holds good for cases where causation is simply regular sequence, which is indeed the case for much of our causal understanding of things in everyday life. When we pursue the

242 "On the Notion of Cause," 396.

analysis of causation more deeply, as Blanshard rightly insists we must, however, we come to recognize that causal processes are continuous. Since in a continuous process there is always another moment between any two moments of it, howevcr, the idea of a cause having its effect as its immediate successor is no longer sound. Separability disappears and so also (again as Blanshard saw) does the argument for the contingency of causation. But as Russell made clear, the argument made by Socrates and the idealists subsequent to him that the separation of cause and effect implied the need for a tie to link cause and effect, an objective necessary connection, also fails, for the causally determined processes that replace simple cause and effect in our thinking have no gaps that need to be bridged. Moreover, again as Russell made clear, for the constant sets of constant differential equations that also replace simple cause and effect regularity in our thinking, there is no reason to suppose that their having the mathematical form that they do is anything but contingent; when they are known, process laws are known inductively, not *a priori*.

In the continuity of the process, the cause, if one still wishes to speak this way, merges into, and, as Blanshard indicates, in a way becomes identical with, the effect. Blanshard uses this as part of his case that the world is shot through with filaments of necessity. These filaments are not the active powers of Socrates and Aristotle and the idealists, but they are still non-empirical necessary connections. Although separability in a way disappears, it does not follow that the idealist necessity returns. To the suggestion that it does, the naturalist or empiricist has a response. In effect, one can still hold that Hume was right. In our world, the world of everyday experience that we know by means of our senses and by inner awareness, the patterns of causation are everywhere contingent and our knowledge of them is everywhere inductive.

9 Not a Conclusion

There is Green's question, "Is a natural science of man possible?" Then there is the question, "What is the nature of causal necessity?" Idealists have answered these questions one way while naturalists or empiricists, disagreeing with these answers and rejecting them, have responded to the idealists with very different answers of their own. In turn, idealists have replied with their own new critiques and their own new answers.

My own view, if it has not been evident, is 1) that a natural science of man is possible; 2) that causation is matter-of-fact regularity, or to put is more precisely, matter-of-fact constancy of differential equations, and 3) that causal patterns are always contingent and any apparent necessity is subjective.

Forms, necessary causal ties, and unanalyzable causal powers, so beloved by the idealists, have no place in any acceptable ontology and certainly no place in any science, especially any science of human being. Accepting such powers invites one to suppose that one knows them, that one has a source of genuine knowledge which is beyond the power of natural science to discover, perhaps obtained merely by sitting in one's armchair in one's college rooms or by marching at the head of the working class. As there are no such forms, ties, or powers, however, one's claims to know them is simply illusion, or, worse, ideology which transforms one's wishes or one's hatreds into apparent objective facts. Ideology thus construed is illusion which has become dangerous. It has no place in any reasonable approach to the world in which we all live.

Yet the history of philosophy imposes on me a certain sense of caution. In the past, idealists have responded in often convincing ways to the naturalist or empiricist position. I have little doubt that they will do so again. That will require yet another response, defending once again the idea of a natural science of human being and the idea that, ontologically, the only world we know is the world of sense experience, a world which is everywhere contingent and in which there are neither objective necessities that are inaccessible to sense nor pure reason that gives us knowledge of them.

Such is the history of philosophy. Such is what we can learn from it by induction. But, then, the empiricist that I am will insist, "How else can we learn?" I anticipate that there is more that remains to be done.

References

Althusser, L. and Balibar, Ė., trans. B. Brewster, *Reading Capital*, London: Verso Editions, 1979.

Bergmann, G. , *Philosophy of Science*, Madison, WI: University of Wisconsin Press, 1956.

Blanshard, Brand, *Reason and Analysis*, London: Allen and Unwin, 1962.

Bosanquet, Bernard, "Science and Philosophy," in his *Science and Philosophy*, London: Allen and Unwin, 1927, 15–33.

Bradley, F.H., *Appearance and Reality*, 2nd ed. (with Appendix), Oxford: Oxford University Press, 1897.

Bradley, F.H., *Collected Essays*, 2 vols., Oxford: Oxford University Press, 1935.

Bradley, F.H., *Essays on Truth and Reality*, Oxford: Clarendon Press, 1914.

Bradley, F.H., *The Principles of Logic*, 2nd ed., revised, 2 vols., London: Oxford University Press, 1922.

Carnap, Rudolph, "Testability and Meaning," *Philosophy of Science*, 3 (1936), 419–471, and *Philosophy of Science* 4 (1937), 1–40.

Chisholm, R., "Law Statements and Counterfactual Inference," *Analysis*, 15 (1957), 97–105.

Coleridge, Samuel Taylor, "On the Principles of Political Knowledge," Essay III (from *The Friend*, First Section), in *The Portable Coleridge*, ed. I.A. Richards, New York: Viking Press, 1961, 332–338.

Epicurus, "Letter to Menoeceous," in *The Philosophy of Epicurus*, translated with Introduction and Commentary by George K. Strodach, Evanston, Ill.: Northwestern University Press, 1963, 178–195.

Ewing, A.C., *Fundamental Questions of Philosophy*, London: Routledge and Kegan Paul, 1951.

Green, T.H. , "Can There be a Natural Science of Man?," *Mind*, Vol.7: No. 25–27 (1882), 1–29 (No. 25); 161–185 (No. 26); 321–348 (No. 27).

Green, T.H. , "Introduction to *Treatise of Human Nature*," in Volume I of *David Hume: The Philosophical Works*, ed., T.H. Green and T.H. Grose, London: 1886; reprint, Aalen: ScientiaVerlag, 1964.

Green, T.H., *Prolegomena to Ethics*, Oxford: Clarendon Press, 1906.

Harrison, J., *Hume's Moral Epistemology*, London: Oxford University Press, 1976.

Hollander, Samuel, *The Economics of John Stuart Mill*, 2 vols., Toronto: University of Toronto Press, 1985.

Hume, David, *An Enquiry concerning the Human Understanding, in his Enquiries concerning the Human Understanding and concerning Morals*, ed. L.A. Selby-Bigge, 2nd ed., Oxford: Oxford University Press, 1902.

Hume, David, *A Treatise of Human Nature*, ed. L.A.,Selby-Bigge (1888); 2nd ed. P.H. Nidditch, Oxford: Clarendon Press, 1978.

Hume, David, "Of the Standard of Taste," in. T.H. Green and T.H. Grosse, eds., *Hume: The Philosophical Works*, 4 vols., London: Oxford University Press, 1882–1886; reprinted Aalen, Scientia, 1964, 3: 246–284.

Huxley, T.H., *Methods and Results*, London: Macmillan, 1892.

James, William, "On Some Hegelisms," *Mind*, o.s. 7 (1882), 186–208.

Kant, Immanuel, *Critique of Pure Reason*, trans. Norman Kemp Smith, 1929; unabridged edition, New York: St. Martin's Press, 1965.

Kirk, G.S., and Raven, J.E., *The Presocratic Philosophers: A Critical History with a Selection of Texts*, Cambridge: Cambridge University Press, 1957.

Lamprecht, S.P., *The Metaphysics of Naturalism*, New York: Appleton-Century-Crofts, 1967.

Levi-Strauss, Claude, *Elementary Structures of Kinship*, 2nd ed., trans. J.H. Bell, J.R. von Sturmer, and R. Needham, Boston: Beacon, 1969.

Lévi-Strauss, Claude, *Structural Anthropology*, trans. C. Jacobson and B.G. Schoepf, Garden City, NY.: Doubleday [Anchor], 1967.

Lucretius, *On the Nature of the Universe*, prose translation by R.E. Latham, Harmondsworth, Middlesex: Penguin, 1951.

Madden, E., *Philosophical Problems of Psychology*, New York: Odyssey Press, 1962.

Mill, J.S. , *An Examination of Sir William Hamilton's Philosophy*, in *The Collected Works of John Stuart Mill*, ed. John Robson, Vol. 9, Toronto: University of Toronto Press, 1979.

Mill, J.S., *System of Logic*, in *The Collected Works of John Stuart Mill*, ed. J. Robson, Vols. 7 and 8, Toronto: University of Toronto Press, 1973–1974.

Mill, J.S., "Utilitarianism," in *The Collected Works of John Stuart Mill*, ed. J. Robson, Vol. 10, Toronto: University of Toronto Press, 1969.

Moore, G.E., "The Refutation of Idealism," *Mind*, New Series, Vol. 12: No. 48 (October 1903), 433–453; reprinted in Moore's *Philosophical Papers*, London: Kegan Paul, Trench, Trubner and Co., 1922, 23–44.

Moore, R.J., "John Stuart Mill at East India House," *Historical Studies*, 20 (1983), 497–519.

Plato, *Phaedo*, in *Plato, Five Dialogues*, trans. G.M.A. Grube, Indianapolis, IN: Hackett, 1981.

Plotinus, *Enneads*, trans. S. McKenna, 4th edition revised by B.S. Page, Foreword by E.R. Dodds, Introduction by Paul Henry, London: Faber and Faber, 1959.

Richard, Thomas J., "Hume's Two Definitions of 'Cause'," in V.C. Chappell, ed., *Hume*, Garden City, NY: Anchor Books, 1966, 148–161.

Robinson, J.A., "Hume's Two Definitions of 'Cause'," in V.C. Chappell, ed., *Hume*, Garden City, NY: Anchor Books, 1966, 129–147.

Robinson, J.A., "Hume's Two Definitions of 'Cause' Reconsidered," in V.C. Chappell, ed., *Hume*, Garden City, NY: Anchor Books, 1966, 162–168.

Russell, Bertrand, "On the Notion of Cause, with Applications to the Free-Will Problem," in H. Feigl and M. Brodbeck, *Readings in the Philosophy of Science*, New York: Appleton-Century-Crofts, 1953, 387–407.

Russell, Bertrand, *Principles of Mathematics*, 2nd ed., London: Allen and Unwin, 1937.

Russell, Bertrand, "The Monistic Theory of Truth," in *Philosophical Essays*, Revised Edition, London: Allen and Unwin, 1966, 150–169.

Spencer, Herbert, *Principles of Psychology*, 2 vols., New York: D. Appleton, 1902.

Stroud, Barry, *Hume*, London: Routledge and Kegan Paul, 1977.

Turnbull, R.G., "Aristotle's Debt to the 'Natural Philosophy' of the *Phaedo*," *Philosophical Quarterly*, 8 (1963), 131–143.

Vlastos, G., "Reasons and Causes in the *Phaedo*," *Philosophical Review*, 78 (1960), 291–325.

Wilson, F., *Acquaintance, Ontology and Knowledge*, Frankfurt: Ontos Verlag, 2007.

Wilson, F., "Bareness, as in 'Bare Particular': Its Ubiquity," in H. Hochberg and K. Mulligan, eds., *Relations and Predicates*, Frankfurt: Ontos Verlag, 2004, 81–112.

Wilson, F., *Body, Mind and Self in Hume's Critical Realism*, Frankfurt: Ontos Verlag, 2008.

Wilson, F., "Burgersdijck, Coleridge, Bradley, Russell, Bergmann, Hochberg: Six Philosophers on the Ontology of Relations," in F. Wilson, *Acquaintance, Ontology and Knowledge*, Frankfurt: Ontos Verlag, 2007, 287–328.

Wilson, F. , "Dispositions: Defined or Reduced?," *Australasian Journal of Philosophy*, 47 (1969), 184–204.

Wilson, F., "Dummett's History," in F. Wilson, *Acquaintance, Ontology, and Knowledge*, Frankfurt: Ontos Verlag, 2007, 683–725.

Wilson, F., *Empiricism and Darwin's Science*, Dordrecht, Holland: D. Reidel, 1991.

Wilson, F., *Hume's Defence of Causal Inference,* Toronto: University of Toronto Press, 1997.

Wilson, F., "Hume's Theory of Mental Activity," in D. F. Norton *et al.*, eds., *McGill Hume Studies*, Los Angeles: Austin Hill Press, 1976, 101–120.

Wilson, F., "Mill on Justice," in *J.S. Mill on Justice*, ed. L. Kahn, London: Palgrave MacMillan Press, 2011.

Wilson, F., "Placing Bergmann," in Laird Addis, Greg Jesson, and Erwin Teigtmeir, eds., *Ontology and Analysis: Essays and Recollections about Gustav Bergmann*, Frankfurt: Ontos Verlag, 2007, 185–276.

Wilson, F., *Psychological Analysis and the Philosophy of John Stuart Mill*, Toronto: University of Toronto Press, 1990.

Wilson, F., "Science and Religion: No Irenics Here," *Metaphysica*, 7 (2006), 159–170.

Wilson, F., *Socrates, Lucretius, Camus*, Lewiston, N.Y.: The Edwin Mellen Press, 2001.

Wilson, F., "The Aboutness of Thought," in T.M. Lennon, ed., *Cartesian Views: Papers Presented to Richard A. Watson*, Leiden: Brill, 2003, 151–164.

Wilson, F., *The External World and Our Knowledge of It: Hume's Critical Realism,* Toronto: University of Toronto Press, 2008.

Wilson, F., "The Impact on Psychology of Bradley's Humean View of the Self," *Bradley Studies*, 5 (1999), 5–44.

Wilson, F., *The Logic and Methodology of Science and Pseudoscience*, Toronto: Canadian Scholars Press, 2000.

Wilson, F., *The Logic and Methodology of Science in Early Modern Thought: Seven Studies*, Toronto: University of Toronto Press, 1999.

CHAPTER 8

Bradleyan Idealism and Philosophical Materialism

K.M. Ziebart

F.H. Bradley is known to history as a member of the late-Victorian philosophical movement called British Idealism, whose followers were proponents of a species of Hegelian-inspired absolute idealism. This reputation had profound consequences for the reception of Bradley's philosophy in the twentieth century, as it set Bradley in seemingly diametrical opposition to the philosophical materialism central to analytic philosophy. The following will reconsider the question of the relation of Bradley's metaphysics to philosophical materialism—in which ways this relation is one of rejection and to what extent it is an idealist rejection. As we shall see, the matter is rather more complicated than the historical classifications would give us to believe.

There are two main issues to consider in this connection. Since one of the main materialist criticisms brought against Bradley's metaphysics (*via* Russell) is the alleged radical monism that results from his rejection of external relations, we will first seek to determine more precisely what sort of monist Bradley actually is, and whether the position he defends is as extreme as has been maintained. Since Bradley's metaphysics has historically also been regarded as what we might call "irrealist" (to the extent that realism is typically equated with materialism), we will then seek to more narrowly circumscribe Bradley's philosophical commitments with respect to realism and idealism.

1 Bradley and Monism

A sampling of encyclopaedia entries shows that materialism and idealism are typically defined through their opposition to one another, such that idealism *per se* is an antimaterialist doctrine and vice versa.[1] Therefore, if Bradley

1 Cf., e.g., article "Idealism" in the *New Catholic Encyclopedia*, Vol. VII, New York: Catholic University of America/McGraw Hill, 1967, 340–345: "Idealism is opposed to materialism, denying that mind originates from or is reducible to matter" (340). Also, G. Stack, "Materialism," in *Routledge Encyclopedia of Philosophy*, E. Craig, ed., London: Routledge, 1998, 171–173: "Materialism is … directly antithetical to a philosophical idealism that denies the existence of matter" (171). Also, article Idealismus in J. Ritter et al., eds., *Historisches Wörterbuch der*

 | DOI:10.1163/9789004415270_010

is indeed an idealist philosopher (despite his own reluctance to call himself one),[2] it might be expected that his metaphysics rests on principles fundamentally opposed to those of philosophical materialism. However, materialism and idealism are not as simply opposed as they appear at first blush. Bradley's philosophy was attacked for being monist, but in fact idealism and materialism both have monist and pluralist variations.[3] The sort of materialism pitted by Russell against "Bradleyan monism" was a pluralist, atomistic materialism. However, monism *per se* is not an anti-materialist doctrine, as even a material pluralist is monist with respect to the belief that the ultimate constituents of reality are all material in nature. Some refinement of these terminological distinctions is thus required.

Monism is commonly divided into two types, usually called "substantival" and "attributive" monism respectively. Materialist monism is usually identified as attributive[4] while idealist monism is sometimes classed as substantival and sometimes as attributive. Substantival monism is always considered the stronger of the two positions, although opinion on the precise commitments of each, as well as the identity of their historical representatives, varies considerably. For this reason it is best to view these types as describing a spectrum of possible monisms, ranging from the view that (a) all of reality is numerically

Philosophie, Bd. 4, Basel: Schwabe & Co., 1976. 30–34: In polemischer Bedeutung ist [Idealismus] zunächst Gegenbegriff zu Materialismus, später auch zu Realismus (30).

2 Bradley did not explicitly identify himself as an idealist. Near the end of *Appearance and Reality*, he wrote: "The conclusion which we have reached, I trust, the outcome of no mere compromise, makes a claim to reconcile extremes. Whether it is to be called Realism or Idealism I do not know, and I have not cared to inquire. It neither puts ideas and thoughts first, nor again does it permit us to assert that anything else by itself is more real" (F.H. Bradley, *Appearance and Reality* (hereafter, AR), 2nd ed. (with an Appendix), Oxford: Clarendon Press, 1897, 485).

3 In fact, materialism is often defined as a monist doctrine: "It is a form of monism in that it holds that everything in existence is reducible to what is material or physical in nature" (Stack, "Materialism," 171). The possibility of pluralist idealisms is mentioned in T.L.S. Sprigge's article on "Idealism" in *Routledge Encyclopedia of Philosophy*, 662–669: "The idealist may be a monist or a pluralist about the mind(s) from which the physical is derivative" (663). According to another author, Russell's own early idealism was pluralist; cf. N. Griffen, *Russell's Idealist Apprenticeship*, Oxford: Clarendon Press, 1991, 327.

4 According to the *Stanford Encyclopedia of Philosophy* (where substantival monism is identified by the term "existence" monism), contemporary proponents of substantival materialist monism do exist, be it in small numbers: "Among contemporary philosophers, Horgan and Potrč are probably the leading, and perhaps the only, existence monists." However, in general, analytic philosophers seem to dismiss the strongest form of monism entirely: "Not for nothing is existence monism widely dismissed as a crazy metaphysic" (J. Schaffer, "Monism," *Stanford Encyclopedia of Philosophy*. http://plato.stanford.edu/entries/monism/).

one and any appearance of plurality in the universe is therefore mere illusion, to the position that (b) all of reality is one substance admitting of a plurality of modes or aspects ultimately derivative of this one substance, to the position that (c) there is ultimately only one category of being within which exists a plurality of substances.[5] Russell's criticism of Bradley assumes that the latter defends the strongest substantival monist position, denying the reality of any sort of plurality. Russell also tends to speak as though monism itself were an exclusively idealist doctrine: "The empirical person would naturally say, there are many things. The monistic philosopher attempts to show that there are not. I should propose to refute his *a priori* arguments."[6]

Russell similarly viewed his rejection of Bradley's relational theory as a rejection of monism itself: "I came to disbelieve Bradley's arguments against relations, and to distrust the logical bases of monism."[7] It is furthermore apparent that he considered Bradleyan monism to be an enemy of the "common-sense" intuition that the universe consists of many things: "When I say that my logic is atomistic, I mean that I share the common-sense belief that there are many separate things."[8] It would seem that Russell felt monism in general was simply incapable of doing justice to the more discontinuous aspects of reality:

> Academic philosophers, ever since the time of Parmenides, have believed that the world is a unity ... The most fundamental of my intellectual beliefs is that this is rubbish. I think that the universe is all spots and jumps, without unity, without continuity, without coherence or orderliness or any of the other properties that governesses love.[9]

5 One opinion which seems unanimous is that the paradigmatic representative of the strongest form of monism is Parmenides, in that he maintained the immutability, indivisibility, and undifferentiatedness of the universe. For example, cf. R. Hall, "Monism," *The Encyclopedia of Philosophy*, Vol. 5, New York: Macmillan, 1967, 363–365; 363; also, E. Craig, "Monism," *Routledge Encyclopedia of Philosophy*, Vol. 6, 474–475; 474.

6 Bertrand Russell, "The Philosophy of Logical Atomism" in *The Philosophy of Logical Atomism*, ed. David Pears, La Salle: Open Court, 1985, 35–155; 48. Moore similarly speaks of idealism as though it were synonymous with the very strongest form of monism, which he dismisses with "common sense" appeals to real plurality, such as "Here is one hand ... and here is another" (in *G. E. Moore: Selected Writings*, ed. T. Baldwin, London: Routledge, 1993, 166).

7 Bertrand Russell, *The Philosophy of Bertrand Russell, The Library of Living Philosophers*, ed. P.A. Schlipp, Vol. v, Evanston, Ill: Northwestern University Press, 1944, 11–12.

8 Bertrand Russell, *The Philosophy of Logical Atomism* (1918); reprinted in his *Logic and Knowledge*, New York: Macmillan, 1956, 178.

9 Bertrand Russell, *The Scientific Outlook*, New York: W.W. Norton, 1931, 98.

It is evident that Russell associates monism with an overly rationalistic, *a priori* style of philosophizing which is hostile both to empiricism and to common-sense experience of reality. But is this a fair assessment of Bradley?

It is plain from Bradley's discussion of relations in Chapters II and III of *Appearance and Reality* that his attack is aimed equally at external and internal relations. Both positions are fraught with contradiction, according to Bradley, and therefore neither can be a true description of reality. He approaches this conclusion by way of an analysis of things and their properties. A lump of sugar, he relates, is a substance with the qualities 'white', 'hard', and 'sweet'. The lump of sugar *is* these qualities and yet the character of this *is* becomes immediately problematic upon inspection. The lump of sugar is not any one of its qualities, considered separately, nor is it simply the conjunction of those separate properties; rather it is all of them together in a certain unity. However, the unity of the lump of sugar must be all of those qualities in such a way that preserves the distinctness of each; for the whiteness in the lump of sugar is not the same as its hardness, etc. In this way we are driven to consider that the lump of sugar must be something beyond its qualities. And yet, what could it be other than its qualities? The unity that is the lump of sugar, Bradley explains, can be found neither within nor without the properties of the thing.[10]

This causes us to turn instead to the relations between the qualities: perhaps, he reasons, it is simply the particular *relation* of the qualities 'white', 'hard', and 'sweet' which makes it a lump of sugar. But here again similar problems arise. As before, the qualities 'white', 'hard', and 'sweet' are distinct; thus, in the lump of sugar, a relation must be asserted of each of them. Where does the relation stand? Bradley rejects the notion that the relation could belong to one of the terms. When we assert "A is related to B," he insists, we do not mean "A *is* 'related to B'." It may then seem better to speak of something *having* a relation to something else, rather than *being* related to something else, but this does not help for then we must explain the meaning of 'have'.[11] The problem, according to Bradley, is that subject-predicate logic cannot adequately express the difference-in-unity that is the lump of sugar. The statement, "The lump of sugar is white," either attributes something (whiteness) to the subject (lump of sugar) that is different from it (in which case the statement is false) or it

10 AR, 16.

11 "No, we should reply, the relation is not identical with the thing. It is only a sort of attribute which inheres or belongs. The word to use, when we are pressed, should not be *is*, but only *has*. But this reply comes to very little. The whole question is evidently as to the meaning of *has*; and apart from metaphors not to be taken seriously, there appears really to be no answer" (AR, 17).

predicates something which is not different from it (in which case the statement tells us nothing).[12]

Having rejected the notion that the relation between two things could be an attribute of the subject, he then goes on to consider the possibility that a relation could be something independent of its relata. However, here we immediately encounter the problem of external relations. For if two terms A and B stand in a relation *R* in such a way that the relation is different from both A and B and is no longer predicated of them, then, according to Bradley, there will have to be something on either side of *R* which relates *R* to A on the one side, and *R* to B on the other side. But if between A and *R* stands a further relation *R1*, which is similarly not ascribed to either term, then *R1* will in turn require further relations, *R2* and *R3*, to relate it to both A and *R*, and those relations will similarly require further relations, and so on into infinite regress. Bradley concludes:

> Thus the problem is not solved by taking relations as independently real. For, if so, the qualities and their relation fall entirely apart, and then we have said nothing. Or we have to make a new relation between the old relation and the terms; which, when it is made, does not help us. It either itself demands a new relation, and so on without end, or it leaves us where we were, entangled in difficulties.[13]

The notion of independently real relations standing alongside their terms is a "failure" and a "delusion" according to Bradley. For relations to work, there has to be "a substantial foundation" in the relata which holds their differences

12 "… if the predicate makes no difference, it is idle; but, if it makes the subject other than it is, it is false" (AR, 17).

13 AR, 18. Bradley mounted essentially the same argument against external relations already in his *Principles of Logic*, in the context of a discussion about experience in terms of units of feeling: "If units have to exist together, they must stand in relation to one another; and, if these relations are also units, it would seem that the second class must also stand in relation to the first. If A and B are feelings, and if C, their relation, is another feeling, you must either suppose that component parts can exist without standing in relation with one another, or else that there is a fresh relation between C and AB. Let this be D, and once more we are launched on the infinite process of finding a relation between D and C—AB; and so on for ever. If relations are facts that exist between facts, then what comes between the relations and the other facts? The real truth is that the units on one side, and on the other side the relations existing between them, are nothing actual. They are fictions of the mind, mere distinctions within a single reality, which a common delusion erroneously takes for independent facts" (*Principles of Logic* (hereafter, PL), 2nd ed., 2 vols., London: Oxford University Press, 1922, 1:96).

together. Without a whole holding differences together, there could be neither differences nor relation.

Bradley never denies that reality appears to us in the form of qualities and relations, or even that "the arrangement of given facts into relations and qualities may be necessary in practice."[14] His point throughout is that, if the matter at hand is how to think without error about reality—and this is the task of *Appearance and Reality*—then this practice will be found to be "theoretically unintelligible."[15]

This is further explained in Chapter 3, where Bradley considers the argument that, even if qualities cannot be separated from their relations in reality, the fact that we are able conceptually to distinguish them means that they exist. Bradley's response is that the distinction made in our minds about a thing in reality is merely that—a point of view or way of thinking. But for this way of thinking about reality to be correct, it must be true *in* reality. Since it has already been shown that independent relations cannot be real, this cannot be a correct way to think about reality. The mere fact that the mind is capable of performing a certain ideal operation does not justify the application of the result of that operation to reality: "There is an operation which, removing one part of what is given, presents the other part in abstraction. This result is never to be found anywhere apart from a persisting abstraction."[16]

He furthermore attacks the intelligibility of qualities without relations and vice versa. Relation, he maintains, is essential to difference and therefore the notion of quality independent of any relation is a self-contradiction. A can only be A insofar as it is distinguished from other things. The very *coherence* of A relies upon relation. Relations without qualities are equally incoherent, since a relation which does not relate something is but "a phrase without meaning." Bradley finds it meaningless to trade on such abstractions.[17] Relations and qualities taken together are also unintelligible. On the one hand, if a relation is externally connected to a quality, then an infinite number of intervening relations will be required. On the other hand, if the relation is somehow internal to the quality, then the unity of that quality is destroyed "… and its contents are dissipated into an endless process of

14 AR, 21.

15 AR, 21ff.

16 AR, 23.

17 "Without a relation it [a quality] has no meaning; it is a mere word, and not a thought; and no one would take it for a thought if he did not, in spite of his protests, import relation into it. […] Have qualities without relation any meaning for thought? For myself, I am sure they have none" (AR, 25).

distinction."[18] Relations, no matter where they are placed, destroy qualities, even as they are constitutive of them.[19]

The conclusion of Bradley's analysis is thus that there is no coherent theory of relations. He does not advocate internal relations, as has often been supposed, nor does he only take into account the subject-predicate form, nor does he confuse the "is" of existence with the "is" of predication—all accusations brought against him, particularly by Russell.[20] Thought—to think correctly about reality—must be able to take reality as both one and many without contradiction, but thought cannot do this. It finds consistency only by disregarding one of these elements and therefore can at best yield an incomplete description of reality. In other words, thought is limited to appearance:

> The conclusion to which I am brought is that a relational way of thought—any one that moves by the machinery of terms and relations—must give appearance and not truth. It is a makeshift, a device, a mere practical compromise, most necessary, but in the end most indefensible.[21]

Bradley's analysis of relations thus provides the logical grounds for his monism. But is this the radical monism of Parmenides, which regards the universe as one immutable, indivisible, and undifferentiated substance, and would dismiss all difference as illusion? Or is it a more moderate form of monism, which views reality as one substance with different modes of appearance? I think the first may essentially be rejected. Bradley's account of relations is clearly at pains, not simply to argue for the underlying unity of all things, but equally to protect the evident differences between them which are given to us in experience. Bradley is no enemy of common sense. His arguments for the unity of reality are always aimed, on the one hand, at *ultimate* reality, or the

18 "Hence the quality must exchange its unity for an internal relation. But, thus set free, the diverse aspects, because each something in relation, must each be something also beyond. This diversity is fatal to the internal unity of each; and it demands a new relation, and so on without limit. In short, qualities in a relation have turned out as unintelligible as were qualities without one" (AR, 26–27).

19 Herein lies Bradley's rejection of Hegelian dialectic, what he refers to as the "Dialectic Method" at AR, 26.

20 For more detailed examinations of the criticism Bradley faced from Russell, Moore, *et al.* in the early twentieth century, cf. S. Candlish, "The Wrong Side of History: Relations, the Decline of British Idealism, and the Origins of Analytical Philosophy," in Guy Stock, ed., *Appearance versus Reality. New Essays on Bradley's Metaphysics*, Oxford: Clarendon Press, 1998, 111–152, and S. Candlish, *The Russell/Bradley Dispute and its Significance for Twentieth-Century Philosophy*, Houndmills: Palgrave/Macmillan, 2007.

21 AR, 28.

Absolute, and, on the other hand, at the logical *foundations* of thought. Insofar as thought aims to think correctly about ultimate reality (which is the project of metaphysics), it is constrained by the condition of unity. However, between these extremes, in the realm of sensory experience and thought which seeks to describe it, plurality abounds, and Bradley nowhere seeks to suppress or reduce it. Thus the discussion of colours and smells he presents in Chapter 2 of *Appearance and Reality* is not aimed at demonstrating the illusory nature of these qualities, but rather seeks to understand how it is that we experience them as the distinct and yet conjoined properties of things.[22] He uses this example to point out that our experience shows us "... how we are forced to arrange things in a certain manner, but it does not justify that arrangement."[23] It is not experience that is problematized but theory.

Bradley's monism is more akin to the moderate type identified above, which views reality as one substance with a plurality of modes or aspects. The label "attributive" would be inappropriate for it since his discussion of relations is a long argument against the possibility that reality consists of things and their attributes. But appearances, on the Bradleyan model, are real, not illusory, and are thus something more like modes or aspects of reality. What sorts of modes or aspects these are requires a closer examination of what Bradley means by "thought," "appearance," and "experience," which brings us to the second part of our study.

2 Idealism and Realism in Bradley

As mentioned above, Bradley himself was noncommittal on the subject of his idealism, indicating—in a way that suggests he did not particularly like either alternative—that idealism or realism would be equally acceptable names for his philosophy. Russell and Moore certainly would not have affirmed the latter possibility, as they considered idealism and realism antithetical. However, they also tended to blend the views of all idealists together, such that their attacks

22 "Thus colours and smells live together at peace in the reality; for the thing divides itself, and so leaves them merely side by side within itself. But colour collides with colour, because their special identity drives them together. And here again, if the identity becomes relational by help of space, they are outside one another, and are peaceful once more. The 'contrary', in short, consists of differences possessed by that which cannot find the relation which serves to couple them apart. It is a marriage attempted without a *modus vivendi*. But where the whole, relaxing its unity, takes the form of an arrangement, there is co-existence with concord" (AR, 19).

23 AR, 19.

on idealism were aimed at a fabricated, composite subject. G.E. Moore's "Refutation of Idealism" is a good example of this tendency. The main target of this paper is considered to have been Bradley, but the view attacked is essentially Berkeleyan.[24]

However, it was not entirely without reason that Moore thought Bradley to be a defender of some sort of Berkeleyan idealism as it is easy to get this impression from Bradley's equation of reality with experience. In Chapter 14 of *Appearance and Reality*, Bradley writes: "Sentient experience, in short, is reality, and what is not this is not real. We may say, in other words, that there is no being or fact outside of that which is commonly called psychical existence."[25] This does really sound like a version of *esse est percipi*. Yet shortly thereafter, Bradley will flatly deny that he is advocating with this notion a form of subjective idealism whereby the only reality is subjective reality. First, however, he makes his case for the equation of reality and experience *via* one of his 'ideal experiments':

> Find any piece of existence, take up anything that any one could possibly call a fact, or could in any sense assert to have being, and then judge if it does not consist in sentient experience. Try to discover any sense in which you still continue to speak of it, when all perception and feeling have been removed; or point out any fragment of its matter, any aspect of its being, which is not derived from and is not still relative to this source. When the experiment is made strictly, I can myself conceive of nothing else than the experienced. Anything, in no sense felt or perceived, becomes to me quite unmeaning. And as I cannot try to think of it without realizing either that I am not thinking at all, or that I am thinking of it against my will as being experienced, I am driven to the conclusion that for me experience is the same as reality.[26]

The meaning of this experiment has been the source of much debate amongst scholars, in no small part due to the striking resemblance it bears to the "conceivability argument" of Berkeley.[27] It is indeed difficult to grasp precisely

24 Moore's criticism is aimed at the central thesis of Berkelyan idealism, *esse est percipi*, which Moore wrongly attributes to all forms of idealism, and which, we shall argue below, is not a doctrine Bradley supports. Cf. G.E. Moore's "Refutation of Idealism" (originally published in *Mind*, 12 [1903]); in T. Baldwin, ed. *G.E. Moore. Selected Writings*, 23–44.

25 AR, Chapter XIV, "The General Nature of Reality (cont.)," 127.

26 AR, 127–128.

27 The Bishop's argument runs as follows: "But say you, surely there's nothing easier than [for me] to imagine trees, for instance, in a park, or books existing in a closet, and nobody

Bradley's intention here.[28] However, certain things are evident. First, Bradley enjoins the reader to focus, not on the nature of his own mental states, but on the contents or objects of those states (i.e., on the "facts" or things being thought of). Bradley is not here making the error (famously attributed to Berkeley) of confusing a thought with the act of thinking it or a representation with what it represents.[29] Second, in accordance with this objective orientation, Bradley does not ask one to try to conceive of a fact or thing existing independently of a perceiving mind (as does Berkeley) but rather asks what would remain of that fact or thing were all traces of experience removed from it—in other words, how any fact could even remain coherent or thinkable in the total absence of anything related to experience *per se*.

Despite the similarity with Berkeley's argument, therefore, what Bradley is trying to get at is something rather different, and much broader. It should also be pointed out that Bradley presents this ideal experiment not in the context of an antimaterialist argument (as does Berkeley), but in the context of an argument against noumenal reality. Moreover, it is significant that by "sentient experience" and "psychical existence," Bradley means to refer *simpliciter* to all forms or modes of living experience or consciousness, including pre-reflective experiences like volition, feeling, desire, and perception. He wants to argue that, since everything of which we can conceive or speak or be even remotely aware—from the vaguest perception or feeling to the most abstract of notions—falls within the bounds of sentient experience, the notion of some

by to perceive them. I answer, you may so, there is no difficulty in it: but what is all this, I beseech you, more than framing in your mind certain ideas which you call *books* and *trees*, and at the same time omitting to frame the idea of anyone that may perceive them? But do not you yourself perceive or think of them all the while? This therefore is nothing to the purpose: it only shows you have the power of imagining or forming ideas in your mind; but it does not shew that you can conceive it possible, the objects of your thought may exist without the mind ..." (George Berkeley, *A Treatise Concerning the Principles of Human Knowledge* (1710, 2nd ed, 1734); Introduction and Notes by M.R. Ayers, London: Dent, 1975, §23, 83). The similarity between Berkeley's argument and Bradley's ideal experiment has led some commentators (like Richard Wollheim) to assume that Bradley was simply repeating the former, and therefore that he was an advocate of subjective idealism.

28 W.J. Mander lists a number of the interpretations to which Bradley's experiment has been subject over the years. His own evaluation of it seems to lead to no conclusive result, although he rules out a number of mistaken views, including subjectivist interpretations (Mander, *An Introduction to Bradley's Metaphysics*, Oxford: Clarendon Press, 1994, 129–132).

29 Indeed, as Mander also points out, it would be odd for Bradley to commit such an error, since he goes to such lengths to make this distinction in the opening pages of his *Principles of Logic* (Mander, *An Introduction to Bradley's Metaphysics*, 130; cf. PL, 1:2–4).

reality beyond sentient experience is something that can in no way be articulated or coherently thought.[30] Therefore, when we speak of a reality beyond sentient experience, that of which we speak is either a) conditioned and constrained by sentient experience, and thus not truly a concept of that of which it claims to be a concept, or b) is a concept stripped down to what Bradley calls "a vicious abstraction whose existence is meaningless nonsense."[31] In other words, if unexperienced reality is a coherent notion, it is only because "I am thinking of it against my will as being experienced," and if all experience were truly excluded, I would no longer be thinking at all.

Bradley recognizes that the conclusion of his ideal experiment could be misunderstood as endorsing "a common error" which, as he puts it, begins with an illicit dividing of the percipient subject from the universe, who is then supposed incapable of transcending his own subjective states.[32] However, Bradley insists that this is not his position. The dictum "reality is experience" does not imply subjective idealism, he explains, because reality is *the whole*. That which is experienced by a subject is therefore not his own perceptions, but reality itself: "And this is the point on which I insist, and it is the very ground on which I stand, when I urge that reality is sentient experience. I mean that to be real is to be indissolubly one thing with sentience."[33]

The key to distinguishing Bradley's dictum from that of Berkeley's is recognizing that, to Bradley, experience—which he also refers to here as "sentience"—is not subjective experience. He is adamant that what we find in experience is neither a subject nor an object; it is, rather, "a whole in which distinctions can be made, but in which divisions do not exist."[34] It is for this reason that there is no danger of falling into a Berkeleyan subjectivism through Bradley's appeal to experience.

30 The possible objection that there are many things of which no one is currently aware, or even that there may be possible modes of experiencing things which do not fall within our current conception of "sentient experience" does not, I believe, pose a threat to this view. Reality is here so defined as to include any possible experience, whatever form it may take. Since Bradley maintains that all appearances are incomplete aspects of reality, "experience," insofar as Bradley equates it with reality, is to be understood as the totality of all possible experiences, including the innumerable ones we cannot even imagine. Just as, to Bradley, truth is the ideal correlate of reality (and is something never completely attained by thought), so is experience, as it posited in this ideal experiment, the sentient correlate of reality.

31 AR, 128.

32 AR, 128.

33 AR, 128.

34 AR, 128.

Since he is not a subjective idealist, the obvious alternative would seem to be that Bradley is an objective, or absolute, idealist. Certainly, in his own mind, Bradley was no follower of Hegel. In the Preface to his *Principles of Logic*, he confesses to having admiration for the philosopher, but maintains that he could never call himself an Hegelian, due in part to the fact that he rejects the main principle of that philosophy, i.e., the belief that the real is rational.[35] He also mocks the notion of his belonging to a so-called "Hegelian school," which he considers to be purely an invention of his own critics.[36]

Despite such statements, Bradley's philosophy is permeated with many strongly Hegelian elements.[37] However, Bradley's equation of reality and experience is not an equation of the real with the ideal. Thought is but one aspect of experience, one appearance of the real, and moreover is not privileged in this respect above other types of appearance, such as feeling or volition.[38] To Bradley, reality *per se* is neither subjective nor objective, neither material nor ideal, since all such designations presuppose a duality of terms which pertains to the ideal realm only, not to reality itself. Therefore, Bradley's philosophy is not a form of objective idealism.

35 We also saw above how Bradley's criticism of relations is in part an attack on Hegelian dialectic.

36 PL, Preface: "I fear that, to avoid worse understandings, I must say something as to what is called 'Hegelianism'. For Hegel himself, assuredly I think him a great philosopher; but I never could have called myself an Hegelian, partly because I can not say that I have mastered his system, and partly because I could not accept what seems his main principle, or at least part of that principle. I have no wish to conceal how much I owe to his writings; but I will leave it to those who can judge better than myself, to fix the limits within which I have followed him. As for the 'Hegelian School' which exists in our reviews, I know no one who has met with it anywhere else" (1:x).

37 This is true especially in Bradley's presentation of appearances as striving to complete themselves and be absorbed in the higher unity of the Absolute: "For each one-sided aspect, to gain itself, blends with that which seemed opposite, and the product of this fusion keeps the riches of all ... To work out its full nature it has sunk itself into these differences. But in each it longs for that absolute self-fruition which comes only when the self bursts its limits and blends with another finite self" (AR, 161).

38 PL also concludes with the categorical rejection of the equation of thought with reality, which moreover reveals Bradley's fundamental distaste for the abstract "bloodlessness" of objective idealism: "Unless thought stands for something that falls beyond mere intelligence, if 'thinking' is not used with some strange implication that never was part of the meaning of the word, a lingering scruple still forbids us to believe that reality can ever be purely rational. It may come from a failure of my metaphysics, or from a weakness of the flesh which continues to blind me, but the notion that existence could be the same as understanding strikes as cold and ghost-like as the dreariest materialism" (PL, 2:590–591).

If the contents of his philosophy support Bradley's insistence that he is neither a Berkeleyan/subjective nor an Hegelian/objective idealist, then the most likely remaining candidate is that he is some kind of transcendental idealist. That said, this notion faces the immediate objection that Bradley explicitly rejected the existence of the *Ding-an-sich*. Chapter XII of *Appearance and Reality* is devoted to the refutation of the idea that reality in itself is beyond all knowledge and experience, a belief he regards as "absurd."[39] Bradley first asserts a commonly-made criticism against Kant's thing-in-itself, namely, that if it were truly unknowable, then we would be in no position to assert its existence.[40] However, this is not his main attack. As in the ideal experiment examined above, his primary argument against the noumenal is that the very notion of reality existing utterly divorced from appearance is unintelligible. For if, he writes, we consider the relation of appearance to reality so construed, we have already fallen into the error of qualifying the latter with properties that it cannot possibly have (such as relation) or, if we (properly) deny it any properties whatsoever, then we are left with something which is a) clearly not real and, b) of no use as a concept of reality.[41]

Bradley's argument for a) is that reality can be real only by virtue of its possessing some quality; however, the noumenal can have no qualities and therefore is nothing at all. His argument for b) is that appearance, or phenomena, must either be regarded as standing in relation to reality or not. Either way, appearance will be left in a confused state since its attribution to the "other world" of the thing-in-itself provides no information or clarification of its nature. "Everything in our concrete world," Bradley writes, "remains the same, and the separate existence somewhere of this wretched abstraction serves us only as a poor and irrelevant excuse for neglecting our own concerns."[42] It is this aspect of the thing-in-itself which troubles Bradley the most. Appearances

39 AR, Chapter XII, "Things in Themselves," 110–115.

40 "The Unknowable must, of course, be prepared either to deserve its name or not. But, if it actually were not knowable, we could not know that such a thing even existed. It would be much as if we said, 'Since all my faculties are totally confined to my garden, I cannot tell if the roses next door are in flower.' And this seems inconsistent" (AR, 111).

41 AR, 112; cf. "Either our Thing has qualities, or it has not. If it has them, then within itself the same puzzles break out which we intended to leave behind – to make a prey of phenomena and to rest contented with *their* ruin. So we must correct ourselves and assert that the Thing is unqualified. But, if so, we are destroyed with no less certainty. For a Thing without qualities is clearly not real. It is mere Being, or mere Nothing, according as you take it simply for what it is, or consider also that which it *means* to be. Such an abstraction is palpably of no use to us" (AR, 112).

42 AR, 113.

are what we experience, and therefore must somehow be connected to reality.[43] They may be just "a beggarly show," he tells us, but nevertheless most certainly belong to reality.

It remains to be shown *how* it is that appearance belongs to reality. In Chapter XV of *Appearance and Reality*, Bradley seeks to provide a viable alternative to the Kantian doctrine he has rejected. Here he answers the question of how thought can find its necessary connection to reality, given that it cannot think reality as such. He will maintain that we are justified in asserting that there is a whole greater than our thought, which is nonetheless not the thing-in-itself because it is in no way conceivable as separate from thought. The reason we are so justified is precisely because it is analysis of the structure of judgment which makes this evident.

Anything, Bradley writes, has two aspects: a "what" and a "that" or a content (i.e., quiddity) and an existence, for "that anything should be, and should yet be nothing in particular, or that a quality should not qualify and give character to anything, is obviously impossible."[44] An idea is a "what" considered separately from its "thatness" or existence. Judgment, which is the operation of thought, attempts the reunification of these two sides, "provisionally estranged."[45] Truth is the aim of thinking, and is the ideal correlate of reality. Thought's striving for truth is its natural tendency to try to repair this broken state.[46] Nevertheless, as it is precisely in the alienation of these two aspects that ideality consists, if thought were to achieve its aim, it would cease to be thought.[47]

Bradley then addresses the "erroneous idea" that "if reality is more than thought, thought itself is, at least, quite unable to say so," and that to maintain otherwise involves positing the thing-in-itself.[48] He begins by urging that, despite its limitations, "... there is no impossibility in thought's existing as an element, and no self-contradiction in its own judgment that it is less than the universe."[49] The reason this is so lies in the fact that the true subject

43 "... appearances exist. That is absolutely certain, and to deny it is nonsense. And whatever exists must belong to reality. ... What appears, for that sole reason, most indubitably *is*; and there is no possibility of conjuring its being away from it. ... And reality, set on one side and apart from all appearance, would assuredly be nothing. Hence what is certain is that, in some way, these inseparables are joined" (AR, 114–115).

44 AR, 143. This basically summarizes the theory of judgment developed in Book I of PL.

45 AR, 145.

46 As Bradley puts it, truth "... is the effort to heal this disease, as it were, homoeopathically" (AR, 146).

47 Thus Bradley's assertion that "thought is aiming at suicide" (AR, 148).

48 AR, 147.

49 AR, 148.

of every judgment is in fact reality itself.[50] There is in every thought, he explains, something which reaches beyond the boundaries of thought: "In other words, in practice thought is always found with, and appears to demand, an Other."[51]

This becomes clearer upon examination of an actual judgment. Bradley notes two characteristics of the subject of any judgment: what he calls its "sensuous infinitude" and its "immediacy." The former refers to the fact that the subject of any judgment, considered closely, "… has a detail which is unlimited." Of course, he does not mean that it is quantitatively unlimited, as in an unlimited plurality of features, but rather that its given content is in a way illimitable, in that it can be determined only through an infinitely expanding series of relations to other things. It is as though the subject has ragged edges which "imply another existence from which it has been torn, and without which it really does not exist."[52] "Immediacy" refers to the given particularity of the subject, which presents itself as one integral whole; however, this is inconsistent with its "endlessness."

Because the self-dependence and immediacy which the subject claims are not possessed by its content, "… in the attempted self-assertion this content drives the subject beyond its actual limits, and so begets a process which is infinite and cannot be exhausted. Thus thought's attempt wholly to absorb the subject must fail."[53] The outward unravelling of the subject through a never-ending series of external relations is not simply practically unfinishable, it is *per se* endless.[54] Furthermore, even if thought could exhaust the series of relations necessary to complete the subject, the result would still fall short of the reality, for it would be but a "maze of relations," lacking the quality of immediacy.

In thus being driven through its own processes to try to repair its broken state and achieve a completeness which ever evades it, thought is brought quite naturally to confront the "Other" within its very structure. Herein lies Bradley's final answer to Kant:

50 By "subject" here, Bradley does not mean the grammatical subject, but rather the referent of the judgment.

51 AR, 155.

52 AR, 156.

53 AR, 157.

54 "We can neither take the terms with their relations as a whole that is self-evident, that stands by itself, and that calls for no further account; nor, on the other side, when we distinguish, can we avoid the endless search for the relation between the relation and its terms" (AR, 157).

> Thought can understand that, to reach its goal, it must get beyond relations. Yet in its nature it can find no other working means of progress. Hence it perceives that somehow this relational side of its nature must be merged and must include somehow the other side. Such a fusion would compel thought to lose and to transcend its proper self. And the nature of this fusion thought can apprehend in vague generality, but not in detail; and it can see the reason why a detailed apprehension is impossible. Such anticipated self-transcendence is an Other; but to assert that Other is not a self-contradiction. Hence in our Absolute thought can find its Other without inconsistency.[55]

Through an analysis of the evident structure of thought and judgment, Bradley infers their necessary connection to a reality which exceeds thought and yet may be apprehended by it "in vague generality" because it possesses it already in incomplete form. This permits him to ground epistemologically his metaphysical view that reality exceeds, yet encompasses, the ideal as one of its aspects.

The relationship between thought and experience gains further clarification in the Appendix to *Appearance and Reality*. Entitled "The Dilemma of the Relation of Thought to Reality," Bradley writes that, as far as thought is concerned, there are two main features to be noted: 1) that thought/truth is incapable of satisfying its own essential requirements and 2) that if it could, it would be reality (and cease to be thought). To this dilemma, Bradley maintains that the solution lies in "the whole view of Reality." When reality is understood to be "... above thought and every partial aspect of being, yet to include them all, then this whole view is had." This whole view is, moreover, experience: "... we know that the Absolute must be one, because anything experienced is experienced in or as a whole, and because anything like independent plurality or external relations cannot satisfy the intellect.[56]

Unity is the defining nature of our experience. We experience reality directly as a unified, unbroken whole; therefore our reasoning about reality will fail to satisfy the intellect so long as it does not capture this unity. We may thus say that Bradley's monism also has a phenomenological justification. Pluralism fundamentally contradicts our experience of reality, and so the intellect cannot reach a state of rest or satisfaction by such a means. Bradley's ultimate

55 AR, 160.

56 AR, 494–495.

criticism of pluralist philosophies is that they neglect experience by not making it the starting point and index of truth.[57]

It is by now clear that Bradley does not endorse the idealist view that mind is the most basic reality or that the physical world exists only as an appearance to, or as an expression of, mind. If one may be called an "idealist" despite maintaining that both the material and ideal are but aspects of a broader yet encompassing reality, then Bradley is rightly so named; however, it is notable that the term as applied to him does not include many of the beliefs which critics like Russell and Moore attacked under that denomination. Thus while it serves to situate him historically with respect to his philosophical contemporaries, the term "idealist" fits rather awkwardly with Bradley's actual philosophical doctrine, and his own hesitance on the matter is understandable.[58]

What Bradley does assert is the absolute reality of experience—broadly construed so as to include everything that could ever be experienced, including, but not limited to, those things and modes of understanding of which we are presently aware. This scheme is moreover strongly realist (if we dismiss the convention which identifies realism with materialism), since there is absolutely nothing that can fall outside of reality, and even the dimmest appearance is still an appearance of reality. It is also a scheme which accords no priority either to the mental/ideal over the material or vice versa.

57 This is how we should understand Bradley's accusation that his critics have not seriously considered "… what the demands of thought really are, or what their satisfaction involves" (AR, 493).

58 This has some support in the research. Richard Ingardia, who finds fundamental agreement between Bradleyan and Thomistic principles, acknowledges that "Bradley and Aquinas are not idealistic philosophers, if an idealistic philosophy is defined as one which regards idea-objects as the direct, primary and immediate objects that we know" (Richard Ingardia, "Bradley and Aquinas: Empirical Realists," in *Perspectives on the Logic and Metaphysics of F.H. Bradley*, ed. W.J. Mander, Bristol: Thoemmes Press, 1996, 75–106; 78–79). Guy Stock has written on "The Realistic Spirit in Bradley's Philosophy," casting doubt on the usefulness of classing Bradley as either an idealist or a Hegelian (Guy Stock, "The Realistic Spirit in Bradley's Philosophy," in *Appearance versus Reality. New Essays on Bradley's Metaphysics*, ed. Guy Stock, Oxford: Clarendon Press, 1998. 1–18). Mander maintains that Bradley is an idealist, and that his idealism is, moreover, a direct consequence of his doctrine of relations; however he does underline that, although his idealism shares some things in common with both subjective and objective idealism, it would be inaccurate to ascribe it to either of these traditions. He also highlights Bradley's realism: "He may be an idealist, but he is a strongly realist one" (Mander, *An Introduction to Bradley's Metaphysics*, 132). That Bradley's metaphysics is a direct consequence of his view on the unreality of relations is certain; however, what Mander sees as peculiarly idealist about this metaphysics is unclear to me.

3 Conclusion: Bradley and Philosophical Materialism

What may we thus conclude about the relation of Bradley's metaphysics to philosophical materialism? Clearly, insofar as materialism posits the ultimate nature of reality as material, his metaphysics is a rejection of this doctrine. However, we have also seen that Bradley's philosophy is one in which the ideal and the material coexist and even overlap. We will conclude by pointing out one concrete way in which this coexistence is manifested.

Bradley does not accept any essential division between material and ideal processes and, in particular, regards thought as a fundamentally embodied, natural phenomenon. Judgment may be what distinguishes human being from that of other creatures, but it is an operation which has evolved out of the other physical processes, and should therefore not be regarded as a clear or essential dividing line between man and the lower animals. That it is in fact a "late acquisition" of human being shows that the possession of mind is really a matter of degree. In *Principles of Logic*, Bradley sketches his position on the development of the function of judgment in the following way:

> As we have defined it above, judgement does not show itself at all the stages of psychical evolution. It is a comparatively late acquisition of the mind, and marks a period of its upward growth. We should probably be wrong if we took it as a boundary which divides the human from the animal intelligence … It is better to treat the mind as a single phenomenon, progressing through stages, and to avoid all discussion as to whether the lines, by which we mark out this progress, fall across or between the divisions of actual classes of animals. Thus with judgment we are sure that, at a certain stage, it does not exist, and that at a later stage it is found in operation …[59]

Bradley affirms the general association of the distinctions requisite of judgment with the acquisition of language, but stresses too that this is a matter of degree, given the inherent difficulty of saying with certainty where language begins.[60] He finds even amongst animals the employment to some degree of universals without which, he maintains, it is incomprehensible that an animal burnt one day at the kitchen fire will the next day shrink from a lighted match.[61] "But if this is so," he writes, "judgment must come before language,

59 PL, 1:28–29.

60 PL, 1:31.

61 "From the very beginning of soul-life universals are used. It is because the results of experience are fixed in an ideal and universal form, that animals are able, I do not say to

and certainly cannot be distinctively human."[62] Mind, therefore, is not to be regarded as an essentially human, "heaven-sent faculty," but rather as something continuous with overall physical evolution.[63]

Bradley's analysis of mental development is very empirical in approach, although it regards ideal operations as continuous with the physical without reducing them to it. Because our mental capacities are merely a stage in a broader evolution, there is nothing absolute about our knowledge of reality; it is always a work in progress, and we must not mistake the limits of our present degree of mental development for the limits of reality itself. Bradley is in no way hostile to the physical, nor to empiricism in general, and indeed embraces both within his philosophy. His realism is so construed as to enfold both the material and the ideal without doing violence to either—which is to say, it seeks to preserve most of our common-sense notions about our experience both of ourselves and of the world around us. Whether his metaphysics ultimately withstands philosophical scrutiny is a matter for further investigation and debate. However, I hope I have at least succeeded in showing that Bradley's philosophy does not merit some of the labels attached to it by his early materialist critics.

References

Baldwin, T., ed., *G. E. Moore: Selected Writings*, London: Routledge, 1993.

Berkeley, George, *A Treatise Concerning the Principles of Human Knowledge* (1710, 2nd ed, 1734); Introduction and Notes by M.R. Ayers, London: Dent, 1975.

Bradley, F.H., *Appearance and Reality*, 2nd ed. (with an Appendix), Oxford: Clarendon Press, 1897.

Bradley, F.H., *Principles of Logic*, 2nd ed., 2 vols., London: Oxford University Press, 1922.

progress, but to maintain themselves in bare existence ... It is not true that among lower animals universal ideas are never used ... Again, if an animal has been burnt one day at the kitchen fire, the next day it may shrink form a lighted match. But how different are the two. How much more unlike than like. Will you say then that the match can not operate unless it first summons up, and then is confused with the image of a kitchen fire; or will you not rather say that a connection between elements, which are none of them particular, is produced in the mind by the first experience? But if so, from the outset universals are used, and the difference between the fact and the idea, the existence and the meaning, is unconsciously active in the undeveloped intelligence ... Our object here has been, in passing, to show that the symbolic use of ideas in judgment, although no early process of the mind, is a natural result of mental development" (PL 1:34–37).

62 PL, 1:32.

63 PL, 1:34ff.

Candlish, S., *The Russell/Bradley Dispute and its Significance for Twentieth-Century Philosophy*, Houndmills: Palgrave/Macmillan, 2007.

Candlish, S. "The Wrong Side of History: Relations, the Decline of British Idealism, and the Origins of Analytical Philosophy," in Guy Stock, ed., *Appearance versus Reality. New Essays on Bradley's Metaphysics*, Oxford: Clarendon Press, 1998, 111–152.

Craig, E., "Monism," *Routledge Encyclopedia of Philosophy*, Vol. 6, 474–475.

Griffen, N., *Russell's Idealist Apprenticeship*, Oxford: Clarendon Press, 1991.

Hall, R., "Monism," *The Encyclopedia of Philosophy*, Vol. 5, New York: Macmillan, 1967, 363–365.

Ingardia, Richard, "Bradley and Aquinas: Empirical Realists," in *Perspectives on the Logic and Metaphysics of F.H. Bradley*, ed. W.J., Mander, Bristol: Thoemmes Press, 1996.

Mander, W.J., *An Introduction to Bradley's Metaphysics*, Oxford: Clarendon Press, 1994.

Moore, G.E., "Refutation of Idealism" (originally published in *Mind*, 12 [1903]); in T. Baldwin, ed. *G.E. Moore. Selected Writings*, London: Routledge, 1993, 23–44.

Russell, Bertrand, "The Philosophy of Logical Atomism," in *The Philosophy of Logical Atomism*, ed. David Pears, La Salle: Open Court, 1985, 35–155.

Russell, Bertrand, *The Scientific Outlook*, New York: W.W. Norton, 1931.

Schaffer, J., "Monism," *Stanford Encyclopedia of Philosophy*. http://plato.stanford.edu/entries/monism/.

Schlipp, P.A., ed., *The Philosophy of Bertrand Russell, The Library of Living Philosophers*, Vol. 5., Evanston, Ill: Northwestern University Press, 1944.

Sprigge, T.L.S. , "Idealism," in *Routledge Encyclopedia of Philosophy*, Volume 5, 662–669.

Stack, G., "Materialism," in *Routledge Encyclopedia of Philosophy*, E. Craig, ed., London: Routledge, 1998, 171–173.

Stock, Guy, "The Realistic Spirit in Bradley's Philosophy," in *Appearance versus Reality. New Essays on Bradley's Metaphysics*, ed. Guy Stock, Oxford: Clarendon Press, 1998, 1–18.

Index

Printed in the United States
By Bookmasters